SEMICONDUCTOR LASER DYNAMICS

Fundamentals and Applications

SEMICONDUCTOR LASER DYNAMICS
Fundamentals and Applications

BY

ANELE BRADLEY
JACK WOOD

KNOWLEDGE
BAKERS

KNOWLEDGE BAKERS

ISBN: 978-81-9759-576-9 (HB)

Published by : **Knowledge Bakers**
Pune, India
Email: info@knowledgebakers.com

PREFACE

Semiconductor lasers represent a cornerstone of modern photonics technology, with applications ranging from telecommunications and optical sensing to biomedical imaging and materials processing. The dynamical behavior of semiconductor lasers plays a crucial role in determining their performance characteristics and potential applications. Understanding the fundamental principles governing the dynamics of semiconductor lasers is essential for engineers, researchers, and practitioners working in the field of photonics. As semiconductor laser technology continues to advance, driven by innovations in materials science, device design, and fabrication techniques, there is a growing need for comprehensive resources that provide insights into the underlying principles and practical applications of semiconductor laser dynamics. As we embark on this exploration of semiconductor laser dynamics, we delve into a realm where physics, engineering, and technology converge, uncovering the principles, phenomena, and applications that define the dynamic behavior of semiconductor lasers.

This book, titled "Semiconductor Laser Dynamics: Fundamentals and Applications," is crafted to provide a comprehensive overview of the fundamental principles and practical applications of semiconductor laser dynamics. Through collaboration with leading experts, researchers, and practitioners in the field of photonics, we aim to offer readers a multidimensional and insightful journey through the intricacies of semiconductor laser dynamics.

As editors of this volume, we have curated a collection of chapters that cover a wide range of topics in semiconductor laser dynamics, including the physics of semiconductor lasers, rate equations modeling, nonlinear dynamics, chaos, synchronization phenomena, and applications in optical communications, optical sensing, spectroscopy, and more. Each chapter is designed to provide theoretical insights, practical methodologies, experimental techniques, and examples to empower readers to understand and apply semiconductor laser dynamics effectively in their research and applications.

Furthermore, this book seeks to highlight the interdisciplinary nature of semiconductor laser dynamics and its significance in advancing various fields of science and technology. By showcasing examples of how semiconductor laser dynamics are applied to solve practical problems and drive technological innovations, we aim to inspire readers to explore new frontiers in semiconductor laser research and to leverage the potential of semiconductor lasers for addressing societal challenges and advancing scientific knowledge.

As we embark on this enlightening journey through the world of semiconductor laser dynamics, we extend our gratitude to the contributors whose expertise and dedication have enriched this book. We hope that "Semiconductor Laser Dynamics: Fundamentals and Applications" serves as a valuable resource for students, researchers, engineers, and practitioners in the field of photonics, empowering them to deepen their understanding of semiconductor laser dynamics and to harness its transformative potential for advancing science and technology.

Anele Bradley
Jack Wood

CONTENTS

Nonlinear Dynamics of a Single-Mode Semiconductor Laser with Long Delayed Optical Feedback: A Modern Experimental Characterization Approach

Abstract: Semiconductor lasers can exhibit complex dynamical behavior in the presence of external perturbations. Delayed optical feedback, re-injecting part of the emitted light back into the laser cavity, in particular, can destabilize the laser's emission. We focus on the emission properties of a semiconductor laser subject to such optical feedback, where the delay of the light re-injection is large compared to the relaxation oscillations period. We present an overview of the main dynamical features that emerge in semiconductor lasers subject to delayed optical feedback, emphasizing how to experimentally characterize these features using intensity and high-resolution optical spectra measurements. The characterization of the system requires the experimentalist to be able to simultaneously measure multiple time scales that can be up to six orders of magnitude apart, from the picosecond to the microsecond range. We highlight some experimental observations that are particularly interesting from the fundamental point of view and, moreover, provide opportunities for future photonic applications.

Keywords: semiconductor lasers; optical feedback; nonlinear dynamics; long delay

---

1. Introduction

Semiconductor lasers (SL) have evolved considerably since their introduction in the early 1960s [1–4]. Nowadays, SL are small and efficient devices regularly used in a variety of applications such as optical data storage, metrology, spectroscopy, material processing, bio-sensing, the pumping of other lasers and optical telecommunications. In particular, its application to long-haul optical data transmission as a high-speed light source has enabled the worldwide optical fiber communication networks.

A characteristic property of most SL is their nonlinear response to perturbations, which manifests itself in a pronounced sensitivity to, e.g., noise, variations in the injection current, external optical injection or delayed optical feedback. This pronounced sensitivity is particularly relevant, as even small amounts of re-injected light can destabilize the SL emission [5–7] and induce chaotic dynamics. Even the small back-reflection from an optical fiber tip can destabilize a SL, which is a nuisance for applications in which stable emission is required. The corresponding instabilities are usually prevented via the introduction of optical isolators that shield the laser diode from feedback. Adding optical isolators implies, however, additional costs and complicates the design of compact and miniaturized photonic integrated circuits. The onset of dynamical instabilities was one of the first aspects that was addressed in the study of the nonlinear properties of SL subject to delayed optical feedback [8–11].

A specific perspective can be adopted when feedback effects are considered from the point of view of nonlinear dynamics. The different dynamical regimes of a delay-coupled SL depend directly on the pump current and feedback parameters, allowing to target specific dynamical regimes by tuning those parameters. The main feedback parameters are the amount of light re-injected into the cavity (feedback rate), the length of the external cavity and the corresponding feedback phase. A SL subject to delayed optical feedback may exhibit many characteristic high-dimensional dynamical phenomena, including hyper-chaotic regimes [12,13] and chaos synchronization when coupled to other SL [14,15]. Moreover, since the complex dynamics generated in delay-coupled SL are exploited in applications as diverse as encrypted communication with synchronized chaotic lasers [16], ultrafast random bit sequence generation [17], light sources with tunable coherence length [18], chaotic LIDAR [19], neuroinspired computation and ultrafast all-optical signal processing [20], a proper experimental and theoretical characterization and understanding are crucial.

The importance of SL subject to delayed optical feedback goes beyond the particular interest in laser dynamics or their photonics related applications. SL are well-controlled and tunable experimental systems, in which we can study nonlinear dynamics phenomena with very high accuracy. Therefore, SL subject to delayed optical feedback are excellent testbed examples of delay-coupled systems in general, being of fundamental importance in a variety of fields. Some examples of these fields are chaos control [21], neuroscience [22], traffic dynamics [23], population dynamics [24], gene regulatory networks [25,26], generic models [27], and secure communications [28].

In this tutorial, we cover some useful experimental tools to characterize the nonlinear dynamics of single-mode SL subject to optical feedback. We focus on the characterization of the intensity dynamics and the corresponding high-resolution optical spectra in the case where the propagation time of the light in the external cavity, i.e., the delay time, is longer than the characteristic time scale of the solitary laser, namely the period of the relaxation oscillations [29].

2. Materials and Methods

2.1. Experimental Setup

The simplest configuration of delayed optical feedback comprises a SL diode and an optical reflector, which could be provided, e.g., by a mirror. Figure 1 shows a schematic view of such an experimental set-up, where the feedback is implemented as a loop via optical fibers. We consider a SL diode that emits stably in a continuous wave operation when unperturbed. The feedback loop redirects a fraction κ_f of the laser's emitted optical field back into the laser cavity after a delay time τ. The properties of the laser emission depend on the amount of light that is reinjected into the laser [9]. Here, we will show experimental results for Discrete-Mode Laser-Diodes that have a Fabry–Perot structure with etched and longitudinally periodic ridges, which have been designed such that they yield single longitudinal-mode operation for a wide range of operating conditions [30]. The discrete-mode laser diode discussed in the following emits at a wavelength of about 1542 nm. The SL used in the following did not include an optical isolator in its butterfly packaging.

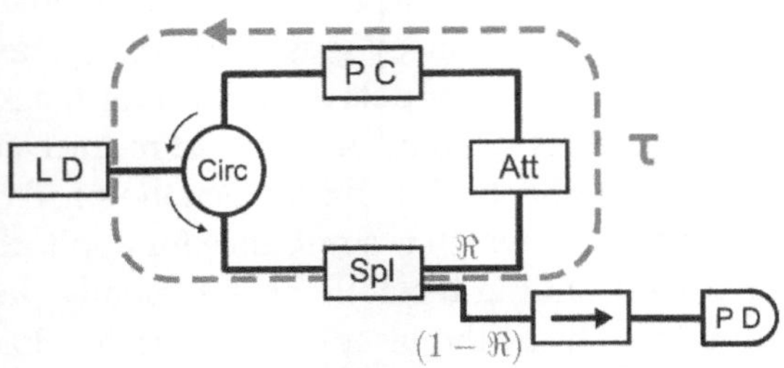

Figure 1. Scheme of our typical experimental setup to study feedback dynamics. LD: laser diode, Circ: optical circulator, PC: polarization controller; Att: optical attenuator, Spl: one by two intensity splitter with $\mathbb{R} = 0.95$ and $(1 - \mathbb{R}) = 0.05$ splitting ratios, $\rightarrow$: optical isolator, and PD: photodiode.

The first and most straightforward consequence of delayed optical feedback that can be measured in SL is its effect on the output power. Coherent optical feedback typically reduces the lasing threshold as compared to solitary operation. This threshold reduction is due to photons being re-injected into the laser cavity, reducing the total optical losses. Figure 2 displays two different power-current characteristics of the same laser under different experimental conditions. The blue curve corresponds to the case of the SL without feedback, in comparison to which the presence of delayed optical feedback reduces the SL's threshold current, see the orange curve. By means of measuring the relative losses through the successive elements in the external cavity, a maximum value of the feedback rate can be estimated. The maximum feedback rate, i.e., the fraction of intensity reflected back to the laser diode for this particular setup is 54%, excluding the laser and fiber-coupling losses. The kink in the feedback power-current characteristics is due to the appearance of low frequency fluctuations dynamics close to the solitary lasing threshold.

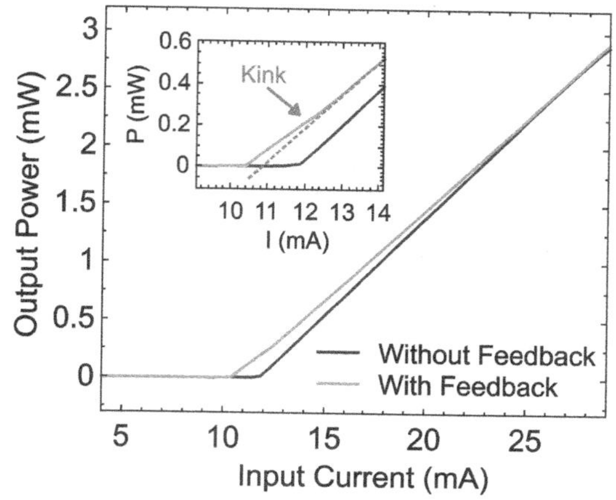

Figure 2. Power-current characteristics of a single-mode laser under different operation conditions: in solitary emission (blue line) and when subject to delayed optical feedback (orange line). The inset is a magnification of the region around the two distinct laser thresholds.

2.2. Spectral Characterization of Feedback-Induced Dynamics

Since the lasing threshold is reduced, temporally averaged input–output characteristics draw a rather positive image of delayed optical feedback. From the viewpoint of the intensity dynamics of the laser, however, delayed feedback has a much more ambiguous effect. In fact, since the early days of SL, feedback was considered a clear nuisance for their operation as it created power instabilities and increased the laser's spectral bandwidth [5–7], i.e., reducing its coherence.

Figure 3 illustrates this optical feedback-induced broadening on the optical and radio-frequency power spectra. Figure 3a depicts the optical spectra as measured with a high-resolution optical spectrum analyzer (resolution 10 MHz). Different colors correspond to different feedback rates, with the orange curve plotting the solitary laser spectrum and the blue color plotting the highest feedback rate. In this figure, the higher the value of the feedback rate, the broader the optical spectrum becomes, mostly broadening towards lower frequencies, that is, longer wavelengths. For the highest feedback rates on this particular device, the broadening of the spectrum already covers more than 20 GHz. This well known phenomenon is frequently referred to in the literature as coherence collapse [8]. For an even higher feedback rate, the SL can eventually become stable again with a reduced linewidth [31].

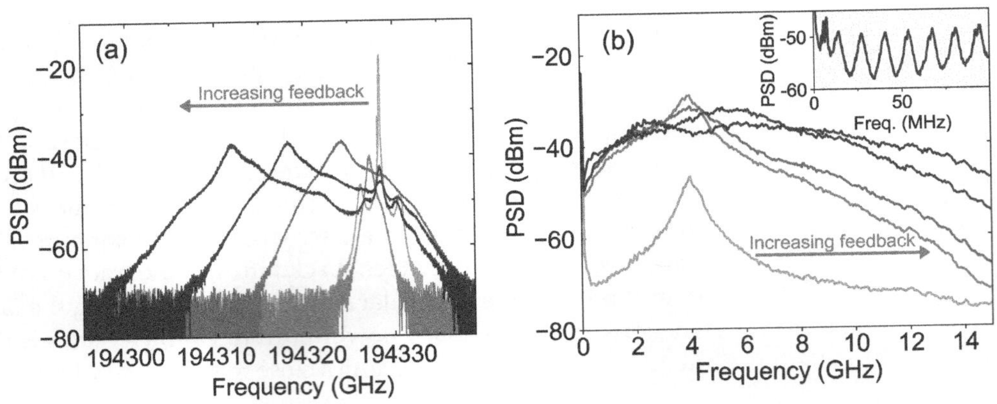

Figure 3. Experimental signatures of delayed optical feedback on the optical and radio-frequency (RF) spectra of a semiconductor laser. (**a**) High-resolution optical spectra of the laser without feedback (orange) and with different amounts of feedback. (**b**) Effect of delayed optical feedback on the RF power spectral density. The RF spectrum is depicted over a 15 GHz frequency span for increasing feedback conditions. RF spectra are shown for the solitary laser emission (orange line) and four successively higher values of the feedback rate (from red towards blue). Inset: Zoom of the first 100 MHz for the maximum feedback rate. The external cavity frequency resonances are here clearly visible as equidistant peaks separated by 13.29 MHz. The arrow indicates the broadening direction of the optical spectrum and the overall bandwidth increase of the RF spectrum when the feedback rate increases, respectively.

A complementary spectral characterization tool for the experimentalist is the radio frequency (RF) power spectrum of the emitted light intensity. Figure 3b depicts the impact of delayed optical feedback on the RF power spectrum, providing additional information complementary to that present in the optical domain. On a multi-GHz scale, one can observe a clear broadening of the RF spectrum due to the feedback. The relaxation oscillations frequency is visible for the solitary laser around 4 GHz (frequency with the highest power density for the orange curve), a feature that is typically observed in the relative intensity noise spectrum of class B lasers [32]. Increasing the feedback rate results in a broadening of the RF spectrum with its maximum shifting towards higher frequencies. Colors for the different feedback scenarios correspond to those of the optical spectra, with blue representing the highest feedback rate. The inset in Figure 3b depicts the fingerprint of the delay time in the RF power spectrum. The visible resonances in the RF power spectrum are separated roughly by the inverse of the light flight time in the external cavity. The resonance separation is here 13.29 MHz, yielding an approximate value of the external cavity delay, here ($\tau \approx 75$ ns) [33]. An experimental method to measure the delay time more precisely is to induce an optical pulse with a steep slope (small rise time) and to measure the time intervals between multiple reflections in the external cavity [33].

The linewidth enhancement factor (α), characteristic for SL, plays a crucial role in the destabilization of the laser emission via delayed optical feedback that ultimately leads to the collapse of the optical coherence. This parameter accounts for the coupling between light intensity and optical phase [34]. As a consequence, any intensity fluctuation in the laser's emission will be fed-back via the delay path, re-enter the gain medium, and affect intensity and optical phase.

2.3. *Temporal Characterization of Feedback-Induced Dynamics*

Figure 4 illustrates the impact of delayed optical feedback on the temporal emission characteristics of SL. The solitary emission is depicted in Figure 4a, where the AC-coupled intensity time series shows small fluctuations around the origin. Those intensity fluctuations are the combination of amplified spontaneous emission and detection noise, particularly when the laser is biased close to the threshold. The corresponding optical linewidth depends on the optical power and for the operating conditions in Figure 4a is approximately 1 MHz. In contrast, in Figure 4b, the irregular emission behavior under delayed optical feedback is illustrated. The intensity dynamics exhibits chaotic pulsations on a sub-ns time scale. The optical spectrum corresponding to such dynamics expands the SL's linewidth from $\sim$1 MHz to tens of GHz. The detection bandwidth plays a fundamental role in the study of the intensity dynamics of the SL subject to optical feedback. The fast intensity pulsations, which remained mostly unresolved until first measured with a streak camera [35], can now be characterized in detail due to oscilloscopes with several tens of GHz analog bandwidth in combination with fast photodetectors.

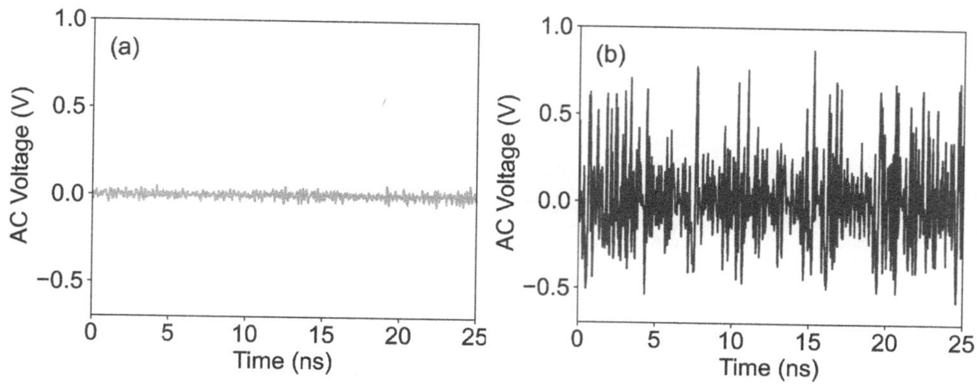

Figure 4. Fingerprint of the coherence collapse phenomenon on the intensity dynamics. The AC-coupled (average power is subtracted by a photodiode, with a low cut-off frequency of a few kHz) time traces are plotted in (**a**) for the solitary laser and (**b**) for the laser subject to delayed feedback. Colors correspond to the experimental conditions depicted in blue and orange colors in Figure 3. In (**b**), the tens of GHz laser's optical linewidth is reflected in fast and chaotic oscillations of the corresponding time trace.

While it is, meanwhile, possible to fully resolve the complex intensity dynamics exhibited by delayed feedback lasers, a direct inspection of chaotic time series is of restricted help for understanding how the dynamics depends on the different system parameters. Here, choosing an appropriate representation of the measured time series is key. Figure 5 illustrates two useful methods to represent features of the time-resolved intensity dynamics in a comprehensive and informative way. Figure 5a depicts the autocorrelation function (ACF) of the intensity dynamics. The most characteristic signature of delayed dynamics is present in the revival peaks located at multiples of the delay time. The inset in Figure 5a zooms on the first ACF peak at a lag time of $\sim\tau$. The spatio-temporal representation of the intensity dynamics is shown in Figure 5b. In this two-dimensional representation, inspired by the two main timescales in the ACF, delay time intervals are plotted as a pseudo-space variable [36]. Here, the abscissa denotes the time offset in a delay interval of length τ, while the vertical axis denotes the ordinal of the current delay interval. In this manner, the intensity time series is divided in temporal segments of length τ and consecutive segments are then plotted on top of each other. For instance, the bottom line in Figure 5b corresponds to time $0 \leq t < \tau$ in the original time-series, the line directly above corresponds to time $\tau \leq t < 2\tau$, and so on. The intensity of the laser is encoded in grey-scale. From such a representation, specific information can be extracted from a long time series, in particular if the plotted dynamics is irregular or chaotic [37].

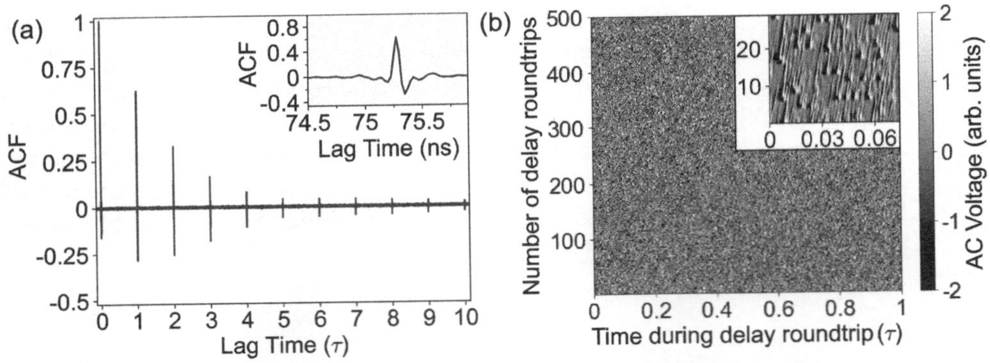

Figure 5. Different representations of the chaotic intensity dynamics of delayed feedback lasers. (**a**) The autocorrelation function (ACF) corresponding to the same feedback conditions as depicted in the intensity dynamics of Figure 4b. The ACF was calculated over a time series of 2 million samples recorded with a sampling rate of 40 GS/s. (**b**) The spatio-temporal representation of the same chaotic time series. The abscissa denotes the time offset in a delay interval of length τ, while the vertical axis stacks the time series segments of length τ. The magnitude of the intensity fluctuations is color-coded.

2.4. Theoretical Background

The results reported in this tutorial have been obtained in experiments using quantum well single mode laser diodes. Even though these results are model-independent, it is of major interest to link these findings with existing numerical models. A SL subject to moderate optical feedback can in many cases be described by the Lang–Kobayashi rate equations as follows [38]:

$$\frac{dE(t)}{dt} = \frac{1+\mathrm{i}\alpha}{2}\left[G_{\mathrm{N}}\left(N(t)-N_o\right)-\frac{1}{\tau_p}\right]E(t)+\kappa E(t-\tau)e^{-\mathrm{i}2\pi f_{th}\tau}, \tag{1}$$

$$\frac{dN(t)}{dt} = \frac{I}{e}-\frac{N(t)}{\tau_s}-G_{\mathrm{N}}\left(N(t)-N_o\right)|E(t)|^2, \tag{2}$$

where E and N are the complex electric field amplitude and the carrier number, respectively. The main feedback parameters, namely the delay time τ and the feedback rate κ, can be seen in Equation (1). Other parameters in Equation (1) are the linewidth enhancement factor α, the differential optical gain G_{N}, the carrier number at transparency N_o, the photon lifetime τ_p, and the laser solitary frequency at lasing threshold f_{th}. In Equation (2), I is the pump current, e the electron charge, and τ_s is the carrier lifetime.

For a detailed analysis of the Lang–Kobayashi model, we refer the reader to the existing literature [10,39]. Here, we will focus on the steady-state solutions of Equations (1) and (2) since they provide the backbone of the observed dynamical phenomena.

2.4.1. External Cavity Modes

The relative equilibria solutions for the SL subject to optical feedback can be obtained by introducing $P(t) = |E(t)|^2 = P_s$, $N(t) = N_s$, and $\phi_s(t) = 2\pi(f_s - f_{th})t$ in Equations (1) and (2). Here, $\phi(t)$ denotes the phase of the electric field $E(t)$, while P_s, N_s and f_s are constant. Accordingly, the steady-state solutions read:

$$N_s = N_{th} - 2\frac{\kappa}{G_N}\cos\left(2\pi f_s\tau\right), \tag{3}$$

$$2\pi(f_s - f_{th}) = \kappa\sqrt{1+\alpha^2}\sin\left(2\pi f_s\tau + \arctan(\alpha)\right), \tag{4}$$

$$P_s = \frac{\frac{I}{e}-\frac{N_s}{\tau_s}}{G_N(N_s-N_o)}, \tag{5}$$

where $N_{th} = N_o + 1/(\tau_p G_N)$. The solutions in Equations (3)–(5) are rotating waves, which lie on an ellipse in the (f_s, N_s) plane [40,41]. Figure 6 shows such ellipses of relative equilibria obtained for different parameter values of the feedback rate, delay time, and linewidth enhancement factor. The external cavity modes (ECM) are the solutions that originate from constructive interference between the laser field and the delayed feedback, blue circles in Figure 6, while the external cavity antimodes are the ones corresponding to destructive interference and are shown as orange triangles in Figure 6.

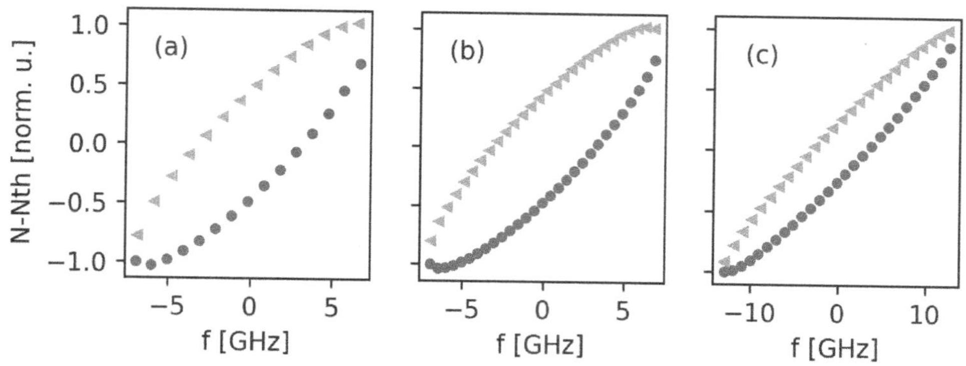

Figure 6. Ellipse of steady-state solutions for the Lang–Kobayashi model. External cavity modes are depicted as blue circles and antimodes as orange triangles. (**a**) Parameters: $\kappa = 20$ ns^{-1}, $\tau = 1$ ns, and $\alpha = 2$. (**b**) Parameters: $\kappa = 20$ ns^{-1}, $\tau = 2$ ns, and $\alpha = 2$. (**c**) Parameters: $\kappa = 20$ ns^{-1}, $\tau = 1$ ns, and $\alpha = 4$.

As illustrated in Figure 6, the maximum feedback-induced frequency shift in the Lang–Kobayashi model is approximately $\Delta f_{fb} = \alpha\kappa/2\pi$ (extremes of the ellipse) and the frequency separation between ECM is proportional to τ^{-1}, where $f_{EC} = \tau^{-1}$ is the frequency associated to the external cavity. Thus, the extent of the ellipse is $\Delta f_{fb} \sim 6.35$ GHz in Figure 6a,b, while $\Delta f_{fb} \sim 12.7$ GHz in Figure 6c, respectively. In turn, the separation between the ECMs is $\tau^{-1} \sim 1$ GHz in Figure 6a,c, while $\tau^{-1} \sim 0.5$ GHz in Figure 6b, respectively. The eccentricity of the ellipse of external cavity modes and antimodes is determined by α.

To understand the feedback-induced laser dynamics, the relaxation oscillation frequency of the solitary laser is of great relevance. The relaxation oscillation frequency of the solitary laser, i.e., Equations (1) and (2) with $\kappa = 0$, is given by $f_{RO} = \frac{1}{2\pi}\sqrt{G_N(I - I_{th})/e}$, where I_{th} is the pump current at the lasing threshold. As will be shown in Section 3, the relationship between Δf_{fb} and f_{RO} plays a major role in the observed dynamics of the SL subject to optical feedback since we are dealing with an experimental system with a large delay ($\tau \approx 75$ ns).

3. Experimental Results: Dynamical Regimes

Spectral and dynamical phenomena in delayed feedback SL depend on the combination of both, laser and feedback parameters. It is possible, however, to group these intricate dependencies in different scenarios or feedback regimes. Here, three characteristic experimental scenarios are analyzed, corresponding to different degrees of influence of feedback on the single laser behavior. In the following, feedback rate and bias current are the two parameters varying between low, intermediate and strong feedback scenarios.

As illustrated in [42], feedback regimes depend mainly on the relation between the system's three principal frequencies: the relaxation oscillations frequency f_{RO}, the external cavity frequency f_{EC}, and the feedback-induced frequency shift Δf_{fb}, respectively. The dynamical characteristics, shown for a given combination of feedback and laser parameters, are similar to any other combination of parameters that keep the ratios $f_{RO}/\Delta f_{fb}$ and f_{RO}/f_{EC} constant. As discussed in Section 2.4.1, f_{RO} depends on bias current I and Δf_{fb} on feedback rate κ. The particular values of the three bias currents and feedback rates in the following were chosen to best illustrate the dynamic and spectral characteristics of feedback-induced semiconductor laser dynamics in the low, intermediate and strong feedback regimes, respectively. By working in the limit of long external cavities, with $f_{RO} \gg f_{EC}$, the ratio $f_{RO}/\Delta f_{fb}$ dominates the global dynamics, while the ratio f_{RO}/f_{EC} determines if the temporal dynamics is contracting or expanding [42].

The experimentally measured transition between the three feedback conditions shown here does not necessarily visit discrete feedback regimes, but rather features continuous transitions between these. A description of such dependence and their relation with the system's time scales can be found in [42]. For a detailed study of the bifurcation-cascade diagrams in delayed feedback lasers we refer, e.g., to this previous work [43].

3.1. Low Feedback Rate

The first illustrated regime corresponds to the weakest feedback condition, the onset region of dynamical instabilities. In the selected example, the bias current I = 18 mA is approximately $\simeq 1.5I_{th}$ and the feedback rate has been strongly attenuated by 31 dB from its maximum value. This combination of bias current and feedback rate leads to the condition $f_{RO} > \Delta f_{fb}$. In particular, the chosen parameter values correspond to undamped relaxation oscillations dynamics [44], occurring just after the destabilization of the solitary laser emission.

Figure 7a shows the optical spectrum, which is characterized by the presence of sharp well defined peaks at left and right sides of the solitary laser mode (f_{SOL}). The frequency peak at its immediate left side corresponds to the feedback-induced frequency shift (Δf_{fb}). In addition, one can observe several side peaks that are shifted by multiples of the relaxation oscillation frequency (f_{RO}). This is the signature of the undamping of the relaxation oscillations. Such resonances are clearly identifiable in the RF spectrum depicted in Figure 7b. As shown in Figure 7a, Δf_{fb} is smaller than the relaxation oscillations frequency f_{RO} for this weak feedback regime.

The ACF is depicted in Figure 7c for an interval of 10τ. The overall ACF decay along multiple delay echoes is very small and the system contains, on average, persistent τ-feedback memory of its past states. The inset in Figure 7c depicts the ACF oscillations around the first delay echo. These oscillations are the signature of the undamped relaxation oscillations in the ACF and are dominated by f_{RO}, as visible in the RF spectrum. In the broader view of the ACF, a particular lobe structure modulates the ACF envelope amplitudes in between delay echoes. Only a slight modification of any of the feedback parameters can significantly modify the lobe structure, but their influence on the ACF height will be minor.

Figure 7d illustrates the spatio-temporal representation of the intensity dynamics. This representation clearly elicits the high correlation of the local dynamics with the delay time τ. For a time window of 500τ (more than 30 μs), two distinct and separate dynamical behaviors—namely stable emission and an oscillatory state—coexist. The region magnified in the inset allows the identification of oscillations of constant frequency f_{RO} that characterize the oscillatory region and a sharp transition to the stable regime. The coexistence of stable and unstable dynamics is a common feature of SL with low optical feedback [43,45,46].

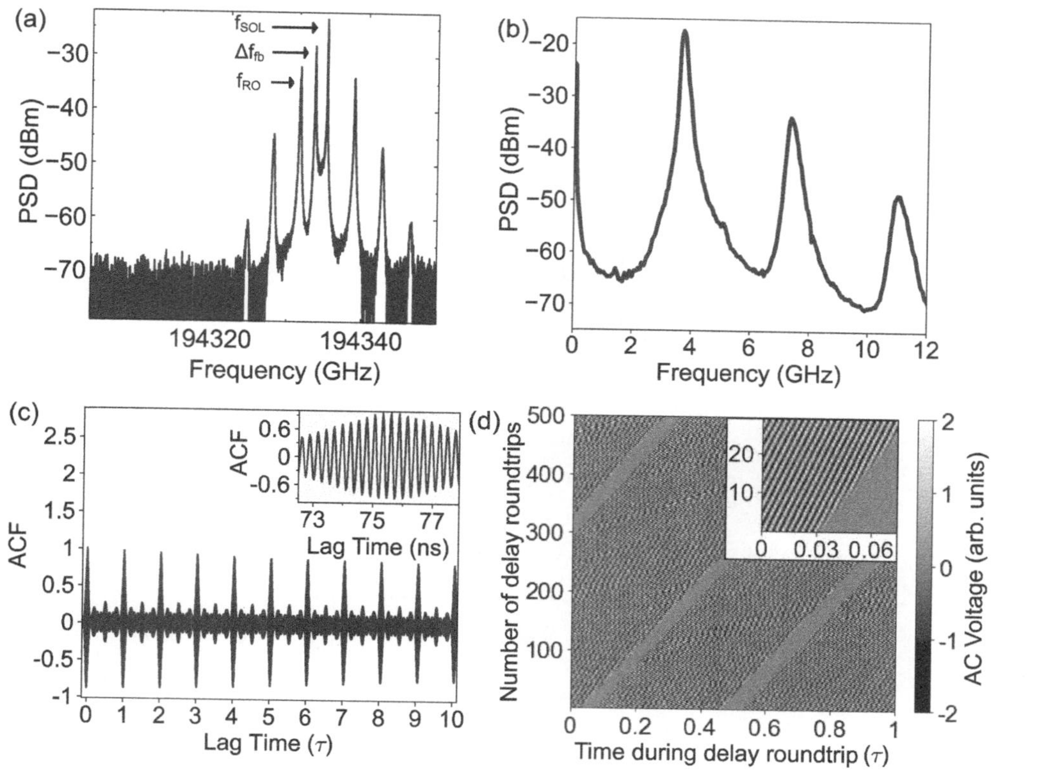

Figure 7. Characteristics for undamped relaxation oscillation dynamics at a bias current of I = 18 mA and a feedback rate that has been attenuated by 31 dB from its maximum value. (**a**) Optical spectrum, (**b**) RF power spectrum, (**c**) Autocorrelation function (ACF) of the intensity time series with inset around the first delay echo, (**d**) Spatio-temporal representation of the intensity time series with inset for magnification of the intensity dynamics around the origin.

3.2. Intermediate Feedback Rate

Here, we show results for a bias current of I = 16 mA ($\simeq 1.3I_{th}$) and a feedback rate that has been attenuated by 13 dB from its maximum value. These parameter choices correspond to the intermediate feedback condition $f_{RO} \approx \Delta f_{fb}$.

The intermediate feedback region can be associated with a strong chaos regime in the weak–strong chaos paradigm for delayed systems [47,48]. In this regime, all correlations in dynamic and spectral signatures are strongly reduced as a result of the enhanced nonlinear mixing between f_{RO} and Δf_{fb}. This feature can be identified in Figure 8a, with a broad optical spectrum covering more than 20 GHz at -20 dB height. No sharp peaks are present neither in the optical spectrum nor in the RF power spectrum shown in Figure 8b.

The envelope of the ACF in Figure 8c illustrates the fast fading memory of the dynamics at successive delay echoes for these experimental conditions. The ACF height at the first delay echo is below 0.3, as visible in the inset. This low value of the ACF at multiples of τ is a signature of strong chaos in delayed feedback lasers [48]. Figure 8d depicts the spatio-temporal representation for such time series. The spatio-temporal plot shows differentiated structures with little correlation between them, disappearing rapidly and losing their correlation within a few delay times.

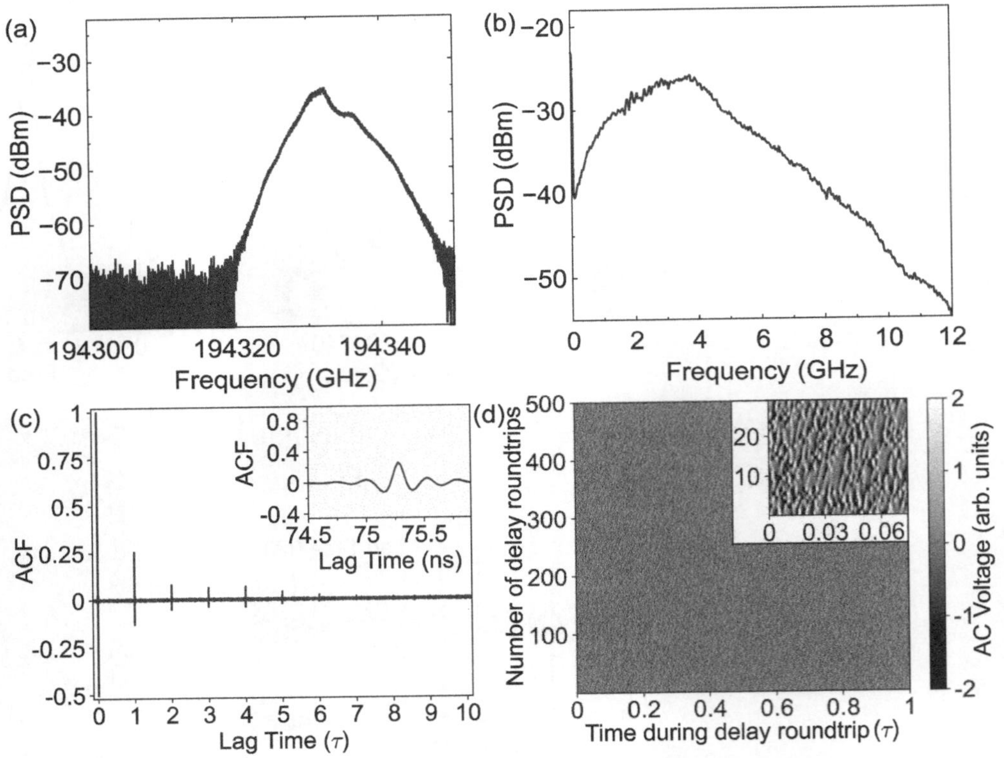

Figure 8. Exemplary dynamics at intermediate feedback levels. The bias current is I = 16 mA and the feedback rate has been attenuated by 13 dB from its maximum value. (**a**) Optical spectrum; (**b**) RF power spectrum; (**c**) Autocorrelation function (ACF) of the intensity time series with inset around the first delay echo, (**d**) Spatio-temporal representation of the intensity time series with inset for magnification around the origin.

3.3. High Feedback Rate

Strong feedback conditions are depicted in Figure 9. In this case, the bias current is I = 13 mA ($\simeq 1.1 I_{th}$) and the feedback rate is the maximum value allowed in the experimental set-up, yielding the condition $f_{RO} < \Delta f_{fb}$.

The optical spectrum in Figure 9a shows a two-peaked structure with a distance of ~20 GHz between the peaks. The highest peak, at lower optical frequencies, corresponds to the spectral signature of the maximum feedback-induced frequency shift (Δf_{fb}), while the peak at the right hand side of the spectrum is the signature of the dynamics in the region of the solitary laser mode. The RF power spectrum in Figure 9b illustrates that the chaotic bandwidth broadens when the feedback rate is increased, with the particularity that the power for low frequencies significantly increases (see inset of Figure 9b). This increase of the RF power density at lower frequencies is due to the appearance of the so-called low frequency fluctuations (LFF) in the intensity dynamics. The LFF dynamics, which appear for currents close to the lasing threshold and strong feedback, is characterized by irregularly distributed dropouts in the intensity dynamics [49–51]. The LFF power dropouts are distinguishable in Figure 9d as dark horizontal lines that typically last a τ-interval and repeat irregularly every $\sim 20\tau$. Here, LFFs are observed in the strong feedback case since the laser is operated close to the lasing threshold $I \simeq 1.1 I_{th}$.

The high feedback rate conditions lead to large delay-induced correlation peaks, as shown in detail at the inset in Figure 9c. Despite the fast chaotic dynamics and a broad RF spectrum, the slow decay of the ACF envelope hints at the fact that this dynamics correspond to a weak chaos regime [48]. The long-term vertical propagation of the structures visible in the spatio-temporal representation of Figure 9d (and magnified in the inset) supports this interpretation. In delayed feedback SL, strong chaos does not occur when feedback is strongest, but rather in the region where the feedback induced frequency shift approaches the frequency of the relaxation oscillations of the solitary laser [42]. It is worth noting that, when the feedback is strong enough, the laser again enters the weak chaos regime and, eventually, for particularly strong feedback and bias currents not far from threshold, the laser intensity can be stabilized again. This latter scenario leads the laser to a particularly high coherent state [31].

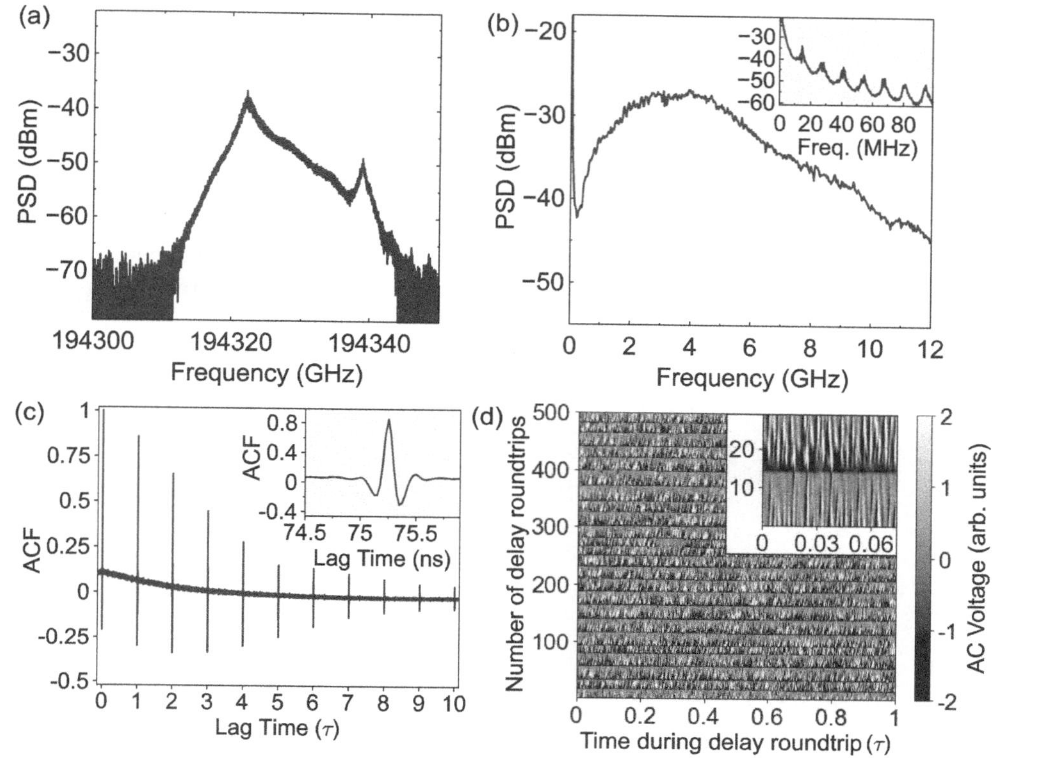

Figure 9. High feedback rate dynamics. The bias current is I = 13 mA and the feedback rate is the maximum value allowed in the experimental set-up. (**a**) Optical spectrum, (**b**) RF power spectrum, (**c**) Autocorrelation function (ACF) of the intensity time series with inset around the first delay echo, (**d**) Spatio-temporal representation of the intensity time series with inset for magnification around the origin.

4. Discussion and Conclusions

The complex dynamical behavior of SL subject to optical feedback leaves characteristic fingerprints that can be observed in time and frequency domains. In this tutorial, we presented a basic overview of the main dynamical phenomena and the feedback-induced features in optical spectra and laser intensity measurements of single-mode SL. The impact of optical feedback on multi-mode lasers is a further, interesting aspect that has not been covered here. We have focused instead on the conditions where long delayed optical feedback yields a transition between stable dynamics and undamped relaxation oscillations, as well as in two conditions that lead to two chaotic regimes of different natures. In all these cases, the interplay between the relaxation oscillations frequency f_{RO} and the maximum feedback-induced frequency shift Δf_{fb} is crucial for the dynamics that can be observed in the laser. The particularities of the dynamical features, such as the shape of the optical spectrum or the structures in the spatio-temporal representation, do, however, depend on the parameters of the considered device. In particular, lasers with a higher linewidth enhancement factor α are typically more sensitive to feedback.

By directly detecting the laser emission with a photodiode, one gets information about the intensity dynamics. Characterization of the phase dynamics is more challenging but also possible through, e.g., heterodyne measurements [52]. For the heterodyne technique, the field of the delayed-feedback laser is coherently added to the field of a narrow linewidth laser of a similar optical frequency, and the mixed fields are then detected by a fast photodetector. In this manner, a signal proportional to the optical frequency dynamics can be measured in the radio-frequency domain. By simultaneously measuring intensity and optical frequency dynamics, along with the carrier dynamics, with high temporal resolution, one obtains a complete picture of the dynamical evolution of the laser along the three dynamical dimensions of the corresponding rate equation model [53]. The full knowledge of the dynamics can allow for tailoring the response of the laser for applications by performing a proper phase space engineering.

In the last few years, the study of SL with optical feedback has been extended beyond the single, fixed delay case. From the fundamental point of view, two notable extensions include the realization of state-dependent delays [54] or 2D spatio-temporal dynamics [55] in laser systems with two delays. From the application point of view, SL subject to optical feedback and optical injection have been demonstrated to allow for information processing in the context of photonic reservoir computing [20,56]. We envision that the complex dynamics of SL with optical feedback described in this tutorial will still prove to be a valuable resource for further research and innovation.

Abbreviations

The following abbreviations are used in this manuscript:

AC	Alternating Current
ACF	Autocorrelation Function
ECM	External Cavity Mode
LFF	Low Frequency Fluctuations
PSD	Power Spectral Density
RF	Radio-Frequency
SL	Semiconductor Laser

References

1. Hall, R.; Fenner, G.; Kingsley, J.; Soltys, T.; Carlson, R. Coherent Light Emission From GaAs Junctions. *Phys. Rev. Lett.* **1962**, *9*, 366–368. [CrossRef]
2. Nathan, M.I.; Dumke, W.P.; Burns, G.; Dill, F.H.; Lasher, G. Stimulated Emission of Radiation from GaAs p-n Junctions. *Appl. Phys. Lett.* **1962**, *1*, 62. [CrossRef]
3. Quist, T.M.; Rediker, R.H.; Keyes, R.J.; Krag, W.E.; Lax, B.; McWhorter, A.L.; Zeigler, H.J. Semiconductor Maser of GaAs. *Appl. Phys. Lett.* **1962**, *1*, 91. [CrossRef]
4. Coleman, J.J. The development of the semiconductor laser diode after the first demonstration in 1962. *Semicond. Sci. Technol.* **2012**, *27*, 090207. [CrossRef]
5. Broom, R. Self modulation at gigahertz frequencies of a diode laser coupled to an external cavity. *Electron. Lett.* **1969**, *5*, 571. [CrossRef]
6. Broom, R.; Mohn, E.; Risch, C.; Salathe, R. Microwave self-modulation of a diode laser coupled to an external cavity. *IEEE J. Quantum Electron.* **1970**, *6*, 328–334. [CrossRef]
7. Risch, C.; Voumard, C. Self-pulsation in the output intensity and spectrum of GaAs-AlGaAs cw diode lasers coupled to a frequency-selective external optical cavity. *J. Appl. Phys.* **1977**, *48*, 2083. [CrossRef]
8. Lenstra, D.; Verbeek, B.; den Boef, A. Coherence collapse in single-mode semiconductor lasers due to optical feedback. *IEEE J. Quantum Electron.* **1985**, *21*, 674–679. [CrossRef]
9. Tkach, R.; Chraplyvy, A. Regimes of feedback effects in 1.5-um distributed feedback lasers. *J. Light. Technol.* **1986**, *4*, 1655–1661. [CrossRef]

10. Mørk, J.; Tromborg, B.; Mark, J. Chaos in semiconductor lasers with optical feedback: Theory and experiment. *IEEE J. Quantum Electron.* **1992**, *28*, 93–108. [CrossRef]
11. Heil, T.; Fischer, I.; Elsäßer, W. Stabilization of feedback-induced instabilities in semiconductor lasers. *J. Opt. B Quantum Semiclassical Opt.* **2000**, *2*, 413–420. [CrossRef]
12. Ahlers, V.; Parlitz, U.; Lauterborn, W. Hyperchaotic dynamics and synchronization of external-cavity semiconductor lasers. *Phys. Rev. E* **1998**, *58*, 7208. [CrossRef]
13. Vicente, R.; Daudén, J.; Colet, P.; Toral, R. Analysis and characterization of the hyperchaos generated by a semiconductor laser subject to a delayed feedback loop. *IEEE J. Quantum Electron.* **2005**, *41*, 541–548. [CrossRef]
14. Heil, T.; Fischer, I.; Elsässer, W.; Mulet, J.; Mirasso, C. Chaos Synchronization and Spontaneous Symmetry-Breaking in Symmetrically Delay-Coupled Semiconductor Lasers. *Phys. Rev. Lett.* **2001**, *86*, 795–798. [CrossRef] [PubMed]
15. Locquet, A.; Masoller, C.; Mirasso, C.R. Synchronization regimes of optical-feedback-induced chaos in unidirectionally coupled semiconductor lasers. *Phys. Rev. E* **2002**, *65*, 056205. [CrossRef] [PubMed]
16. Argyris, A.; Syvridis, D.; Larger, L.; Annovazzi-Lodi, V.; Colet, P.; Fischer, I.; García-Ojalvo, J.; Mirasso, C.R.; Pesquera, L.; Shore, K.A. Chaos-based communications at high bit rates using commercial fibre-optic links. *Nature* **2005**, *438*, 343–346. [CrossRef]
17. Uchida, A.; Amano, K.; Inoue, M.; Hirano, K.; Naito, S.; Someya, H.; Oowada, I.; Kurashige, T.; Shiki, M.; Yoshimori, S.; et al. Fast physical random bit generation with chaotic semiconductor lasers. *Nat. Photonics* **2008**, *2*, 728–732. [CrossRef]
18. Peil, M.; Fischer, I.; Elsäßer, W.; Bakić, S.; Damaschke, N.; Tropea, C.; Stry, S.; Sacher, J. Rainbow refractometry with a tailored incoherent semiconductor laser source. *Appl. Phys. Lett.* **2006**, *89*, 091106. [CrossRef]
19. Lin, F.-Y.; Liu, J.-M. Chaotic LIDAR. *IEEE J. Sel. Top. Quantum Electron.* **2004**, *10*, 991–997. [CrossRef]
20. Brunner, D.; Soriano, M.C.; Mirasso, C.R.; Fischer, I. Parallel photonic information processing at gigabyte per second data rates using transient states. *Nat. Commun.* **2013**, *4*, 1364. [CrossRef]
21. Pyragas, K. Continuous control of chaos by self-controlling feedback. *Phys. Lett. A* 1992, 170, 421–428. [CrossRef]
22. Stepan, G. Delay effects in brain dynamics. Introduction. *Philos. Trans. Ser. Math. Phys. Eng.* **2009**, *367*, 1059–1062.
23. Orosz, G.; Wilson, R.E.; Szalai, R.; Stépán, G. Exciting traffic jams: Nonlinear phenomena behind traffic jam formation on highways. *Phys. Rev. E* **2009**, *80*, 046205. [CrossRef] [PubMed]
24. Mackey, M.C.; Glass, L. Oscillation and chaos in physiological control systems. *Science* **1977**, *197*, 287–289. [CrossRef] [PubMed]
25. Elowitz, M.B.; Leibler, S. A synthetic oscillatory network of transcriptional regulators. *Nature* **2000**, *403*, 335–338. [CrossRef]
26. Chen, L.; Aihara, K. Stability of genetic regulatory networks with time delay. *IEEE Trans. Circuits Syst. Fundam. Appl.* **2002**, *49*, 602–608. [CrossRef]
27. Yeung, M.; Strogatz, S. Time Delay in the Kuramoto Model of Coupled Oscillators. *Phys. Rev. Lett.* **1999**, *82*, 648–651. [CrossRef]
28. Keuninckx, L.; Soriano, M.C.; Fischer, I.; Mirasso, C.R.; Nguimdo, R.M.; van der Sande, G. Encryption key distribution via chaos synchronization. *Sci. Rep.* **2017**, *7*, 43428. [CrossRef]
29. Soriano, M.C.; García-Ojalvo, J.; Mirasso, C.R.; Fischer, I. Complex photonics: Dynamics and applications of delay-coupled semiconductors lasers. *Rev. Mod. Phys.* **2013**, *85*, 421–470. [CrossRef]
30. Kelly, B.; Phelan, R.; Jones, D.; Herbert, C.; O'Carroll, J.; Rensing, M.; Wendelboe, J.; Watts, C.; Kaszubowska-Anandarajah, A.; Guignard, C.; et al. Discrete mode laser diodes with very narrow linewidth emission. *Electron. Lett.* **2007**, *43*, 1282. [CrossRef]
31. Brunner, D.; Luna, R.; Latorre, A.D.i.; Porte, X.; Fischer, I. Semiconductor laser linewidth reduction by six orders of magnitude via delayed optical feedback. *Opt. Lett.* **2017**, *42*, 163–166. [CrossRef]
32. Baili, G.; Alouini, M.; Malherbe, T.; Dolfi, D.; Sagnes, I.; Bretenaker, F. Direct observation of the class-B to class-A transition in the dynamical behavior of a semiconductor laser. *Europhys. Lett.* **2009**, *87*, 44005. [CrossRef]
33. Porte, X.; D'Huys, O.; Jüngling, T.; Brunner, D.; Soriano, M.C.; Fischer, I. Autocorrelation properties of chaotic delay dynamical systems: A study on semiconductor lasers. *Phys. Rev. E* **2014**, *90*, 052911. [CrossRef]
34. Henry, C. Theory of the linewidth of semiconductor lasers. *IEEE J. Quantum Electron.* **1982**, *18*, 259–264. [CrossRef]
35. Fischer, I.; van Tartwijk, G.; Levine, A.; Elsässer, W.; Göbel, E.; Lenstra, D. Fast Pulsing and Chaotic Itinerancy with a Drift in the Coherence Collapse of Semiconductor Lasers. *Phys. Rev. Lett.* **1996**, *76*, 220–223. [CrossRef] [PubMed]
36. Arecchi, F.; Giacomelli, G.; Lapucci, A.; Meucci, R. Two-dimensional representation of a delayed dynamical system. *Phys. Rev. A* **1992**, *45*, R4225–R4228. [CrossRef]
37. Masoller, C. Spatiotemporal dynamics in the coherence collapsed regime of semiconductor lasers with optical feedback. *Chaos* **1997**, *7*, 455–462. [CrossRef] [PubMed]
38. Lang, R.; Kobayashi, K. External optical feedback effects on semiconductor injection laser properties. *IEEE J. Quantum Electron.* **1980**, *16*, 347–355. [CrossRef]
39. Petermann, K. External optical feedback phenomena in semiconductor lasers. *IEEE J. Sel. Top. Quantum Electron.* **1995**, *1*, 480–489. [CrossRef]
40. Sano, T. Antimode dynamics and chaotic itinerancy in the coherence collapse of semiconductor lasers with optical feedback. *Phys. Rev. A* **1994**, *50*, 2719–2726. [CrossRef] [PubMed]
41. Henry, C.; Kazarinov, R. Instability of semiconductor lasers due to optical feedback from distant reflectors. *IEEE J. Quantum Electron.* **1986**, *22*, 294–301. [CrossRef]
42. Porte, X.; Soriano, M.C.; Fischer, I. Similarity properties in the dynamics of delayed-feedback semiconductor lasers. *Phys. Rev. A* **2014**, *89*, 023822. [CrossRef]

43. Kim, B.; Li, N.; Locquet, A.; Citrin, D.S. Experimental bifurcation-cascade diagram of an external-cavity semiconductor laser. *Opt. Express* **2014**, *22*, 2348–2357. [CrossRef]
44. Mørk, J.; Mark, J.; Tromborg, B. Route to chaos and competition between relaxation oscillations for a semiconductor laser with optical feedback. *Phys. Rev. Lett.* **1990**, *65*, 1999–2002. [CrossRef] [PubMed]
45. Masoller, C.; Abraham, N.B. Stability and dynamical properties of the coexisting attractors of an external-cavity semiconductor laser. *Phys. Rev. A* **1998**, *57*, 1313–1322. [CrossRef]
46. Dong, J.-X.; Ruan, J.; Zhang, L.; Zhuang, J.P.; Chan, S.-C. Stable-unstable switching dynamics in semiconductor lasers with external cavities. *Phys. Rev. A* **2021**, *103*, 053524. [CrossRef]
47. Heiligenthal, S.; Dahms, T.; Yanchuk, S.; Jüngling, T.; Flunkert, V.; Kanter, I.; Schöll, E.; Kinzel, W. Strong and Weak Chaos in Nonlinear Networks with Time-Delayed Couplings. *Phys. Rev. Lett.* **2011**, *107*, 234102. [CrossRef]
48. Heiligenthal, S.; Jüngling, T.; D'Huys, O.; Arroyo-Almanza, D.A.; Soriano, M.C.; Fischer, I.; Kanter, I.; Kinzel, W. Strong and weak chaos in networks of semiconductor lasers with time-delayed couplings. *Phys. Rev. E* **2013**, *88*, 012902. [CrossRef]
49. Mørk, J.; Tromborg, B.; Christiansen, P.L. Bistability and low-frequency fluctuations in semiconductor lasers with optical feedback: A theoretical analysis. *IEEE J. Quantum Electron.* **1988**, *24*, 123–133. [CrossRef]
50. Sukow, D.W.; Gardner, J.R.; Gauthier, D.J. Statistics of power-dropout events in semiconductor lasers with time-delayed optical feedback. *Phys. Rev. A* **1997**, *56*, R3370. [CrossRef]
51. Heil, T.; Fischer, I.; Elsäßer, W. Coexistence of low-frequency fluctuations and stable emission on a single high-gain mode in semiconductor lasers with external optical feedback. *Phys. Rev. A* **1998**, *58*, R2672–R2675. [CrossRef]
52. Brunner, D.; Porte, X.; Soriano, M.C.; Fischer, I. Real-time frequency dynamics and high-resolution spectra of a semiconductor laser with delayed feedback. *Sci. Rep.* **2012**, *2*, 732. [CrossRef]
53. Brunner, D.; Soriano, M.C.; Porte, X.; Fischer, I. Experimental Phase-Space Tomography of Semiconductor Laser Dynamics. *Phys. Rev. Lett.* **2015**, *115*, 053901. [CrossRef] [PubMed]
54. Martínez-Llinàs, J.; Porte, X.; Soriano, M.C.; Colet, P.; Fischer, I. Dynamical properties induced by state-dependent delays in photonic systems. *Nat. Commun.* **2015**, *6*, 7425. [CrossRef] [PubMed]
55. Yanchuk, S.; Giacomelli, G. Pattern formation in systems with multiple delayed feedbacks. *Phys. Rev. Lett.* **2014**, *112*, 174103. [CrossRef]
56. Van der Sande, G.; Brunner, D.; Soriano, M.C. Advances in photonic reservoir computing. *Nanophotonics* **2017**, *6*, 561–576. [CrossRef]

Nonlinear Dynamics of Interband Cascade Laser Subjected to Optical Feedback

Abstract: We present a theoretical study of the nonlinear dynamics of a long external cavity delayed optical feedback-induced interband cascade laser (ICL). Using the modified Lang–Kobayashi equations, we numerically investigate the effects of some key parameters on the first Hopf bifurcation point of ICL with optical feedback, such as the delay time (τ_f), pump current (I), linewidth enhancement factor (LEF), stage number (m) and feedback strength (f_{ext}). It is found that compared with τ_f, I, LEF and m have a significant effect on the stability of the ICL. Additionally, our results show that an ICL with few stage numbers subjected to external cavity optical feedback is more susceptible to exhibiting chaos. The chaos bandwidth dependences on m, I and f_{ext} are investigated, and 8 GHz bandwidth mid-infrared chaos is observed.

Keywords: interband cascade laser; mid-infrared chaos; optical feedback; nonlinear dynamics

1. Introduction

As a mid-infrared semiconductor laser, the interband cascade laser (ICL) has made significant progress in the last two decades [1–6]. The RAND Corporation reports that mid-infrared lasers in the 3–5 μm band of the atmospheric transmission window have good atmospheric transmission characteristics, lower transmission losses than other bands, and are less susceptible to weather factors [7]. In addition, the mid-infrared band covers the absorption peaks of many atoms and molecules [8]. Therefore, ICL can be used in applications such as gas detection [9,10], clinical respiratory diagnosis [11] and free-space optical communication [12].

In contrast to the quantum cascade laser (QCL), the ICL is a bipolar device, with the electronic transition of the ICL occurring between the conduction and the valence sub-bands [13]. Therefore, the carrier lifetime of the ICL is in the nanosecond order like in more conventional semiconductor lasers. Furthermore, in recent experimental reports the linewidth enhancement factor of the ICL was found to be about 2.2, which is much higher than that of QCL [14,15]. Both of these characteristics suggest that when subjected to external perturbation, the ICL will exhibit rich nonlinear dynamics. Wang et al.'s recent experiments confirm that with external optical feedback the ICL exhibits periodic oscillations and weak chaos [16]. 450 MHz low frequency oscillation (detector bandwidth limited) chaos was observed. However, the route to chaos and the identification of the means for obtaining strong chaos are open for detailed study. A recent report shows that based on a QCL with external optical feedback, a generated mid-infrared low frequency chaotic oscillation was used to achieve 0.5 Mbit/s message private free-space optical communication [17]. To realize a much higher-speed message secure transmission, strong broadband chaos is essential.

In this paper, modified Lang–Kobayashi equations are used to investigate the dynamics of the ICL subject to external optical feedback. ICLs with short external cavities have been used to affect wavelength tuning [18]. In contrast, and with a view to performing experiments with discrete devices [19–25], we focus on the case of optical feedback from longer external cavities, wherein the external feedback delay time is larger than the oscillation relaxation time of the ICL [26]. The pump current, feedback strength, stage number and linewidth enhancement factor effects on the stability of the ICL with optical feedback are analyzed, and the influence of the stage number, pump current and feedback strength on the bandwidth of chaos are investigated.

2. Theoretical Model

Figure 1a presents an ICL structure having three stages. Initially, the electron transition in the ICL occurs between the first conduction sub-band and the first valence sub-band, as indicated as E_e (blue potential well) and E_h (red potential well) in the left upper corner in Figure 1a [27]. After the first electron transition, the electron reaches the second stage conduction sub-band through interband tunneling and then repeats the electron transition in the second stage and then in the third stage. Figure 1b is the schematic diagram of an ICL subjected to external mirror feedback, where τ_f is the feedback time delay.

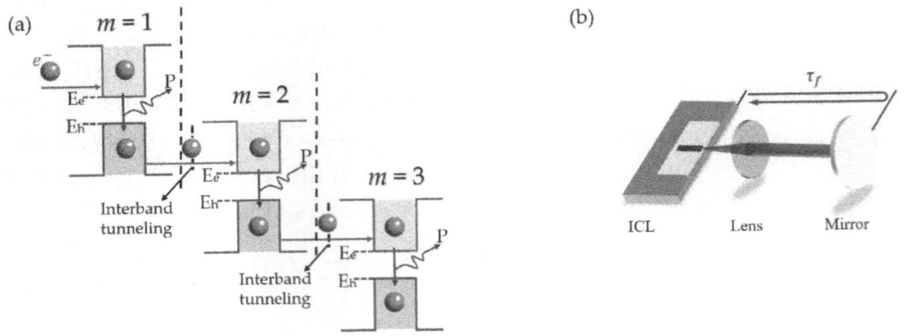

Figure 1. (**a**) 3-stage number structure of interband cascade laser(ICL); (**b**) Schematic diagram of ICL with optical feedback structure.

Using appropriately modified Lang–Kobayashi equations, the rate equations for the ICL with optical feedback are as follows [28,29]:

$$\frac{dN(t)}{dt} = \eta\frac{I}{q} - \Gamma_p v_g g S - \frac{N(t)}{\tau_{sp}} - \frac{N(t)}{\tau_{aug}} \quad (1)$$

$$\frac{dS(t)}{dt} = \left[m\Gamma_p v_g g - \frac{1}{\tau_p}\right]S(t) + m\beta\frac{N(t)}{\tau_{sp}} + 2k\sqrt{S(t)S(t-\tau_f)}cos\theta(t) \quad (2)$$

$$\frac{d\varphi(t)}{dt} = \frac{\alpha_H}{2}\left[m\Gamma_p v_g g - \frac{1}{\tau_p}\right] - k\sqrt{\frac{S(t-\tau_f)}{S(t)}}sin\theta(t) \quad (3)$$

where $N(t)$, $S(t)$ and $\varphi(t)$ respectively represent the per stage carrier number, total photon number of all gain stages and phase of the electric field. M is the number of gain stages, τ_{sp} is the spontaneous radiation lifetime, and τ_{aug} is the Auger recombination lifetime. Since the Auger recombination lifetime τ_{aug} in the ICL is smaller than the spontaneous radiation lifetime τ_{sp}, τ_{aug} must be considered in the current-carrying dynamics. In principle, the Auger coefficient has a carrier density dependence, as in ref. [30]. In this work, we follow [29] in assuming a constant value for the Auger lifetime. η is the current injection efficiency, Γ_p is the optical confinement factor per gain stage, v_g is the group velocity of light, g is the material gain per stage which is given by $g = a_0[N(t) - N_{tr}]/A$. τ_p is the photon lifetime, and k is the feedback coefficient which is given by $k = 2C_l\sqrt{f_{ext}}/\tau_{in}$, where τ_{in} is the internal cavity roundtrip time, f_{ext} is the feedback strength which is defined as the power ratio between the feedback light and the laser output, and C_l is an external coupling coefficient. The external coupling coefficient can be expressed as $C_l = (1 - R)/2\sqrt{R}$, with R being the reflection coefficient of the laser front facet facing the external mirror.

The steady-state solutions for the ICL operating above the threshold current are as follows [29]:

$$N = \frac{1}{m}\frac{A}{\Gamma_p v_g a_0 \tau_p} + N_{tr} \tag{4}$$

$$S_0 = m\eta\tau_p \frac{I - I_{\text{th}}}{q} \tag{5}$$

$$I_{\text{th}} = \frac{q}{\eta}\left(\frac{1}{m}\frac{A}{\Gamma_p v_g a_0 \tau_p} + N_{tr}\right)\left(\frac{1}{\tau_{sp}} + \frac{1}{\tau_{aug}}\right) \tag{6}$$

The descriptions of other symbols and their values used in the simulation are given in Table 1, as taken from [14,27,29–32]. The integration time step in the simulation is 0.1 ps.

Table 1. ICL parameters used in the simulations.

Parameter	Symbol	Value
Cavity length	L	2 mm
Cavity width	W	4.4 μm
Group velocity of light	v_g	8.38×10^7 m/s
Wavelength	λ	3.7 μm
Active area	A	8.8×10^{-9} m^2
Facet reflectivity	R	0.32
Refractive index	n_r	3.58
Optical confinement factor	Γ_p	0.04
Stage number	m	3–20
Injection efficiency	H	0.64
Photon lifetime	τ_p	10.5 ps
Spontaneous emission time	τ_{sp}	15 ns
Auger lifetime	τ_{aug}	1.08 ns
Threshold current	I_{th}	19.8 mA (m = 5)
Feedback strength	f_{ext}	0~30%
Time delay	τ_f	1.5~5.0 ns
Differential gain	a_0	2.8×10^{-10} cm
Transparent carrier number	N_{tr}	6.2×10^7
Spontaneous emission factor	β	1×10^{-4}
Linewidth enhancement factor	α_H	2.2

3. Numerical Results

We calculate the carrier number and photon number for an increasing bias current, as shown in Figure 2a,b, respectively. It is found that the number of stages m has little effect on the carrier number (in Figure 2a) but that it influences the photon number (in Figure 2b). For relatively large numbers of stages such as m = 10 (red dashed curve), the output power is much higher than in the case of m = 5 (black solid curve), as shown in Figure 2b.

3.1. Route to Chaos

Figure 3 shows the output of the ICL with external optical feedback as the feedback strength increases, for the case of m = 5. The ICL output is stable when the feedback strength f_{ext} ranges from 0 to 0.019%. Without feedback, that is f_{ext} = 0, the relaxation oscillation frequency f_R can be observed from the RF spectrum in Figure 3(d-i) to be 1.02 GHz. This is in accordance with the value 1.035 GHz obtained by calculating the relaxation oscillation frequency via the relation $f_R = (G_0 S_0/\tau_p)^{-1/2}/(2\pi)$, where $G_0 = \Gamma_p v_g a_0/A$ and S_0 are found from Equation (5). As the feedback strength increases, the ICL enters into period-1 dynamics (ii), quasi-periodic dynamics (iii), weak chaos oscillations (iv), and then displays strong chaos (v). The frequency of period-1 oscillations is 1.03 GHz, as shown in Figure 3(d-ii), which is a little larger than the relaxation oscillation frequency shown in Figure 3(a-ii). As the feedback strength increases, a quasi-periodic oscillation is found, which is confirmed by the RF spectrum and phase diagram; that is, more frequencies are induced in Figure 3(d-iii) and more loops are found in the phase in Figure 3(c-iii), respectively. An irregular laser intensity oscillation is observed where the ICL enters into weak chaos, as shown in Figure 3(a-iv). Although the highest peak in the RF spectrum is still at 1.03 GHz, more frequency components appear, as presented in Figure 3(d-iv). For a further increase in the feedback strength, more complex nonlinear dynamics are introduced, hence achieving strong mid-infrared optical chaos, as shown in Figure 3(a-v–e-v). In the optical spectra of the chaos shown in Figure 3(e-v), many external cavity modes with a frequency interval around 410 MHz are found. This can be confirmed from the auto-correlation functions shown in Figure 3(b-v), where the sidelobe peak is at 2.4 ns, corresponding to a cavity length of 36 cm.

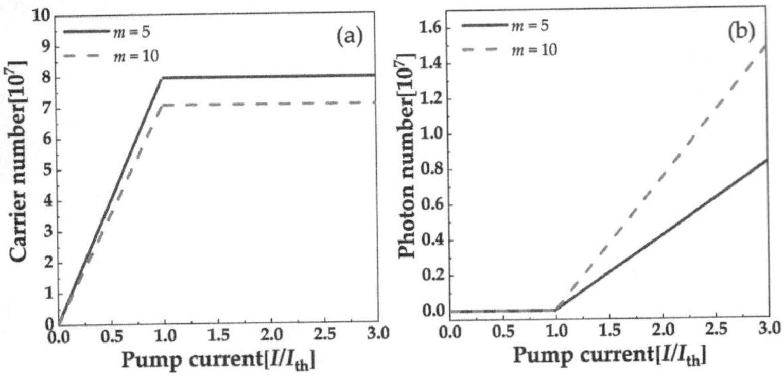

Figure 2. (**a**) Carrier number and (**b**) photon number vs. pump current. Black solid and red dashed curves represent stage numbers $m = 5$ and $m = 10$, respectively.

By using bifurcation diagrams, the dynamics of the ICL for increasing feedback strength can be obtained, as shown in Figure 4(a-i) with $m = 5$ and Figure 4(b-i) with $m = 10$. Maximum Lyapunov exponents [33,34] can be used to determine whether the ICL output with external optical feedback is chaotic (red dots) or not (blue dots), as shown in Figure 4(a-ii,b-ii). As shown in Figure 4(a-i), the route to chaos for a 10-stage ICL with external optical feedback is from stable (S) to period-1 (P1), then quasi-periodic (QP) and then chaos (C). This route is different from that of a 10-stage ICL where multiple-periodic (MP) oscillations are observed. The number of stages m has an effect on the photon number and phase, as presented in Equations (2) and (3), which results in different routes to chaos.

3.2. *Hopf Bifurcation Analysis*

In this section, we explore the stability of ICLs with external optical feedback. We first compare two stage numbers, which are $m = 5$ and $m = 10$, to reveal the effects of the time delay, bias current and linewidth enhancement factor on the Hopf bifurcation points. Then, we ascertain how the Hopf bifurcation changes for stage numbers m in the range of 3 to 20.

The pump current is set a little above the threshold current, that is $1.1I_{th}$. Figure 5 shows that the external cavity delay has little effect on the Hopf bifurcation point. As the external cavity delay increases from 1.5 ns to 5.0 ns, the Hopf bifurcation point values, that is the feedback power ratio where the ICL enters into P1, are around 0.02% to 0.33%. As shown in Figure 5, there is a periodic dependence of the Hopf bifurcation point on the external cavity delay time; for a 2.1 ns delay time, the value of the Hopf bifurcation point reaches a maximum and the second peak is at 3.0 ns. As the delay time increases, the differences between the Hopf bifurcation points reduce.

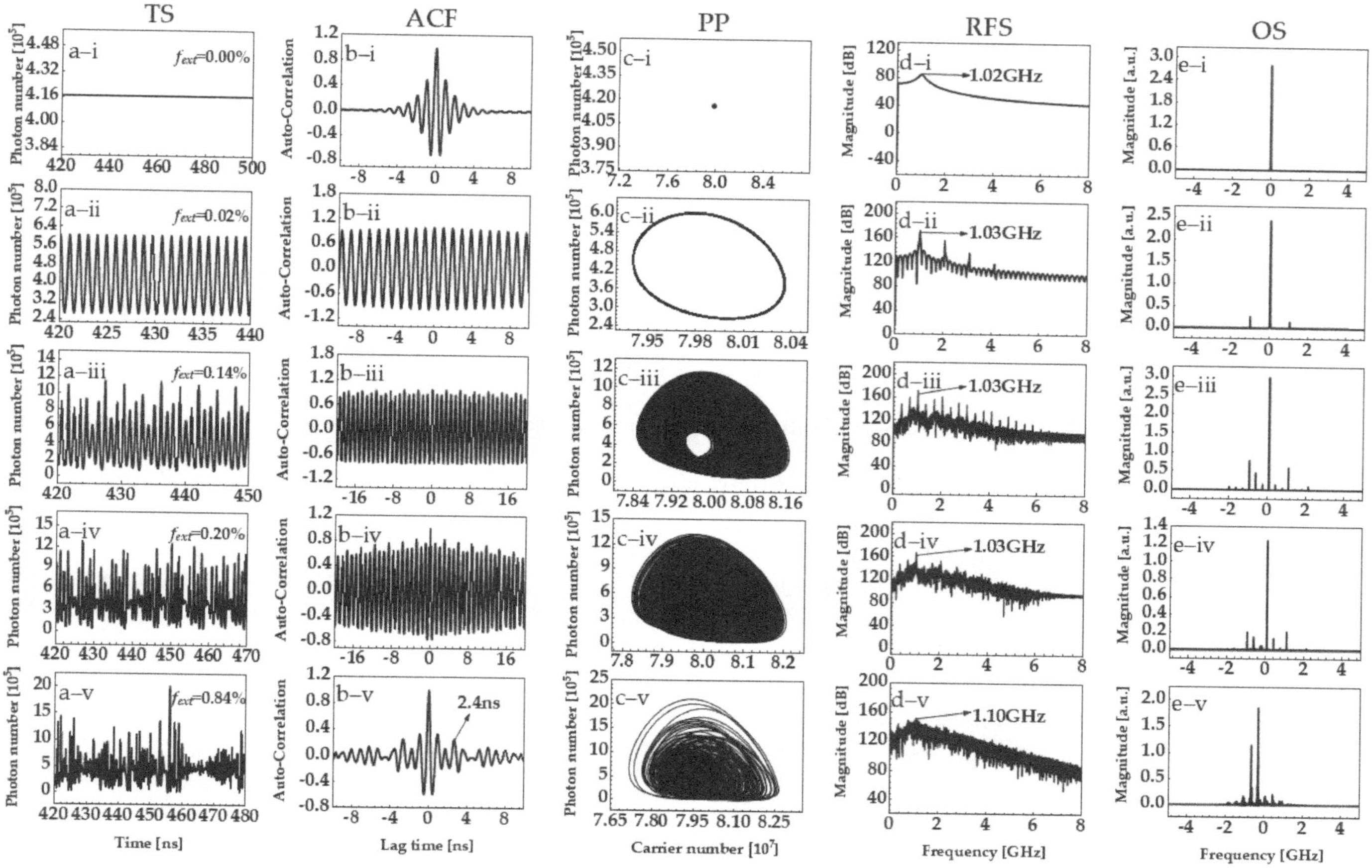

Figure 3. Output of ICL with external optical feedback as feedback strength increases with $m = 5$, $I = 1.1I_{th}$, and $\tau_f = 2.4$ ns; from the top down, i: $f_{ext} = 0.0\%$, ii: $f_{ext} = 0.02\%$, iii: $f_{ext} = 0.14\%$, iv: $f_{ext} = 0.20\%$, v: $f_{ext} = 0.84\%$; columns (**a–e**) are time series (TS), autocorrelation curve functions (ACF), phase portrait (PP), radio-frequency spectrum (RFS) and optical spectra (OS), respectively.

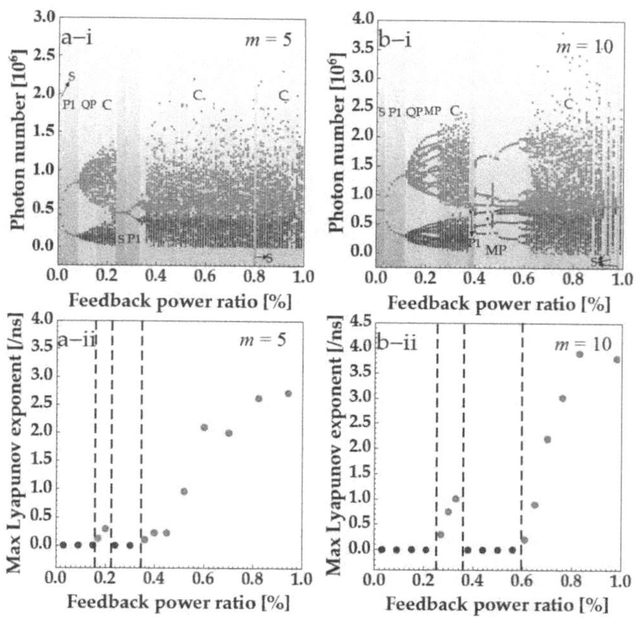

Figure 4. Bifurcations (i) and Maximum Lyapunov exponents (ii) of ICL output with external optical feedback under $I = 1.1I_{th}$, $\tau_f = 2.4$ ns. (**a**) $m = 5$, (**b**) $m = 10$.

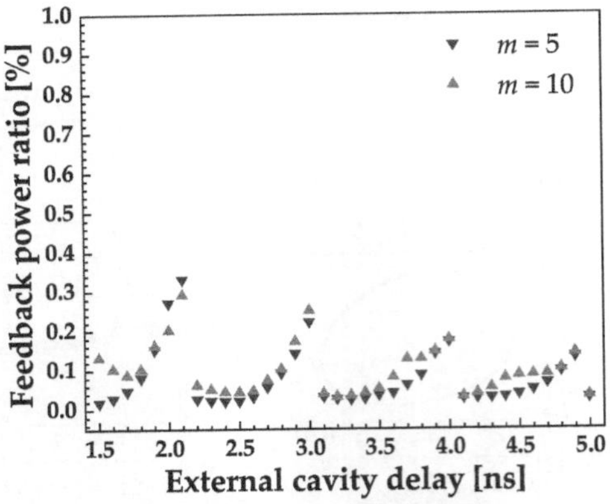

Figure 5. Hopf bifurcation point vs. time delay τ_f for $I = 1.1I_{th}$.

To illustrate the pump current effects, the dependence of the Hopf bifurcation point on bias current is investigated. In Figure 6a,b, two external cavity delays are compared, viz $\tau_f = 2.0$ ns and $\tau_f = 2.4$ ns. As the pump current increases, the Hopf bifurcation point gradually increases for both m = 5 (squares) and m = 10 (circles) cases, as well as for $\tau_f = 2.0$ ns and $\tau_f = 2.4$ ns. Compared with $\tau_f = 2.4$ ns, at $\tau_f = 2.0$ ns there is a need for a larger feedback power ratio to enable the ICL to enter an unstable state, which is around 1.3 times that of τ_f = 2.4 ns for a pump current of $3I_{th}$. Since these two delays show similar trends for the Hopf bifurcation point versus bias current, we focus our attention on $\tau_f = 2.4$ ns in the following results.

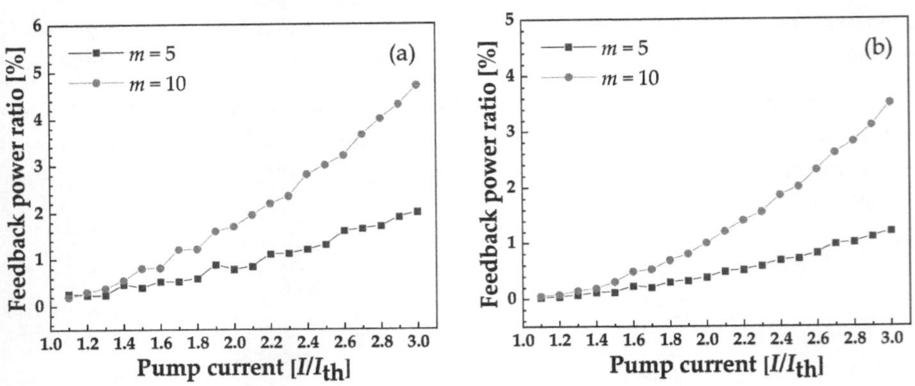

Figure 6. Hopf bifurcation point vs. bias current. (**a**) $\tau_f = 2.0$ ns, (**b**) $\tau_f = 2.4$ ns.

It is well-appreciated that the linewidth enhancement factor, α_H, plays a crucial role in semiconductor laser nonlinear dynamics. For common quantum well laser diodes, α_H is in the range of 2.0 to 5.0 [35,36]. A recent report shows that the below-threshold linewidth enhancement factor of ICL is in the range of 1.1–1.4 [14]. Here, we calculate the Hopf bifurcation point versus linewidth enhancement factor, which ranges from 1 to 5, as shown in Figure 7. As the linewidth enhancement factor increases from 1 to 2, the Hopf bifurcation point value reduces rapidly and then tends to be stable as α_H increases further. Thus, as expected, a small α_H imparts the ICL with considerable dynamic stability.

For ICLs, the number of stages, m, is usually less than 20. Figure 8 shows the Hopf bifurcation point value versus stage number with $I = 1.5I_{th}$, $\tau_f = 2.4$ ns. It is seen that an increased stage number gives rise to an exponential increasing Hopf bifurcation point value, which indicates that ICLs with a large number of stages are more stable.

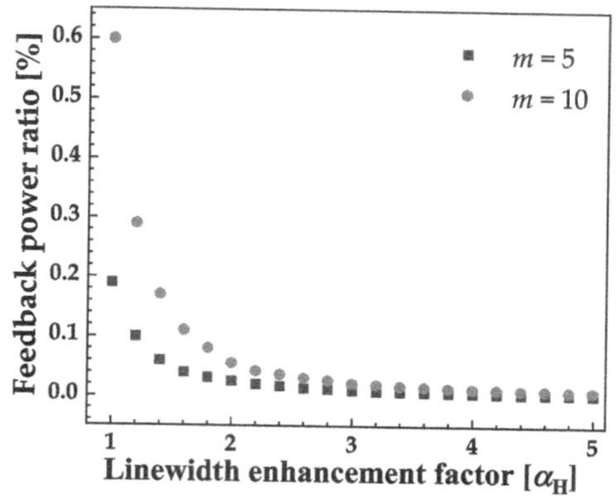

Figure 7. Hopf bifurcation point vs. linewidth enhancement factor under $I = 1.1I_{th}$, $\tau_f = 2.4$ ns.

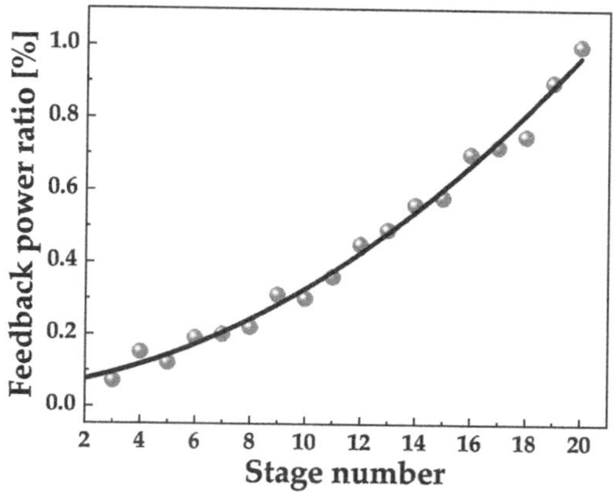

Figure 8. Hopf bifurcation point value vs. stage number m with $I = 1.5I_{th}$, $\tau_f = 2.4$ ns.

3.3. Bandwidth of Chaos

A broadband RF spectrum is one of the significant characteristics of chaos. The bandwidth of chaos determines the range resolution of chaotic lidar [37], the bit rate of random sequence generation [38] and the transmission rate of optical chaos communications [39]. We use a conventional definition of bandwidth of chaotic signals as the frequency span between the DC and the frequency where 80% of the energy is contained [40], and investigate the bandwidth of mid-infrared chaos. Similar to regular quantum well laser diodes, the bandwidth of chaos from ICL with external optical feedback increases as the feedback power ratio increases, as shown in Figure 9.

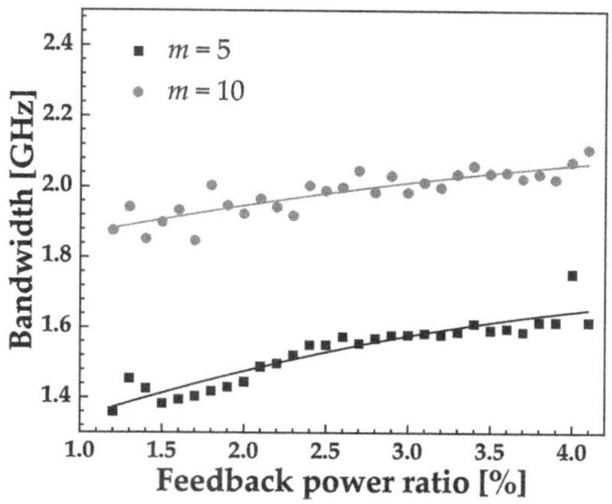

Figure 9. Bandwidth of chaos vs. feedback power ratio under $I = 1.1I_{th}$, $\tau_f = 2.4$ ns, $m = 5$ (black squares) and $m = 10$ (red circles).

Increasing the pump current and hence the relaxation oscillation frequency is expected to enhance the bandwidth of chaos, and this is confirmed in Figure 10. Here, we notice that mid-infrared chaos from ICL with different stage numbers has the same increasing tendency. Once the pump current increases to $2I_{th}$, a 6 GHz chaos bandwidth is obtained when the stage number is 10, and a 8 GHz chaos bandwidth can be achieved with $3I_{th}$, as shown in Figure 10.

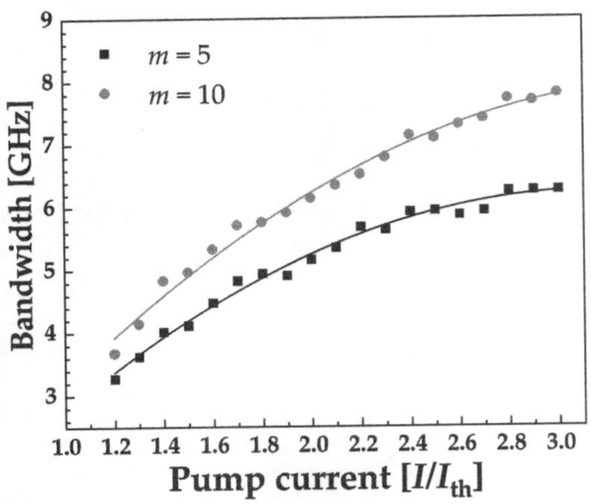

Figure 10. Bandwidth of chaos vs. bias current when f_{ext} = 28% and τ_f = 2.4 ns with m = 5 (black squares), m = 10 (red circles).

Figure 11 respectively presents the time series, auto-correlation, phase diagram, RF spectrum and optical spectral of mid-infrared chaos for m = 5 (in Figure 11a) and m = 10 (in Figure 11b). This indicates that for relatively high number of stages the bandwidth of chaos from ICLs is further enhanced, as shown in Figure 11(a-iv,b-iv). The enhanced bandwidth of the chaos is due to the relaxation oscillation frequency of the ICL increasing with the number of stages.

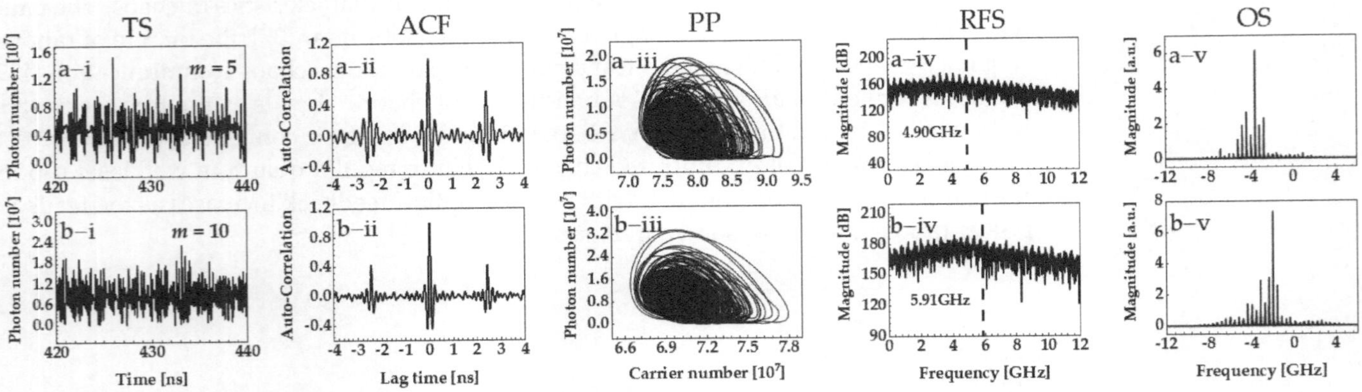

Figure 11. Output chaos of ICL with external optical feedback as stage number increases with $I = 1.9I_{th}$, τ_f = 2.4 ns and f_{ext} = 28%. (**a**) m = 5, (**b**) m = 10. Columns (i–v) are TS, ACF, PP, RFS, OS, respectively.

The stage number effect is presented in Figure 12, where m ranges from 3 to 20 and the pump currents are $1.1I_{th}$ (black squares) and $2I_{th}$ (red circles). Although the increase in bandwidth for the relatively high pump current, $2I_{th}$, is faster than that of the relatively low pump current, $1.1I_{th}$, the tendency of the stage number effects is the same, that is the bandwidth of chaos increases as the stage number increases. This tendency is verified in Figures 9–11.

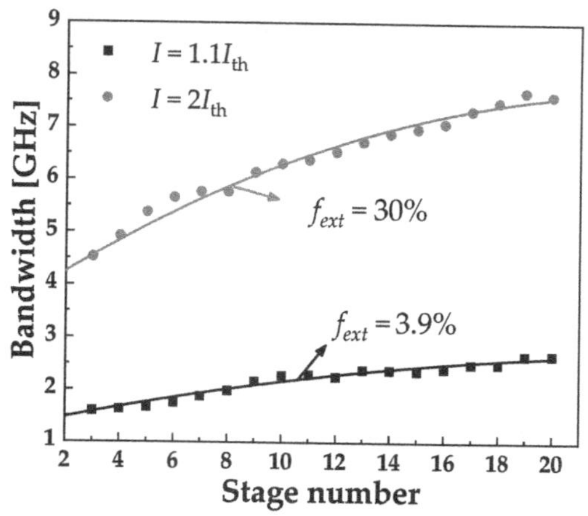

Figure 12. Bandwidth of chaos vs. stage number $I = 1.1I_{th}$ (black squares) and $I = 2I_{th}$ (red circles) with $\tau_f = 2.4$ ns.

By using Maximum Lyapunov exponents, we distinguish chaos from QP, and calculate the 80% energy power bandwidth versus feedback power ratio and pump current, as shown in Figure 13. By using dashed curves, we distinguish QP and C, while the white region represents the stable(S) output. It is found, in Figure 13a, that S is observed for a small number of stages $m = 5$ when the bias current is below $1.2I_{th}$ and the feedback power ratio is in the range of 4% to 30%. Stable output is also found in the top left corner of Figure 13a. For a large stage number, that is $m = 10$, S is observed when the pump current is small, below $1.2I_{th}$, and the feedback power ratio is around 28%, as indicated in the bottom right corner of Figure 13b. S is also observed for $m = 10$ when the bias current is higher than $2.16I_{th}$ and the feedback power ratio is less than 4%, as seen in Figure 13b. These results indicate that an ICL with a relatively high bias current will exhibit S, QP and then enter into C as the feedback power ratio increases. As the stage number increases from 5 to 10 and the ICL is subject to a relatively high bias current, both S and QP regions are extended, as shown towards the left of Figure 13a,b. This confirms that for relatively few stage numbers an ICL with external optical feedback is amenable to exhibiting chaos. The results for both cases confirm that in order to obtain broadband mid-infrared chaos, one needs a relatively high pump current as well as a large feedback power ratio. Furthermore, a 8 GHz bandwidth of mid-infrared chaos can be obtained for $m = 10$, as shown in the top right corner of Figure 13b.

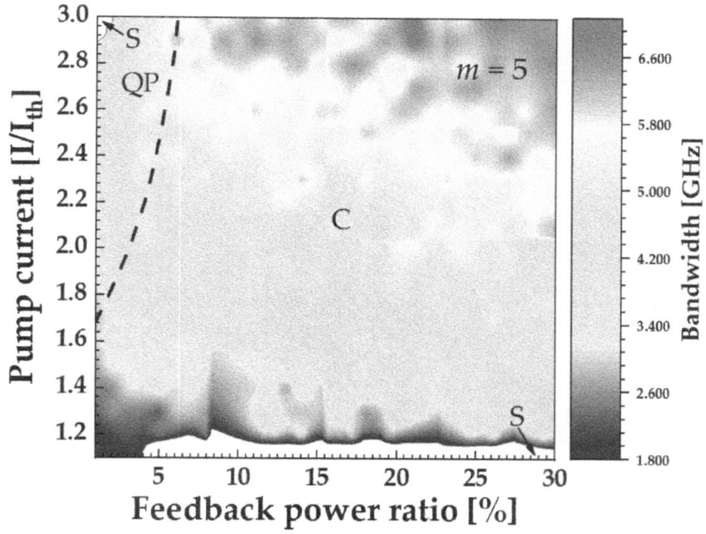

(a)

Figure 13. *Cont.*

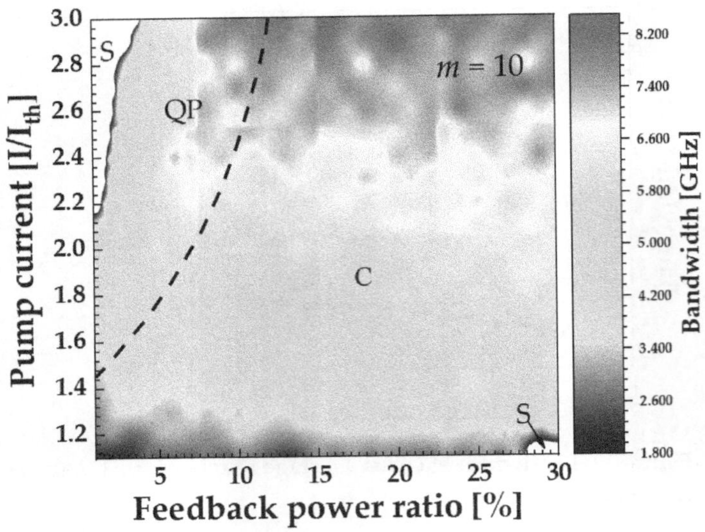

(b)

Figure 13. Bandwidth vs. feedback power ratio and pump current, with τ_f = 2.4 ns. (**a**) m = 5, (**b**) m = 10.

4. Discussion

A mid-infrared chaotic laser will be a promising source for implementing secure free-space optical communication. In this paper, we show how to obtain broadband mid-infrared chaos using an ICL with long-cavity optical feedback. The analysis of the laser stability is effected by identifying the Hopf bifurcation point. From such considerations we find that ICLs with a relatively small number of stages are more unstable and thus susceptible to exhibiting chaos. The chaos bandwidth of the ICL is related to the relaxation oscillation frequency, which increases linearly with the cube root of the stage number and laser pump current. Therefore, to obtain broadband chaos from an ICL there is a need for a high pump current, large number of stages as well as strong optical feedback strength. Our calculations show that 8 GHz bandwidth mid-infrared chaos can be generated using a 10-stage ICL biased at three times the threshold current. With the availability of such broadband mid-infrared chaos, it is expected that of-order Gbit/s secure free-space optical communication is feasible.

Abbreviations

The following abbreviations are used in this manuscript:

ICL	interband cascade laser
LEF	linewidth enhancement factor
QCL	quantum cascade laser
TS	time series
ACF	autocorrelation curves functions
PP	phase portrait
RFS	radio-frequency spectrum
OS	optical spectral
S	stable
P1	period-1
QP	quasi-periodic
MP	multiple-periodic
C	chaos

References

1. Lin, C.-H.; Yang, R.Q.; Zhang, D.; Murry, S.; Pei, S.; Allerman, A.; Kurtz, S. Type-II interband quantum cascade laser at 3.8 μm. *Electron. Lett. Abbrevi.* **1997**, *33*, 598–599. [CrossRef]
2. Yang, R.Q.; Bradshaw, J.L.; Bruno, J.D. Room temperature type-II interband cascade. *Appl. Phys. Lett.* **2002**, *81*, 397–399. [CrossRef]
3. Yang, R.Q.; Hill, C.J.; Yang, B.H.; Wong, C.M.; Muller, R.E.; Echternach, P.M. Continuous-wave operation of distributed feedback interband cascade lasers. *Appl. Phys. Lett.* **2004**, *84*, 3699–3701. [CrossRef]
4. Kim, M.; Canedy, C.L.; Bewley, W.W.; Kim, C.S.; Lindle, J.R.; Abell, J.; Vurgaftman, I.; Meyer, J.R. Interband cascade laser emitting at λ = 3.75 μm in continuous wave above room temperature. *Appl. Phys. Lett.* **2008**, *92*, 191110. [CrossRef]
5. Bagheri, M.; Frez, C.; Sterczewski, L.; Gruidin, I.; Fradet, M.; Vurgaftman, I.; Canedy, C.L.; Bewley, W.W.; Merritt, C.D.; Kim, C.S.; et al. Passively mode-locked interband cascade optical frequency combs. *Sci. Rep.* **2018**, *8*, 3322. [CrossRef] [PubMed]
6. Yang, H.; Yang, R.Q.; Gong, J.; He, J.-J. Mid-Infrared Widely Tunable Single-Mode Interband Cascade Lasers Based on V-Coupled Cavities. *Opt. Lett.* **2020**, *45*, 2700–2702. [CrossRef] [PubMed]
7. Chen, C.C. *Attenuation of Electromagnetic Radiation by Haze, Fog, Clouds, and Rain*; Rand Corp.: Santa Monica, CA, USA, 1975; pp. 1–39.
8. Vurgaftman, I.; Weih, R.; Kamp, M.; Meyer, J.R.; Canedy, C.L.; Kim, C.S.; Kim, M.; Bewley, W.W.; Merritt, C.D.; Abell, J.; et al. Interband cascade lasers. *J. Phys. D Appl. Phys.* **2005**, *48*, 123001. [CrossRef]
9. Horstjann, M.; Bakhirkin, Y.; Kosterev, A.; Curl, R.; Tittel, F.; Wong, C.; Hill, C.; Yang, R. Formaldehyde sensor using interband cascade laser based quartz-enhanced photoacoustic spectroscopy. *Appl. Phys. B* **2004**, *79*, 799–803. [CrossRef]
10. Wysocki, G.; Bakhirkin, Y.; So, S.; Tittel, F.K.; Hill, C.J.; Yang, R.Q.; Fraser, M.P. Dual interband cascade laser based trace-gas sensor for environmental monitoring. *Appl. Opt.* **2007**, *46*, 8202–8209. [CrossRef]
11. Risby, T.H.; Tittel, F.K. Current status of midinfrared quantum and interband cascade lasers for clinical breath analysis. *Opt. Eng.* **2010**, *49*, 111123.
12. Soibel, A.; Wright, M.W.; Farr, W.H.; Keo, S.A.; Hill, C.J.; Yang, R.Q.; Liu, H.C. Midinfrared Interband Cascade Laser for Free Space Optical Communication. *IEEE Photonics Technol. Lett.* **2010**, *22*, 121–123. [CrossRef]
13. Yang, R.Q.; Li, L.; Jiang, Y.C. Interband cascade laser: From original concept to actual device. *Prog. Phys.* **2014**, *34*, 169–190.
14. Deng, Y.; Zhao, B.B.; Wang, C. Linewidth broadening factor of an interband cascade laser. *Appl. Phys. Lett.* **2019**, *115*, 181101. [CrossRef]
15. Green, R.P.; Xu, J.-H.; Mahler, L.; Tredicucci, A.; Beltram, F.; Giuliani, G.; Beere, H.E.; Ritchie, D. Linewidth enhancement factor of terahertz quantum cascade lasers. *Appl. Phys. Lett.* **2008**, *92*, 071106. [CrossRef]
16. Deng, Y.; Fan, Z.F.; Wang, C. Optical feedback induced nonlinear dynamics in an interband cascade laser. *Proc. SPIE* **2021**, *11680*, 116800J.
17. Spitz, O.; Herdt, A.; Wu, J.; Maisons, G.; Carras, M.; Wong, C.-W.; Elsäßer, W.; Grillot, F. Private communication with quantum cascade laser photonic chaos. *Nat. Commun.* **2021**, *12*, 3327. [CrossRef]
18. Caffey, D.; Day, T.; Kim, C.S.; Kim, M.; Vurgaftman, I.; Bewley, W.W.; Lindle, J.R.; Canedy, C.L.; Abell, J.; Meyer, J.R. Performance characteristics of a continuous wave compact widely tunable external cavity interband cascade lasers. *Opt. Express* **2010**, *18*, 15691. [CrossRef]
19. Wang, A.; Li, P.; Zhang, J.; Zhang, J.; Li, L.; Wang, Y. 4.5 Gbps high-speed real-time physical random bit generator. *Opt. Express* **2013**, *21*, 20452–20462. [CrossRef]
20. Wang, A.; Yang, Y.; Wang, B.; Zhang, B.; Li, L.; Wang, Y. Generation of wideband chaos with suppressed time-delay signature by delayed self-interference. *Opt. Express* **2013**, *21*, 8701–8710. [CrossRef] [PubMed]

21. Wang, L.; Wang, D.; Gao, H.; Guo, Y.; Wang, Y.; Hong, Y.; Shore, K.A.; Wang, A. Real-time 2.5-Gb/s correlated random bit generation using synchronized chaos induced by a common laser with dispersive feedback. *IEEE J. Quantum Electron.* **2020**, *56*, 1–8. [CrossRef]
22. Zhang, T.; Jia, Z.; Wang, A.; Hong, Y.; Wang, L.; Guo, Y.; Wang, Y. Experimental observation of dynamic-state switching in VCSELs with optical feedback. *IEEE Photonics Technol. Lett.* **2021**, *33*, 335–338. [CrossRef]
23. Jumpertz, L.; Schires, K.; Carras, M.; Sciamanna, M.; Grillot, F. Chaotic light at mid-infrared wavelength. *Light Sci. Appl.* **2016**, *5*, e16088. [CrossRef]
24. Spitz, O.; Wu, J.; Carras, M.; Wong, C.-W.; Grillot, F. Low-frequency fluctuations of a mid-infrared quantum cascade laser operating at cryogenic temperatures. *Laser Phys. Lett.* **2018**, *15*, 116201. [CrossRef]
25. Jumpertz, L.; Carras, M.; Grillot, F. Regimes of external optical feedback in 5.6 μm distributed feedback mid-infrared quantum cascade lasers. *Appl. Phys. Lett.* **2014**, *105*, 131112. [CrossRef]
26. Panajotov, K.; Sciamanna, M.; Arteaga, M.A.; Thienpont, H. Optical feedback in vertical-cavity surface-emitting lasers. *IEEE J. Sel. Top. Quantum Electron.* **2013**, *19*, 1700312. [CrossRef]
27. Yang, R.Q. Mid-infrared interband cascade lasers based on type-II heterostructures. *Microelectron. J.* **1999**, *30*, 1043–1056. [CrossRef]
28. Lang, R.; Kobayashi, K. External optical feedback effects on semiconductor injection laser properties. *IEEE J. Quantum Electron.* **1980**, *16*, 347–355. [CrossRef]
29. Deng, Y.; Wang, C. Rate Equation modeling of interband cascade lasers on modulation and noise dynamics. *IEEE J. Quantum Electron.* **2020**, *56*, 2300109. [CrossRef]
30. Bewley, W.W.; Lindle, J.R.; Kim, C.S.; Kim, M.; Canedy, C.L.; Vurgaftman, I.; Meyer, J.R. Lifetimes and Auger coefficients in type-II W interband cascade lasers. *Appl. Phys. Lett.* **2008**, *93*, 041118. [CrossRef]
31. Vurgaftman, I.; Canedy, C.L.; Kim, C.S.; Kim, M.; Bewley, W.W.; Lindle, J.R.; Abell, J.; Meyer, J.R. Mid-infrared interband cascade lasers operating at ambient temperatures. *New J. Phys.* **2009**, *11*, 125015. [CrossRef]
32. Vurgaftman, I.; Canedy, C.L.; Kim, C.S.; Kim, M.; Bewley, W.W.; Lindle, J.R.; Abell, J.; Meyer, J.R. Mid-IR type-II interband cascade lasers. *IEEE J. Sel. Top. Quantum Electron.* **2011**, *17*, 1435–1444. [CrossRef]
33. Wolf, A.; Swift, J.B.; Swinney, H.L.; Vastano, J.A. Determining Lyapunov exponents from a time-series. *Phys. D Nonlinear Phenomena.* **1985**, *16*, 285–317. [CrossRef]
34. Wolf, A. Wolf Lyapunov Exponent Estimation from A Time Series. MATLAB Central File Exchange. Available online: https://www.mathworks.com/matlabcentral/fileexchange/48084-wolf-lyapunov-exponent-estimation-from-a-time-series (accessed on 5 January 2021).
35. Wang, D.; Wang, L.; Li, P.; Zhao, T.; Jia, Z.; Gao, Z.; Guo, Y.; Wang, Y.; Wang, A. Bias Current of Semiconductor Laser-An Unsafe Key for Secure Chaos Communication. *Photonics* **2019**, *6*, 59. [CrossRef]
36. Huang, Y.; Zhou, P.; Li, N.Q. Broad tunable photonic microwave generation in an optically pumped spin-VCSEL with optical feedback stabilization. *Opt. Lett.* **2021**, *46*, 3147–3150. [CrossRef]
37. Wang, B.; Wang, Y.; Kong, L.; Wang, A. Multi-target real-time ranging with chaotic laser radar. *Chin. Opt. Lett.* **2008**, *6*, 868–870. [CrossRef]
38. Li, P.; Guo, Y.; Guo, Y.; Fan, Y.; Guo, X.; Liu, X.; Shore, K.A.; Dubrova, E.; Xu, B.; Wang, Y.; et al. Self-balanced real-time photonic scheme for ultrafast random number generation. *APL Photonics* **2018**, *3*, 061301. [CrossRef]
39. Wang, L.; Mao, X.; Wang, A.; Wang, Y.; Gao, Z.; Li, S.-S.; Yan, L. Scheme of coherent optical chaos communication. *Opt. Lett.* **2020**, *45*, 4762. [CrossRef]
40. Lin, F.Y.; Liu, J.M. Nonlinear dynamical characteristics of an optically injected semiconductor laser subject to optoelectronic feedback. *Opt. Commun.* **2003**, *221*, 173–180. [CrossRef]

Dynamics of Semiconductor Lasersunder External Optical Feedback from Both Sides of the Laser Cavity

Abstract: To increase the spectral efficiency of coherent communication systems, lasers with ever-narrower linewidths are required as they enable higher-order modulation formats with lower bit-error rates. In particular, semiconductor lasers are a key component due to their compactness, low power consumption, and potential for mass production. In field-testing scenarios their output is coupled to a fiber, making them susceptible to external optical feedback (EOF). This has a detrimental effect on its stability, thus it is traditionally countered by employing, for example, optical isolators and angled output waveguides. In this work, EOF is explored in a novel way with the aim to reduce and stabilize the laser linewidth. EOF has been traditionally studied in the case where it is applied to only one side of the laser cavity. In contrast, this work gives a generalization to the case of feedback on both sides. It is implemented using photonic components available via generic foundry platforms, thus creating a path towards devices with high technology-readiness level. Numerical results shows an improvement in performance of the double-feedback case with respect to the single-feedback case. In particularly, by appropriately selecting the phase of the feedback from both sides, a broad stability regime is discovered. This work paves the way towards low-cost, integrated and stable narrow-linewidth integrated lasers.

Keywords: laser dynamics; optical feedback; narrow-linewidth lasers; semiconductor lasers; laser stability

1. Introduction

The effect of external optical feedback (EOF) on diode laser dynamics has been extensively studied for the past half century [1,2]. EOF has been proven to affect laser performance, showing regimes that can aid in linewidth reduction [3–6], as well as others responsible for highly unstable behavior, from mode hopping to the case of coherence collapse [7–12]. Methods to improve laser stability thus need to take EOF into account, as even weak feedback can be detrimental. A traditional approach to mitigate its effects is to include an off-chip isolator at the laser output. Yet, this component negatively impacts the dimensions of packaged devices as well as fabrication times and costs. As such, research is ongoing to develop an integrated solution that can minimize the negative effects of EOF. Efforts include adjusting the feedback phase to tune into line-narrowing regimes [13,14], using unidirectional phase modulators [15,16], reducing the linewidth enhancement factor, e.g., using quantum dots [17–19], employing electromagnetic effects [20,21], harnessing the mode propagation properties of ring lasers [22,23], or the extended cavity approach [24–27].

Double external feedback on the same side of the laser cavity has been previously explored for chaos stabilization in Reference [28]. Its results numerically show that, by introducing an additional feedback term, a chaotic regime can be stabilized in terms of output power, where a robust stable region for a wide parameter range can be seen. Furthermore, linewidth is numerically shown to be reduced with respect to the single feedback case. However, it lacks an analytical expression for the linewidth, and it does not explore the effect of the added term on other optical feedback regimes, which limits the scope of the analysis [7]. Dual external cavities have also been explored to maximize the sensitivity of self-mixing interferometers for sensing applications [29–31]. These studies focus on improving the interferometric signal in terms of output power characteristics, yet they do not study the effect EOF has on linewidth, which is crucial for other applications such as coherent communications [32] or frequency metrology [33]. Furthermore, in all works exploring dual external cavities, both external reflections return to the same side of the laser cavity and thus have the same propagation direction. Moreover, phase shifts in the external phase cavity are not accounted for, thus ignoring, e.g., possible phase shifts at the mirrors.

The established line of thought relies on the key assumption that feedback is introduced from only one side of the laser cavity. Current integration technologies make this assumption obsolete, as they allow for arbitrarily complex design geometries with a variety of functionalities, such as tunability and modulation, while maintaining narrow-linewidth performance [34–38]. Consequently, this work aims to extend the theoretical foundations of EOF to the case of feedback coupling into the laser cavity from both sides. This system is studied to obtain and analyze its dynamic rate equations. Furthermore, the frequency noise power spectral density and subsequently the intrinsic linewidth's dependence on feedback is computed. The Lang–Kobayashi approach is used [39], where an additional term to account for the extra feedback term is included. In a similar fashion, to obtain analytical solutions both small-signal and weak feedback conditions are assumed. The obtained equations are then numerically solved. Results show that feedback-insensitivity can be achieved by tuning the feedback parameters. In contrast with previous works, this can be accomplished in a monolithic platform by including a single active component in a straightforward geometric configuration. As such, design complexity is reduced which enables devices with a compact form factor. Furthermore, existing methods for laser fabrication are suitable for realizing the proposed device. In particular, the proposed laser system can be made with mature photonic integrated components available in generic foundry platforms [40,41], thus creating a path towards devices with high technology-readiness level while maintaining low cost and size.

2. Rate Equations Model

This section includes a derivation of the dynamic equations of a laser cavity with EOF coupled into the laser cavity from both sides. Starting from the Lang–Kobayashi model [39], the lasing frequency and threshold gain shifts due to feedback are obtained, as well as an analytical expression for the frequency noise power spectral density from which the change of intrinsic linewidth can be computed.

This work proposes a revised laser system, as shown in Figure 1. The laser cavity of length L is delimited by two mirrors with complex reflection coefficients ρ_1 and ρ_2, respectively. Assuming two interfaces at each side of the main cavity, two additional back-reflections ($\rho_{1,\mathrm{ext}}$ and $\rho_{2,\mathrm{ext}}$) are considered and accounted for by computing effective reflection coefficients. The case of weak feedback is discussed here, for which:

$$|\rho_{j,\mathrm{ext}}| \ll 1. \tag{1}$$

The following parameters are introduced:

$$\kappa_j \equiv \frac{1-\rho_j^2}{\rho_j}\frac{|\rho_{j,\mathrm{ext}}|}{t_{\mathrm{cav}}} \tag{2a}$$

$$\phi_j \equiv \omega_{\mathrm{FB}} t_j + \phi_{\mathrm{m}_j} \tag{2b}$$

$$\delta\omega \equiv \omega_{\mathrm{FB}} - \omega_{\mathrm{ref}} \tag{2c}$$

$$t_{\mathrm{cav}} = 2L/v_{\mathrm{g}}, \tag{2d}$$

with $j = 1, 2$, where κ_j is the coupling coefficient; t_{cav} is the cavity round-trip time, with the group velocity v_{g}; ϕ_j is the phase delay due to the external cavities determined by the external round-trip time t_j, the lasing frequency in the presence of feedback ω_{FB}, and a phase shift at the external mirrors ϕ_{m_j}; and ω_{ref} is the free running laser frequency.

The first step is extracting the lasing conditions of the proposed laser system.

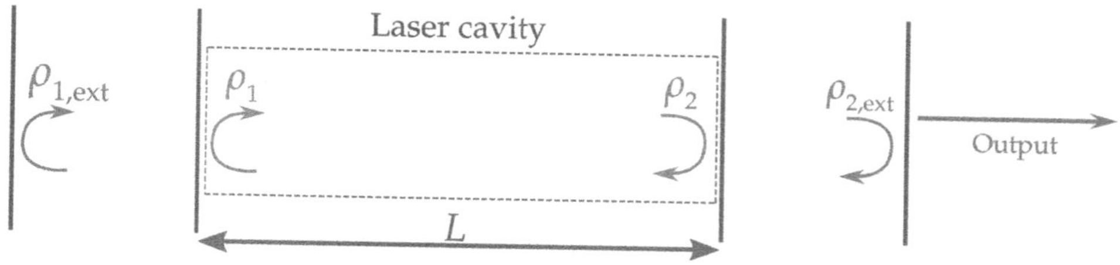

Figure 1. Schematic of a laser cavity affected by external optical feedback from both sides.

2.1. Lasing Conditions

By analyzing the slowly-varying amplitude of the complex electric field, effective reflection coefficients $\left(\rho_j^{\text{eff}}\right)$ can be obtained:

$$\frac{\rho_j^{\text{eff}}}{\rho_j} \approx 1 + \kappa_j t_{\text{cav}} e^{\pm i\phi_j}, \tag{3}$$

where the plus and minus signs corresponds to $j = 1$ (left mirror) and $j = 2$ (right mirror), respectively. The additional reflection influences the lasing condition, as shown in Appendix A, and thus the threshold gain and lasing frequency. To calculate them, the following definitions are convenient:

$$G_{\text{FB}} \equiv \Gamma g_{\text{FB}} v_{\text{g}} \tag{4a}$$

$$G_{\text{ref}} \equiv \Gamma g_{\text{ref}} v_{\text{g}} \tag{4b}$$

$$\Delta\phi_{\text{m}} \equiv \phi_{\text{m}_1} - \phi_{\text{m}_2}, \tag{4c}$$

$$\gamma_{\text{H}} \equiv \sqrt{1 + \alpha_{\text{H}}^2} \tag{4d}$$

$$\theta_{\text{H}} \equiv \arctan(\alpha_{\text{H}}) \tag{4e}$$

where Γ is the confinement factor; g_{FB} and g_{ref} are the threshold gain coefficient with and without feedback, respectively; and α_{H} is the linewidth enhancement factor [42]. From Appendix B, it is possible to obtain:

$$G_{\text{FB}} - G_{\text{ref}} \equiv \delta G \overset{(\text{A16})}{\approx} -2\kappa_2 \cos(\phi_2) - 2\kappa_1 \cos(\phi_1) \tag{5a}$$

$$\delta\omega \overset{(\text{A23})}{\approx} -\gamma_{\text{H}} [\kappa_2 \sin(\phi_2 + \theta_{\text{H}}) + \kappa_1 \sin(\theta_{\text{H}} - \phi_1)]. \tag{5b}$$

The relation between the right-hand terms determines the shift in threshold gain and lasing frequency with feedback. When comparing with previous literature regarding dual external cavity lasers, a sign difference is observed with respect to the lasing frequency shift equations found in References [28,29], arising from the different propagation direction of the additional feedback term. A clearer contrast is present with the lasing frequency shift shown in ref. [30], where a single term accounts for both external reflections. Furthermore, a phase shift at the external mirrors (ϕ_{m_j}) is not considered in the mentioned sources, thus the feedback terms included in both the excess gain and lasing frequency variations are distinct with respect to those in Equation (5).

Given the transcendental form of Equation (5b), a numerical analysis under different feedback conditions is studied in Section 3. Nevertheless, an analytical solution for lasing frequency stability can be found for the condition:

$$\delta\omega = 0. \tag{6}$$

Under this condition, Equation (5b) becomes:

$$\kappa_2 \sin(\phi_2 + \theta_{\mathrm{H}}) \stackrel{(6)}{=} -\kappa_1 \sin(\theta_{\mathrm{H}} - \phi_1), \tag{7}$$

which is determined by the feedback parameters κ_j and ϕ_j, the latter being dependent on the time delay t_j as well as the lasing frequency. Finding a stable solution that does not depend on the lasing frequency is of particular interest, as it can be advantageous for tunable lasers and their numerous applications. With the following assumption:

$$\kappa_2 = \kappa_1 \equiv \kappa, \tag{8}$$

Equation (7) can be rewritten as:

$$\begin{aligned} \sin(\phi_2 + \theta_{\mathrm{H}}) &= -\sin(\theta_{\mathrm{H}} - \phi_1) = \sin(\phi_1 - \theta_{\mathrm{H}}) \\ \Rightarrow \phi_2 + \theta_{\mathrm{H}} &= \phi_1 - \theta_{\mathrm{H}} + 2m\pi. \end{aligned} \tag{9}$$

Without loss of generality, the parameter m is set to $m = 0$, thus:

$$\begin{aligned} 2\theta_{\mathrm{H}} &= \phi_1 - \phi_2 \stackrel{(2b)(4c)}{=} \omega_{\mathrm{FB}}(t_1 - t_2) + \Delta\phi_{\mathrm{m}} \\ \Rightarrow 2\theta_{\mathrm{H}} - \Delta\phi_{\mathrm{m}} &= \omega_{\mathrm{FB}}(t_1 - t_2). \end{aligned} \tag{10}$$

Choosing an equal time delay in both external cavities:

$$t_2 = t_1 \equiv t_{\mathrm{ext}}, \tag{11}$$

sets the right hand term of Equation (10) to zero, so that:

$$2\theta_{\mathrm{H}} = \Delta\phi_{\mathrm{m}}. \tag{12}$$

This result suggests that by tuning the phase in the external cavities, so that condition (12) is met, it is possible to obtain a feedback-insensitive lasing frequency. An active method to tune the phase is however required as α_{H} is dependent on laser parameters, such as carrier density and wavelength [43]. This can be managed by, for example, phase shifters, which are mature and widely used components that can be included on-chip in a laser. The shown stable solution thus requires meeting the conditions (8), (11) and (12), which constrain the feedback parameters of one side of the cavity with respect to those of the other side, but do not restrict their absolute value. Nevertheless, if conditions (8) and (11) are not met, solutions for stable performance become frequency dependent. Such a case would thus only be satisfied for certain lasing frequency values for a given set of feedback parameters, which can potentially yield unstable solutions for other frequencies.

Regarding the variations in threshold gain, under conditions (8) and (11) Equation (5a) becomes:

$$\delta G = -2\kappa[\cos(\omega_{\mathrm{ref}} t_{\mathrm{ext}} + \phi_{\mathrm{m}_1}) + \cos(\omega_{\mathrm{ref}} t_{\mathrm{ext}} + \phi_{\mathrm{m}_2}))], \tag{13}$$

which vanishes if:

$$
\begin{aligned}
&\cos(\omega_{\text{ref}}t_{\text{ext}} + \phi_{\text{m}_1}) = -\cos(\omega_{\text{ref}}t_{\text{ext}} + \phi_{\text{m}_2})\\
&\cos(\omega_{\text{ref}}t_{\text{ext}} + \phi_{\text{m}_1}) = \cos(\omega_{\text{ref}}t_{\text{ext}} + \phi_{\text{m}_2} - \pi)\\
&\Rightarrow \omega_{\text{ref}}t_{\text{ext}} + \phi_{\text{m}_1} = \omega_{\text{ref}}t_{\text{ext}} + \phi_{\text{m}_2} - \pi + 2m\pi \quad m \in \mathbb{Z}\\
&\overset{\text{m=0}}{\Rightarrow} \Delta\phi_{\text{m}} = \pi.
\end{aligned}
\tag{14}
$$

This condition, while different than condition (12), also yields stability regardless of lasing frequency in this case for the threshold gain. Both cases are studied numerically in Section 3.

2.2. Rate Equations

In order to obtain the frequency noise (FN) power spectral density (PSD), and from it the laser intrinsic linewidth, the laser rate equations for the intensity and phase, as well as one for the carrier number, need to be studied. The former two can be extracted from the dynamic equations for the field inside the laser cavity, following the Lang-Kobayashi [39] approach. Its full derivation is shown in Appendix C. Furthermore, Langevin noise terms are included to account for shot noise fluctuations. The following definitions are useful to simplify notation:

$$\mathcal{A}(t) = \sqrt{S(t)}e^{-i\phi(t)} \tag{15a}$$

$$\mathcal{S}_j^{\pm} \equiv \kappa_j\sqrt{S(t \pm t_j)} \tag{15b}$$

$$\phi_{t_j}^{\pm} \equiv \phi(t \pm t_j) \tag{15c}$$

$$\Delta\Phi_1^{+} \overset{(15c)}{=} \phi(t) - \phi_{t_1}^{+} - \phi_1 \tag{15d}$$

$$\Delta\Phi_2^{-} \overset{(15c)}{=} \phi(t) - \phi_{t_2}^{-} + \phi_2 \tag{15e}$$

$$\Delta G \equiv G_{\text{FB}} - \tau_{\text{ph}}^{-1}, \tag{15f}$$

with $j = 1, 2$ relating to the EOF components from the left and right, respectively, and τ_{ph} is the photon decay time, which accounts for cavity and mirror losses. The field amplitude $\mathcal{A}$ is assumed to be slowly varying, where S is the photon number inside the laser cavity and ϕ is the phase of the field. With these definitions, the rate equations of the system can be written as:

$$\dot{S} \overset{(A27)}{=} S\Delta G + 2\mathcal{S}_2^{-}\sqrt{S}\cos(\Delta\Phi_2^{-}) + 2\mathcal{S}_1^{+}\sqrt{S}\cos(\Delta\Phi_1^{+}) + R_{\text{sp}} + F_S \tag{16a}$$

$$\dot{\phi} \overset{(A28)}{=} \frac{\alpha_{\text{H}}\Delta G}{2} - \delta\omega - \frac{\mathcal{S}_2^{-}}{\sqrt{S}}\sin(\Delta\Phi_2^{-}) - \frac{\mathcal{S}_1^{+}}{\sqrt{S}}\sin(\Delta\Phi_1^{+}) + F_\phi \tag{16b}$$

$$\dot{N} \overset{[3]}{=} I - GS(t) - N\tau_{\text{sp}}^{-1} + F_N, \tag{16c}$$

where I is the effective rate of injected current (in electrons), τ_{sp} is the carrier lifetime, and R_{sp} is the spontaneous recombination rate. This system is compatible with those presented in refs. [28,31], by accounting for the sign change in the delayed feedback term corresponding to the left external mirror, due to it having the opposite propagation direction.

The steady state solution of Equation (16) can be seen in Equation (A33). The Langevin noise sources $F_S(t)$, $F_\phi(t)$ and $F_N(t)$ satisfy [44]:

$$\langle F_i(t)\rangle = 0 \tag{17a}$$

$$\langle F_i(t_1)F_j(t_2)\rangle = 2D_{ij}\delta(t_1-t_2) \quad \text{with } i,j = S,\, \phi \text{ or } N, \tag{17b}$$

where:

$$D_{SS} = R_{\text{sp}}S \;\; ; \;\; D_{\phi\phi} = \frac{R_{\text{sp}}}{4S} \;\; ; \;\; D_{NN} = R_{\text{sp}}S + N\tau_{\text{sp}}^{-1} \;\; ; \;\; D_{SN} = -R_{\text{sp}}S, \tag{18}$$

are standard diffusion coefficients. A usual approach for solving the system from Equation (16) involves a small-signal analysis. Small deviations from a steady-state value are assumed:

$$S \simeq S_0 + S_\Delta = S_0 + \int_{-\infty}^{\infty} e^{i\Omega' t} S_{0\text{p}}(\Omega')\text{d}\Omega' \quad \text{with} \quad S_0 \gg S_\Delta \tag{19a}$$

$$\phi \simeq \phi_\Delta = \int_{-\infty}^{\infty} e^{i\Omega' t} \phi_{0\text{p}}(\Omega')\text{d}\Omega' \tag{19b}$$

$$N \simeq N_0 + N_\Delta = N_0 + \int_{-\infty}^{\infty} e^{i\Omega' t} N_{0\text{p}}(\Omega')\text{d}\Omega' \quad \text{with} \quad N_0 \gg N_\Delta, \tag{19c}$$

where the steady state value of the phase is assumed to be zero. The full linearization of the rate equations is shown in Appendix D, which uses the following definitions:

$$\kappa_j^{\text{c}} \equiv \kappa_j t_j \cos(\phi_j) \tag{20a}$$

$$\kappa_j^{\text{s}} \equiv \kappa_j t_j \sin(\phi_j) \tag{20b}$$

$$\text{K}_s \equiv \kappa_2^{\text{s}} + \kappa_1^{\text{s}} \tag{20c}$$

$$\text{K}_c \equiv 1 + \kappa_2^{\text{c}} - \kappa_1^{\text{c}} \tag{20d}$$

$$\zeta_s \equiv R_{\text{sp}}/S_0 \tag{20e}$$

$$a_g = \Gamma v_g a \tag{20f}$$

$$G_i \approx a_g(N_i - N_{\text{tr}}) \tag{20g}$$

$$\tau_{\text{e}}^{-1} \equiv a_g S_0 + \tau_{\text{sp}}^{-1}, \tag{20h}$$

where a linear approximation for the gain has been introduced, with a the differential gain coefficient and N_{tr} the number of electrons at transparency. Applying the Fourier transform to Equation (A36a–c), the following system of equations is obtained in the frequency domain:

$$i\Omega \text{K}_c S_{0\text{p}} \overset{\text{(A36a)}}{=} a_g S_0 N_{0\text{p}} - \zeta_s S_{0\text{p}} - 2i\Omega S_0 \text{K}_s \phi_{0\text{p}} + \widehat{F}_S \tag{21a}$$

$$2i\Omega \text{K}_c \phi_{0\text{p}} \overset{\text{(A36b)}}{=} \alpha_{\text{H}} a_g N_{0\text{p}} + i\Omega \frac{\text{K}_s}{S_0} S_{0\text{p}} + 2\widehat{F}_\phi \tag{21b}$$

$$i\Omega N_{0\text{p}} \overset{\text{(A36c)}}{=} -\tau_{\text{e}}^{-1} N_{0\text{p}} - G_0 S_{0\text{p}} + \widehat{F}_N, \tag{21c}$$

where the unknowns $S_{0\text{p}}$, $\phi_{0\text{p}}$, $N_{0\text{p}}$, and $\widehat{F}_S$, $\widehat{F}_\phi$ and $\widehat{F}_N$ depend on the Fourier angular frequency Ω. These equations are the first step to obtain the FN PSD.

2.3. Power Spectral Density and Laser Intrinsic Linewidth

The next step is to calculate the FN PSD, and from it the laser intrinsic linewidth. By solving the system in Equation (21) it is possible to find an expression for the PSD [45]:

$$S_f^{(1)}(\Omega) = \frac{\Omega^2}{2\pi^2}\langle|\phi_{0p}(\Omega)|^2\rangle, \tag{22}$$

which, using the following definitions:

$$\mathsf{F}_0 \equiv \zeta_s^2 + \mathsf{K}_c^2\tau_e^{-2} - 2\mathsf{K}_c a_g G_0 S_0 \tag{23a}$$

$$\mathsf{F}_1 \equiv \mathsf{K}_s^2 + \mathsf{K}_c^2 \tag{23b}$$

$$\Lambda_4 \equiv 4\mathsf{F}_1 D_{\phi\phi} \tag{23c}$$

$$\mathsf{F}_2 \equiv \mathsf{K}_c\alpha_H + \mathsf{K}_s \tag{23d}$$

$$\mathsf{F}_3 \equiv \mathsf{K}_c - \alpha_H\mathsf{K}_s \tag{23e}$$

$$\Delta_4 \equiv \mathsf{F}_1^2, \tag{23f}$$

and:

$$\Lambda_2 \equiv a_g^2\mathsf{F}_2^2 D_{NN} + 4D_{\phi\phi}\left(\mathsf{K}_s^2\tau_e^{-2} + \mathsf{F}_0\right) - 2\frac{\mathsf{K}_s}{S_0}D_{SS}a_g\left(\tau_e^{-1}\mathsf{F}_2 - \zeta_s\alpha_H - \alpha_H G_0\right) \tag{23g}$$

$$\Lambda_0 \equiv \alpha_H^2 a_g^2\left[D_{SS}\left(\zeta_s^2 + G_0^2 + 2\zeta_s G_0\right) + \zeta_s N\tau_{sp}^{-1}\right] + \left(\tau_e^{-1}\zeta_s + a_g G_0 S_0\right)^2 4D_{\phi\phi} \tag{23h}$$

$$\Delta_2 \equiv \mathsf{K}_c^2\zeta_s^2 + \tau_e^{-2}\mathsf{F}_1^2 - 2\mathsf{F}_1 a_g G_0 S_0 \mathsf{F}_3 \tag{23i}$$

$$\Delta_0 \equiv \left(a_g G_0 S_0 \mathsf{F}_3 + \mathsf{K}_c\zeta_s\tau_e^{-1}\right)^2, \tag{23j}$$

can be written as:

$$4\pi^2 S_f^{(1)} \overset{(23)(A46)}{=\!=} \frac{\Lambda_4\Omega^4 + \Lambda_2\Omega^2 + \Lambda_0}{\Delta_4\Omega^4 + \Delta_2\Omega^2 + \Delta_0}. \tag{24}$$

This is an analytical solution for the FN PSD of the isolated laser system. From the following expression [46] :

$$S_f^{(1)}(\Omega \to 0) = 2\pi\Delta f, \tag{25}$$

which is valid for a Lorentzian lineshape, the intrinsic linewidth can be obtained. As shown in Appendix F:

$$F \equiv \frac{\Delta f}{\Delta f_0(1+\alpha_H^2)} \overset{(A49)}{=} [1 + \gamma_H\kappa_2 t_2\cos(\phi_2 + \theta_H) - \gamma_H\kappa_1 t_1\cos(\phi_1 - \theta_H)]^{-2}, \tag{26}$$

where Δf_0 is the Schawlow–Townes linewidth [47]. The expression found for the intrinsic linewidth has two feedback terms, one contribution from each side, with a sign that depends on ϕ_1 and ϕ_2. Recalling from Equation (2b) that these quantities are a function of t_j and ϕ_{m_j}, a proper design of the laser can yield linewidth stability or a reduction of the intrinsic linewidth with respect to the case of one-sided feedback. This is further explored using a numerical analysis in Section 3. Additionally, it is possible to find an analytical expression for Equation (26) under the conditions for frequency stability, namely conditions (8), (11) and (12). Using this assumptions in Equation (26):

$$F = \{1 + \gamma_H \kappa t_{ext}[\cos(\phi_2 + \theta_H) - \cos(\phi_1 - \theta_H)]\}^{-2}. \quad (27)$$

Taking a closer look at the feedback terms yields:

$$\cos(\phi_2 + \theta_H) \overset{(12)}{=} \omega_{ref} t_{ext} + \phi_{m_1} - 2\theta_H + \theta_H = \cos(\phi_1 - \theta_H). \quad (28)$$

Using Equation (28) in Equation (27) yields a value of $F = 1$ which indicates that, under the assumed conditions, the intrinsic linewidth is insensitive to feedback. This result is significant as under the same condition the frequency is also feedback-insensitive, as shown in Section 2.1, regardless of lasing frequency. It is worth noting however that weak feedback is assumed in this analysis, with which the upper bound for feedback strength under which these equation are valid is not established. Nevertheless, achieving stability in the full frequency domain even under this condition is an improvement with respect to the single feedback case.

3. Numerical Study

Laser stability is studied by numerically evaluating the equations for the shift in lasing frequency, threshold gain and intrinsic linewidth under the revised EOF conditions, namely Equations (5) and (26), under different feedback parameters. Particular attention is given to the previously analyzed case under conditions (8), (11) and (12), which shows feedback-insensitive solutions for the lasing frequency and intrinsic linewidth. System tolerances to these conditions are explored by varying each while keeping the other two fixed. The simulated equations are plotted as a function of the unperturbed laser frequency multiplied by t_2, which represents the phase delay in the right external cavity for the free running laser frequency. It is kept between 0 and 1 (i.e., $\omega_{ref} t_2 \in (0, 2\pi]$) given the periodicity of the numerically solved functions. Furthermore, without loss of generality, ϕ_{m_2} is kept fixed at zero so that the value of $\Delta\phi_m$ is selected by varying ϕ_{m_1}. Additionally, simulations assume $\alpha_H = 3$. This value is compatible with measurements for semiconductor lasers [48], and thus meeting condition (12) requires that $\Delta\phi_m = 2\theta_H \simeq 2.5$. A summary of the values used in the numerical solutions is given in Table 1.

Table 1. Summary of values used in numerical solutions of Equations (5) and (26).

Parameter	Value
$\omega_{ref} t_{ext}$	$(0,2\pi)$
α_H	3
$\kappa_2 t_2$	0.1581 (case 1) 0.3162 (case 2) 0.4111 (case 3)
ϕ_{m_2}	0

Results are compared with the case with feedback from a single side, where:

$$\kappa_1 = 0. \quad (29)$$

In this case, as shown in [7], as feedback strength increases solutions for the lasing frequency become multi-valued. This gives rise to instabilities in the system such as mode hopping or coherence collapse regimes. The separation between single-valued and multi-valued solutions is related to the coefficient:

$$C = \gamma_H \kappa_2 t_2, \quad (30)$$

where $C = 1$ is the critical value that separates both behaviours.

Numerical solutions of the proposed system under conditions (8) and (11) are thus compared to the single feedback case for three cases:

Case 1: $C = 0.5$. This represents the single-feedback case with a single solution, and results for various values of $\Delta\phi_m$ are shown in Figure 2. Column A, B and C show the numerical solutions for the variations in lasing frequency, threshold gain and intrinsic linewidth, given by Equations (5) and (26) , respectively. By plotting the logarithm of the latter, negative values correspond to linewidth narrowing. The blue and orange plots represent the double-feedback and the single-feedback case, respectively, and these labels are maintained throughout the document.

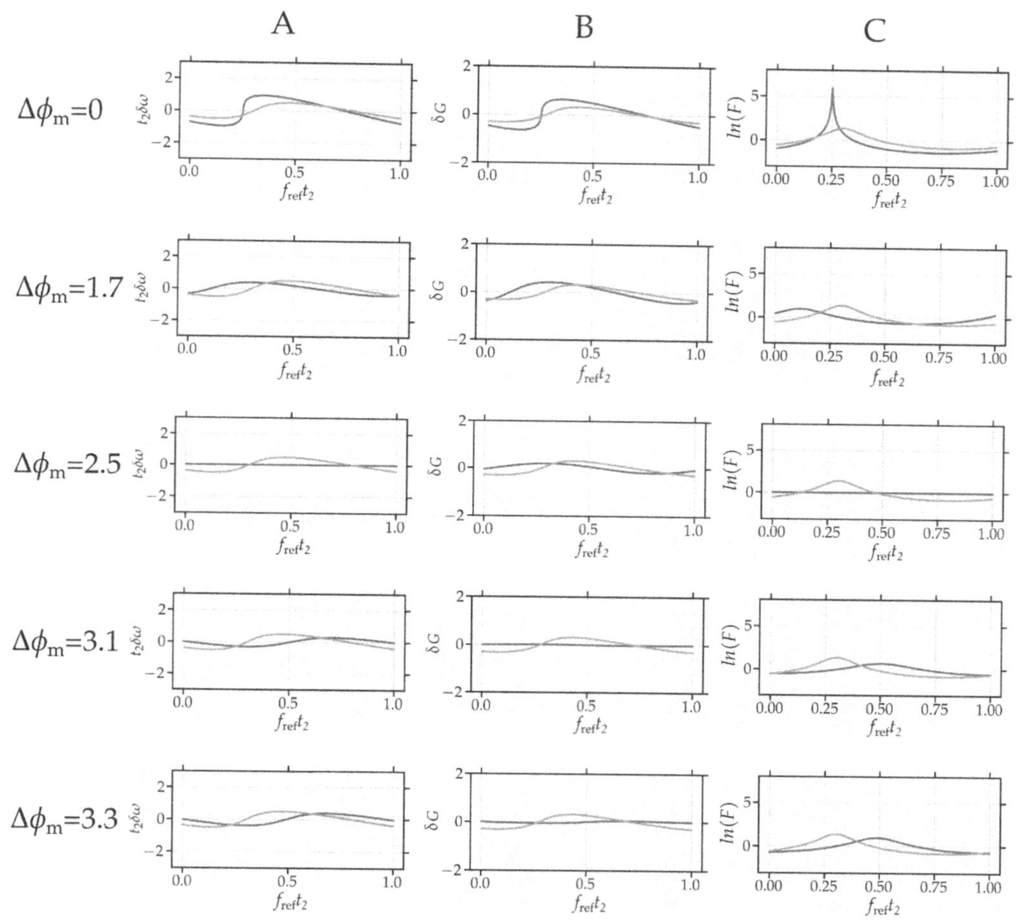

Figure 2. Numerical solutions for $C = 0.5$ under conditions (8) and (11) for different $\Delta\phi_m$. Double-feedback case shown in blue, single feedback case shown in orange. Column (**A**) shows the lasing frequency shift results. Column (**B**) shows the threshold gain shift. Column (**C**) shows the intrinsic linewidth variations.

The upper row shows the case where $\Delta\phi_m = 0$, i.e., additional phase shifts at, for example, the external mirrors (ϕ_{m_j}) are kept at zero. Linewidth narrowing for a wide range of frequencies can be observed, evidenced, as mentioned, by $ln(F) < 1$. As an example, when $f_{ref}t_2 = 0.75$, a 74% and a 42% reduction in linewidth is seen with respect to the free running laser and the single feedback case, respectively. Additionally, the signal is singled valued in the full domain, yet it is close to the critical point where multi-valued solutions arise. As $\Delta\phi_m$ increases, the amplitude of the lasing frequency shift is reduced until becoming zero for all values when condition (12) is met, as expected from previous analysis. Under this condition the intrinsic linewidth does not experience fluctuations either, and the amplitude of the threshold gain fluctuations is lower than in the single feedback case, indicating better stability across the three analyzed parameters with respect to the single feedback case. For the case of $\Delta\phi_m = \pi$, the threshold gain shows no fluctuations as predicted by Equation (14), and while the lasing frequency and intrinsic linewidth fluctuations are no longer zero, they are less pronounced than in the single feedback case. Further increases in $\Delta\phi_m$ show an increase in the fluctuations across all functions, and for $\phi_m > 6$ multi-valued solutions arise.

Case 2: $C = 1$. This represents the limiting case between single and multivalued solutions in the single-feedback case. Results for various values of $\Delta\phi_m$ are shown in Figure 3. As expected, the threshold gain is stable for $\Delta\phi_m = \pi$, and meeting condition (12) results in a stable lasing frequency and intrinsic linewidth. As $\Delta\phi_m$ deviates from these optimal points in either direction, the amplitude of fluctuations increase until reaching multi-valued solutions for $\Delta\phi_m < 1.5$ and $\Delta\phi_m > 3.5$. Comparing these results with the previous case shows that as feedback increases, the single valued solutions become more sensitive to the value of $\Delta\phi_m$.

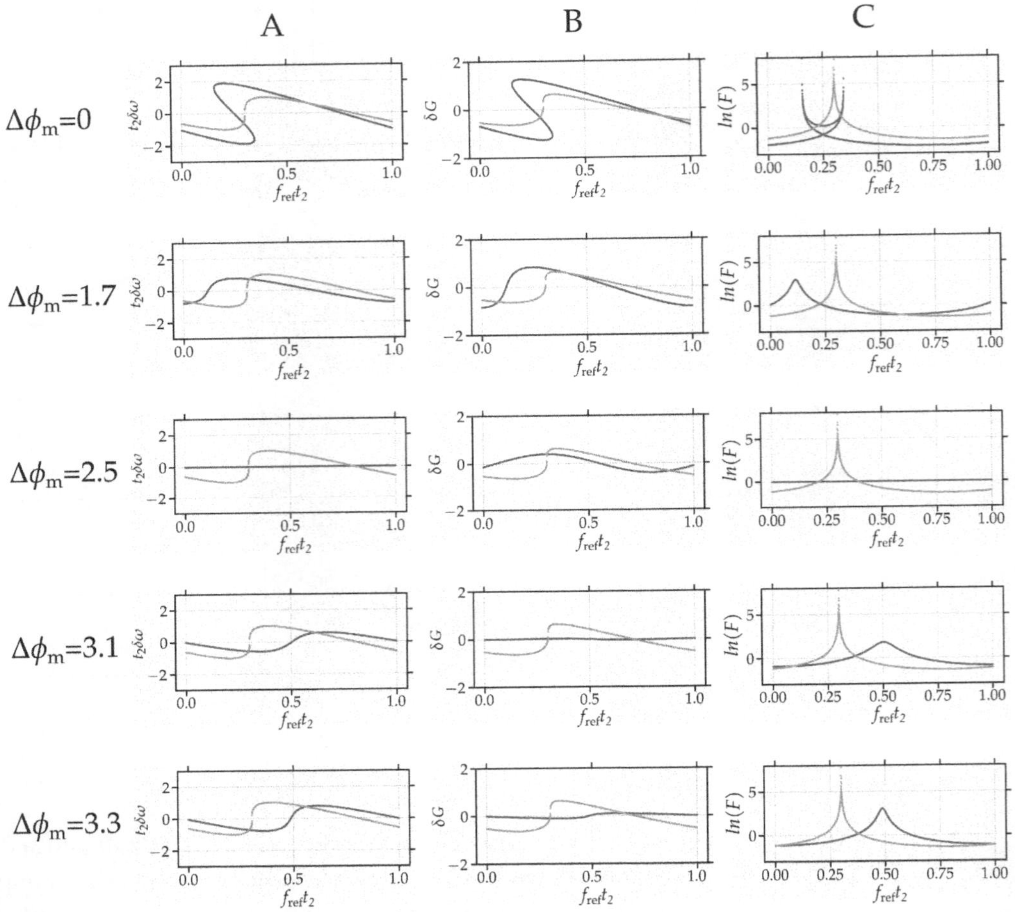

Figure 3. Numerical solutions for $C = 1$ under conditions (8) and (11) for different $\Delta\phi_m$. Double-feedback case shown in blue, single feedback case shown in orange. Column (**A**) shows the lasing frequency shift results. Column (**B**) shows the threshold gain shift. Column (**C**) shows the intrinsic linewidth variations.

Case 3: $C = 1.3$. This represents the single-feedback case with multi-valued solutions. Results for various values of $\Delta\phi_m$ are shown in Figure 4. In the single feedback case, multi-valued solutions are present in a given frequency range, and this span increases with increasing feedback strength. The multi-valued characteristics are evidenced experimentally with unstable regimes characterized by mode hopping and eventually coherence collapse for sufficiently high feedback. In contrast, the system proposed in this work shows that by tuning the value of $\Delta\phi_m$ to meet condition (12), even with increasing feedback it is possible to achieve stable performance regardless of frequency. In the case shown in

Figure 4 for $C = 1.3$, single valued solutions can be found for $\Delta\phi_m \in (1.5, 3.3)$ which is equivalent to a phase variation of more than 90°. Still, comparing with previous cases it is possible to see that as feedback increases, the single valued solutions tolerance with respect to $\Delta\phi_m$ decreases. Nevertheless, it is an improvement with respect to the single feedback case which shows no single value solutions across all frequencies for $C > 1$.

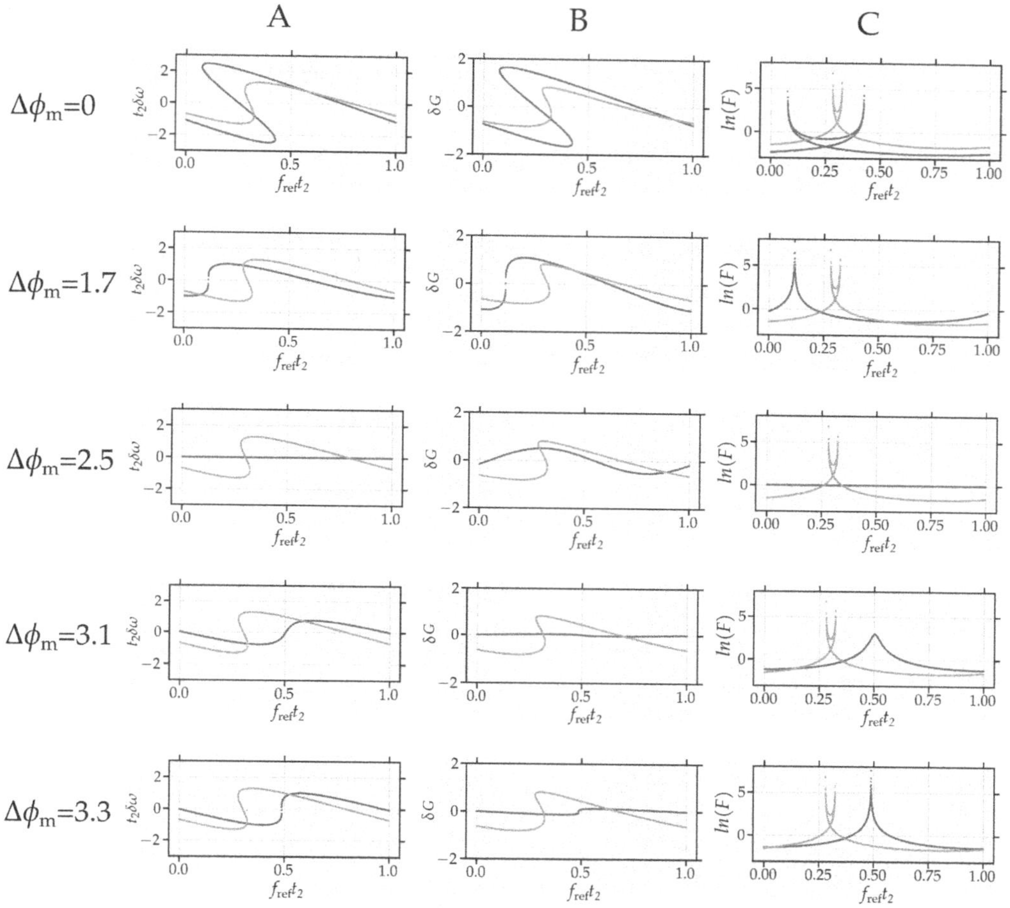

Figure 4. Numerical solutions for $C = 1.3$ under conditions (8) and (11) for different $\Delta\phi_m$. Double-feedback case shown in blue, single feedback case shown in orange. Column **(A)** shows the lasing frequency shift results. Column **(B)** shows the threshold gain shift. Column **(C)** shows the intrinsic linewidth variations.

Taking all cases into account, it can be seen that linewidth narrowing regions are present for all cases of $\Delta\phi_m$ analyzed. The only exception are the stable cases when meeting the three conditions (8), (11) and (12), yet the feedback-insensitivity provided by this case is also beneficial. Selecting an appropriate $\Delta\phi_m$ can thus be used to harness linewidth narrowing properties at a desired frequency.

Furthermore, system tolerances to conditions (8) and (11) are studied, while maintaining condition (12). Results for up to a 20% deviation are shown in Figures A2–A4 for cases 1, 2 and 3, respectively, where single valued solutions are observed in all cases in the full frequency span. Results show a high tolerance with respect to feedback strength. Looking at case 3, single valued solutions are obtained for $\kappa_1/\kappa_2 \in (0.2, 1.7)$. While the system is no longer feedback-insensitive, results evidence single-valued solutions that are robust with respect to condition (8). The system is more sensitive with respect to (11), with single valued solutions achieved within $t_1/t_2 \in (0.47, 1.2)$. Nevertheless, tolerances become once again stricter as feedback increases for both parameters, thus laser design is of paramount importance and should focus on meeting the discussed stability conditions. In particular, choosing equal lengths for both external cavities should suffice to meet condition (11). Fabrication tolerances in foundry processes are the main limiting factor for time delay accuracy. To meet condition (8), a possible approach is to merge the output of the two external cavities into a single one using a coupler, which can be included in the laser on-chip. Finally, as mentioned before, condition (12) can be met by using a phase shifter, which is a mature component in active platforms.

All in all, results demonstrate that, with proper design of the laser cavity, conditions (8), (11) and (12) can be met, with which it is possible to obtain lasing frequency and intrinsic linewidth insensitivity to feedback.

4. Discussion

The existing mature photonic integration fabrication processes are very flexible with respect to device geometry. They are however limited by a lack of commercially available on-chip isolators, and thus new approaches are required to minimize the effect feedback has on laser stability. The current work proposes a theoretical extension of laser dynamics under EOF by considering two external reflections, one from each side of the cavity. The proposed analysis yields new laser dynamic equations. These are numerically solved, which show the existence of a stable regime with high feedback tolerances. Similar results have been seen in previous double external cavity schemes [28], with EOF on the same side of the laser cavity, which showed robust stabilization of the chaotic regime by tuning the external feedback parameters. Yet other feedback regimes are not explored, and neither analytical stability conditions nor dependence with free laser running frequency are reported. The present analysis shows that feedback-insensitivity is achievable for a wide range of feedback strengths, under conditions that can be met with current laser fabrication processes and components, such as phase shifters for meeting condition (12) and couplers for condition (8).

It has previously been shown, for the single feedback case, that tuning the feedback phase can result in linewidth narrowing [13], however this still requires low feedback levels and a precise phase shift which can suffer variations due to external parameters, such as temperature or driving currents. In the approach proposed in this work, the stability conditions do not require specific values, instead relating the feedback parameters from one side to those from the other side. For example, condition (11) only implies equal round-trip times at the external cavities, regardless of value. This allows for additional flexibility in the feedback parameters and gives versatility to the device. Furthermore, this method allows for feedback-insensitivity across the full spectra, which is not seen in the single feedback case. This is of particular importance for tunable lasers, as all lasing frequencies are thus equally affected. Additionally, it relaxes the need for an isolator, reducing the cost and size of packaging processes. Another significant improvement of the proposed method with respect to the single feedback case is the increase in feedback tolerance: by choosing feedback parameters close to the stability conditions of Equations (11) and (12), higher levels of feedback strength are allowed without seeing multi-valued solutions, which results in experimentally seen mode-hopping. As a weak feedback approximation is used, the upper bound for feedback tolerance cannot be extracted from this analysis. Despite this, even under this approximation, tolerances are higher than that of the single feedback case. Furthermore, this system has a high tolerance to deviations from the optimal stability conditions as analyzed in the previous section. Linewidth narrowing can be achieved in these cases for certain frequency values, which can be tuned by selecting appropriate feedback parameters, as was the case for single feedback conditions, while maintaining stable solutions.

Finally, while the dynamics under consideration are complex, the laser system itself involves a straight-forward configuration using widely used on-chip components, which are available in generic foundry platforms. Previously studied methods to reduce feedback sensitivity include resourceful yet intricate designs. The proposed system is, in contrast, potentially easier to design, fabricate and characterize. An experimental study of this laser system is essential to validate the obtained results, and more importantly to explore the limitations of the model, and is the next step for a more comprehensive understanding of the proposed system. It is worth noting that, for the single feedback case, the Lang-Kobayashi approach yields results compatible with experimental data [49–51], which suggests that future experimental realization of the method here presented will be compatible with its theoretical predictions.

5. Conclusions

This work explores an extension of the theoretical background of EOF. By assuming that feedback couples into the laser cavity from both sides, new dynamic equations are found for the lasing frequency, the threshold gain and the intrinsic linewidth. These are numerically evaluated to analyze laser stability. Solutions with linewidth reduction are observed, where tuning of the feedback parameters yielded a 74% and a 42% reduction with respect of the free running laser and the single feedback case, respectively. Results also show the existence of a stable solution, with feedback-insensitive lasing frequency and intrinsic linewidth, regardless of the lasing frequency. This case is obtained by tuning the phase of the feedback field, for external cavities with equal lengths and coupling factors. Furthermore, the feedback-insensitive case exists regardless of the feedback strength, within a weak feedback approximation, which is an major improvement with respect to the single feedback case. Additionally, the stability conditions show good tolerances with respect to all feedback parameters, albeit they become stricter as the feedback strenght increases. Choosing feedback parameters close to the feedback-insensitive conditions ensures stable solutions that are feedback tolerant. Finally, the proposed system relies on few components in straight-forward configurations, and the stable conditions can be met with mature components available in generic foundry platforms. This enables close to market, low cost, feedback-tolerant semiconductor lasers which have direct applications in multiple fields that rely on stable laser sources, such as coherent communications and spectroscopy.

Appendix A. Amplitude and Phase Conditions

To obtain the revised lasing conditions resulting from the additional feedback component, extracting the amplitude and phase of the effective reflection coefficients is needed. In polar notation:

$$\rho_j^{\text{eff}} = \left|\rho_j^{\text{eff}}\right| e^{i\varphi_j}, \tag{A1}$$

where the magnitude is computed as:

$$\begin{aligned} \left|\rho_j^{\text{eff}}\right|^2 \Big/ \rho_j^2 &\overset{(3)}{=} \left[1 + \kappa_j t_{\text{cav}} \cos(\phi_j)\right]^2 + \left[\kappa_j t_{\text{cav}} \sin(\phi_j)\right]^2 \\ &= 1 + 2\kappa_j t_{\text{cav}} \cos(\phi_j) + \kappa_j^2 t_{\text{cav}}^2. \end{aligned} \tag{A2}$$

Assuming condition (1) of weak external feedback, the last term on the right-hand side of Equation (A2) is neglected, resulting in:

$$\left|\rho_j^{\text{eff}}(\omega_{\text{FB}})\right| / \rho_i = \sqrt{1 + 2\kappa_j t_{\text{cav}} \cos(\phi_j)} \overset{(1)}{\approx} 1 + \kappa_j t_{\text{cav}} \cos(\phi_j). \tag{A3}$$

The phase of the effective reflection coefficient is extracted from:

$$\varphi_j = \arctan\left[\left(\rho_j^{\text{eff}}\right)'' / \left(\rho_j^{\text{eff}}\right)'\right] \overset{(3)}{=} \arctan\left[\frac{\pm\kappa_j t_{\text{cav}} \sin(\phi_j)}{1 + \kappa_j t_{\text{cav}} \cos(\phi_j)}\right]$$

$$\overset{(1)}{\approx} \arctan[\pm\kappa_j t_{\text{cav}} \sin(\phi_j)] \overset{(1)}{\approx} \pm\kappa_j t_{\text{cav}} \sin(\phi_j). \tag{A4}$$

The fields traveling forward and backward in the laser cavity, $\mathcal{E}_{\text{f}}$ and $\mathcal{E}_{\text{b}}$ shown in Figure A1, can now be related by the effective reflection coefficients:

$$\mathcal{E}_{\text{f}}(z = 0) = \rho_1^{\text{eff}}\, \mathcal{E}_{\text{b}}(z = 0) \tag{A5a}$$

$$\mathcal{E}_{\text{b}}(z = L) = \rho_2^{\text{eff}}\, \mathcal{E}_{\text{f}}(z = L). \tag{A5b}$$

Using the propagation constant:

$$\beta \equiv n\omega/c\,, \tag{A6}$$

with n the effective refractive index of the lasing mode and c the speed of light in vacuum, the fields can be written as:

$$\mathcal{E}_{\text{f}} = A_f e^{-i\beta z + \frac{1}{2}(\Gamma g - \alpha)z} \tag{A7a}$$

$$\mathcal{E}_{\text{b}} = A_b e^{-i\beta(L-z) + \frac{1}{2}(\Gamma g - \alpha)(L-z)}, \tag{A7b}$$

where g is the gain coefficient and α is the attenuation coefficient. Replacing Equation (A5) into Equation (A7):

$$\mathcal{E}_{f0} \overset{(A1)}{=} \left|\rho_2^{\text{eff}}\right| e^{i\varphi_2} A_b e^{-i\beta L + (\Gamma g - \alpha)L/2} \tag{A8a}$$

$$\mathcal{E}_{b0} \overset{(A1)}{=} \left|\rho_1^{\text{eff}}\right| e^{i\varphi_1} A_f e^{-i\beta L + (\Gamma g - \alpha)L/2}, \tag{A8b}$$

and inserting Equation (A8a) into Equation (A8b) results in:

$$1 = \left|\rho_1^{\text{eff}}\right| e^{i\varphi_1} \left|\rho_2^{\text{eff}}\right| e^{i\varphi_2} e^{-2i\beta L + (\Gamma g - \alpha)L} = \left|\rho_1^{\text{eff}} \rho_2^{\text{eff}}\right| e^{-i(2\beta L - \varphi_1 - \varphi_2)} e^{(\Gamma g - \alpha)L}. \tag{A9}$$

Once lasing has been established, the gain assumes its threshold value:

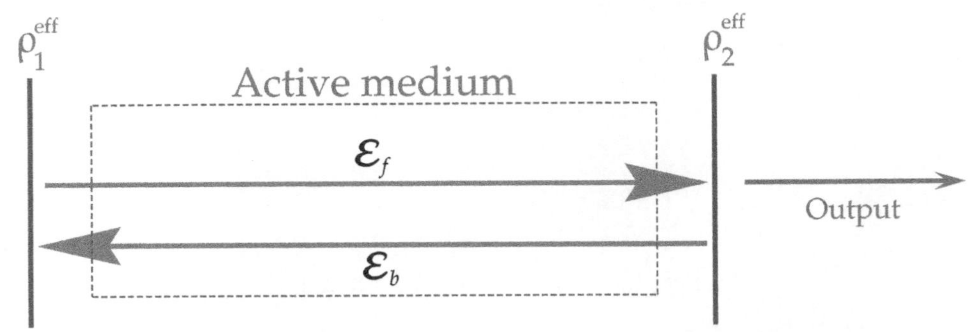

Figure A1. Schematic of the effective cavity of the laser, resulting from calculating effective reflection coeffitients.

$$g = g_{\text{FB}}, \tag{A10}$$

where g_{FB} is the threshold gain with feedback. Thus, Equation (A9) yields a lasing condition for the amplitude:

$$1 = \left|\rho_1^{eff}\rho_2^{eff}\right| e^{(\Gamma g_{FB}-\alpha)L}$$

$$\overset{(A10)(A3)}{\approx} \rho_2\rho_1[1+\kappa_2 t_{cav}\cos(\phi_2)][1+\kappa_1 t_{cav}\cos(\phi_1)]e^{(\Gamma g_{FB}-\alpha)L}, \tag{A11}$$

and the phase:

$$2\pi m = 2\beta L - \varphi_2 - \varphi_1 \overset{(A4)}{=} 2\beta L + \kappa_2 t_{cav}\sin(\phi_2) - \kappa_1 t_{cav}\sin(\phi_1) \quad , \quad m \in \mathbb{Z}. \tag{A12}$$

where the influence of feedback gives rise to two terms in Equations (A11) and (A12), one from each side. The new lasing conditions result in a variation of the lasing frequency and threshold gain of the system, and thus have to be studied to determine the laser dynamics.

Appendix B. Threshold Gain Reduction and Lasing Frequency Shift

Under feedback from both sides of the laser cavity, new lasing conditions are found which subsequently result in a shift of the laser threshold gain and lasing frequency with respect to the case without feedback, in which:

$$\kappa_j = 0. \tag{A13}$$

In this case, the amplitude condition from Equation (A11) becomes:

$$1 \overset{(A13)}{=} \rho_2\rho_1 e^{(\Gamma g_{th}-\alpha)L}. \tag{A14}$$

Using the expansion:

$$\ln(1+x) \simeq x, \tag{A15}$$

the threshold gain reduction due to feedback can be found by computing the ratio between Equations (A11) and (A14):

$$1 = \frac{\rho_2\rho_1[1+\kappa_2 t_{cav}\cos(\phi_2)][1+\kappa_1 t_{cav}\cos(\phi_1)]e^{(\Gamma g_{FB}-\alpha)L}}{\rho_2\rho_1 e^{\Gamma g_{ref}-\alpha L}}$$

$$= [1+\kappa_2 t_{cav}\cos(\phi_2)][1+\kappa_1 t_{cav}\cos(\phi_1)]e^{\Gamma(g_{FB}-g_{ref})L}$$

$$\overset{(2)}{\Leftrightarrow} G_{FB} - G_{ref} \approx -2\kappa_2\cos(\phi_2) - 2\kappa_1\cos(\phi_1). \tag{A16}$$

The relation between the right hand terms determines the threshold gain reduction, as discussed in Section 2.1.

The phase lasing condition from Equation (A12) yields the lasing frequency shift equation. Consider the following definitions related to the effective refractive index [52]:

$$n \equiv n' + in'' \tag{A17a}$$

$$n_g \equiv n + \omega\frac{\partial n}{\partial \omega} \tag{A17b}$$

$$n'' \equiv -\frac{cG}{2\omega v_g}, \tag{A17c}$$

$$\alpha_H \equiv \Delta n' / \Delta n'' \tag{A17d}$$

$$\frac{\partial n}{\partial N} = \frac{\partial n}{\partial n''}\frac{\partial n''}{\partial N}$$

$$\overset{(A17a)}{=}\overset{(A17c)}{=}\overset{(A17d)}{=} -\frac{\partial G}{\partial N}\frac{\alpha_H c}{2\omega v_g}, \tag{A17e}$$

where n_g is the group refractive index. To find the lasing frequency shift equation, calculating the change in β is first needed:

$$c\delta\beta \overset{(A6)}{=} \delta(n\omega) = \omega\delta n + n\delta\omega \overset{(2c)}{=} \omega\left[\frac{\partial n}{\partial N}(N - N_{th}) + \frac{\partial n}{\partial \omega}\delta\omega\right] + n\delta\omega$$

$$\overset{(A17b)}{=}\overset{(A17e)}{=} -\frac{G_{FB} - G_{th}}{2v_g}\alpha_H c + n_g\delta\omega \overset{(5a)}{=} [\kappa_2\cos(\phi_2) + \kappa_1\cos(\phi_1)]\frac{\alpha_H c}{v_g} + n_g\delta\omega, \tag{A18}$$

with which:

$$2L\delta\beta \overset{(5a)}{=} [\kappa_2\cos(\phi_2) + \kappa_1\cos(\phi_1)]\alpha_H t_{cav} + t_{cav}\delta\omega. \tag{A19}$$

Furthermore, using:

$$\sin[\arctan(x)] = \frac{x}{\sqrt{1+x^2}} \tag{A20a}$$

$$\cos[\arctan(x)] = \frac{1}{\sqrt{1+x^2}} \tag{A20b}$$

$$\sin(x \pm y) = \sin(x)\cos(y) \pm \cos(x)\sin(y) \tag{A20c}$$

$$\cos(\theta_H) \overset{(4e)}{=}\overset{(A20b)}{=}\overset{(4d)}{=} 1/\gamma_H, \tag{A20d}$$

the following can be computed:

$$\begin{aligned}\alpha_H\cos(\phi_j) &\overset{(A20a)}{=}\overset{(4d)}{=}\overset{(4e)}{=} \gamma_H\sin(\theta_H)\cos(\phi_j)\\ &\overset{(A20c)}{=} \gamma_H\left[\sin(\theta_H \pm \phi_j) \mp \cos(\theta_H)\sin(\phi_j)\right]\\ &\overset{(A20d)}{=} \gamma_H\left[\sin(\theta_H \pm \phi_j) \mp \sin(\phi_j)/\gamma_H\right]\end{aligned} \tag{A21}$$

$$\Leftrightarrow \alpha_H\cos(\phi_j) \pm \sin(\phi_j) = \gamma_H\sin(\theta_H \pm \phi_j).$$

Finally, using the phase condition in Equation (A12), and assuming without generality loss that $m = 0$, it is possible to compute:

$$2\pi m \overset{(A19)}{=} t_{cav}\delta\omega + \kappa_2 t_{cav}[\alpha_H\cos(\phi_2) + \sin(\phi_2)] + \kappa_1 t_{cav}[\alpha_H\cos(\phi_1) - \sin(\phi_1)]$$

$$\overset{m=0}{\Leftrightarrow} \delta\omega = -\kappa_2[\alpha_H\cos(\phi_2) + \sin(\phi_2)] + \kappa_1[\alpha_H\cos(\phi_1) - \sin(\phi_1)] \tag{A22}$$

$$\delta\omega \overset{(A21)}{=} -\gamma_H[\kappa_2\sin(\phi_2 + \theta_H) + \kappa_1\sin(\phi_1 - \theta_H)],$$

which describes the lasing frequency shift as a function of the feedback parameters $\kappa_1, \kappa_2, \phi_1$ and ϕ_2, Similarly to the threshold gain reduction, the interaction between the two right hand terms determines the lasing frequency stability which is discussed in Section 2.1.

Appendix C. Deriving the Rate Equation for the Intensity and Phase

To further inspect the laser dynamics, the rate equations for the intensity and phase must be studied. Considering Equation (15), and a slowly varying electric field given by:

$$\mathcal{E}(t) = \mathcal{A}(t)e^{-i\omega_{\rm FB}t}, \tag{A23}$$

and following the approach from Lang and Kobayashi [39], the laser field equation which considers EOF from both sides of the laser cavity can be written as:

$$\dot{\mathcal{E}} = \left(-i\omega_{\rm ref} + \Delta G\frac{1-i\alpha_{\rm H}}{2}\right)\mathcal{E}(t) + \kappa_2\mathcal{E}(t-t_2) + \kappa_1\mathcal{E}(t+t_1)$$

$$\overset{(A23)}{\Leftrightarrow} \frac{d\left[\mathcal{A}(t)e^{-i\omega_{\rm FB}t}\right]}{dt} \overset{(15)}{=} e^{-i\omega_{\rm FB}t}\left[\left(-i\omega_{\rm ref} + \Delta G\frac{1-i\alpha_{\rm H}}{2}\right)\mathcal{A}(t) + \kappa_2\mathcal{A}(t-t_2)e^{i\phi_2} + \kappa_1\mathcal{A}(t+t_2)e^{-i\phi_1}\right] \tag{A24}$$

$$\Leftrightarrow \dot{\mathcal{A}}(t) \overset{(2c)(2b)}{=\!=} \left(i\delta\omega + \Delta G\frac{1-i\alpha_{\rm H}}{2}\right)\mathcal{A}(t) + \kappa_2\mathcal{A}(t-t_2)e^{i\phi_2} + \kappa_1\mathcal{A}(t+t_2)e^{-i\phi_1}.$$

The last two right-hand terms appear as a result of the imposed feedback conditions, each term to account for feedback on each side of the cavity. In the case without feedback the lasing frequency become $\omega_{\rm FB} = \omega_{\rm ref}$ and $\kappa_j = 0$, thus recovering the no-feedback field Equation [42]. The slowly varying field amplitude $\mathcal{A}$ can be modeled as Equation (15a), and thus the rate equations for the photon number S and phase ϕ can be found using:

$$\dot{S} = \frac{d[\mathcal{A}\mathcal{A}^*]}{dt} = \mathcal{A}\dot{\mathcal{A}}^* + \mathcal{A}^*\dot{\mathcal{A}} \tag{A25a}$$

$$\dot{\phi} = -\frac{1}{S}\Im(\mathcal{A}^*\dot{\mathcal{A}}). \tag{A25b}$$

Replacing Equations (15a) and (A24) into Equation (A25a) the photon rate equation reads:

$$\dot{S} \overset{(15d)(15e)}{=\!=} S\Delta G + 2S_2^-\sqrt{S}\cos(\Delta\Phi_2^-) + 2S_1^+\sqrt{S}\cos(\Delta\Phi_1^+). \tag{A26}$$

In the case of the phase, its rate equation comes from replacing Equations (15a) and (A24) into Equation (A25b):

$$\dot{\phi} \overset{(15)}{=} \frac{\Delta G}{2}\alpha_{\rm H} - \delta\omega - \frac{S_2^-}{\sqrt{S}}\sin(\Delta\Phi_2^-) - \frac{S_1^+}{\sqrt{S}}\sin(\Delta\Phi_1^+). \tag{A27}$$

Equations (A26) and (A27) are the amplitude and phase rate equations for the laser system proposed in this work. These are the starting point to compute the frequency noise PSD, and extract the intrinsic linewidth.

Appendix D. Small-Signal Analysis

To find the FN PSD, the system shown in Equation (16) is to be solved. This is done using the small-signal analysis proposed in Equation (19). Assuming a narrow-linewidth laser, i.e., a long coherence time with respect to the external cavity lenghts:

$$t_{\rm ext} < t_{\rm coh}, \tag{A28}$$

the following approximation is valid:

$$\Omega t_{\rm ext} \ll 1. \tag{A29}$$

By linearizing the following expressions:

$$\sqrt{\frac{S(t\pm t_j)}{S(t)}} \overset{(A28)}{\approx} = \sqrt{1\pm\frac{\dot{S}}{S}t_j} \overset{(19a)}{\approx} \sqrt{1\pm\frac{i\Omega' S_\Delta}{S_0}t_j} \overset{(A29)}{\approx} 1\pm\frac{i\Omega' S_\Delta}{2S_0}t_j \tag{A30a}$$

$$\phi-\phi(t\pm t_j) \overset{(A28)}{\approx} \phi-\phi\mp t_j\dot{\phi} \overset{(19b)}{=} \mp i t_j\Omega'\phi_\Delta, \tag{A30b}$$

it is possible to rewrite Equation (16) as:

$$i\Omega' S_\Delta \overset{(A30)}{=} S\Delta G + R_{\text{sp}} + 2\frac{\kappa_2}{t_{\text{cav}}}S\left(1-\frac{i\Omega' S_\Delta}{2S_0}t_2\right)\cos\left(i\Omega'\phi_\Delta t_2+\phi_2\right)$$

$$+2\kappa_1 S\left(1+\frac{i\Omega' S_\Delta}{2S_0}t_1\right)\cos\left(i\Omega'\phi_\Delta t_1+\phi_1\right)+F_S$$

$$i\Omega'\phi_\Delta \overset{(A30)}{=} \alpha_{\text{H}}\frac{\Delta G}{2}-\delta\omega-\frac{\kappa_2}{t_{\text{cav}}}\left(1-\frac{i\Omega' S_\Delta}{2S_0}t_2\right)\sin\left(i\Omega'\phi_\Delta t_2+\phi_2\right) \tag{A31a}$$

$$+\kappa_1\left(1+\frac{i\Omega' S_\Delta}{2S_0}t_1\right)\sin\left(i\Omega'\phi_\Delta t_1+\phi_1\right)+F_\phi \tag{A31b}$$

$$i\Omega' N_\Delta \overset{(19)(A30)}{=} I-G_{\text{FB}}S-N\tau_{\text{sp}}^{-1}+F_N. \tag{A31c}$$

Solving Equation (A31) requires the steady-state solution of Equation (16). Under stationary conditions:

$$\dot{S}=0\Rightarrow S(t)=S(t\pm t_j)=S_0 \tag{A32a}$$

$$\dot{\phi}=0\Rightarrow \phi(t)=\phi(t\pm t_j) \tag{A32b}$$

$$\dot{N}=0\Rightarrow N(t)=N(t\pm t_j)=N_0, \tag{A32c}$$

the steady-state equations are:

$$\tau_{\text{ph}}^{-1}=G_0+2\kappa_2\cos(\phi_2)+2\kappa_1\cos(\phi_1)+\frac{R_{\text{sp}}}{S_0} \tag{A33a}$$

$$\delta\omega=\alpha_{\text{H}}\frac{G_0-\tau_{\text{ph}}^{-1}}{2}-\kappa_2\sin(\phi_2)+\kappa_1\sin(\phi_1) \tag{A33b}$$

$$I=G_0S_0+N_0\tau_{\text{sp}}^{-1}, \tag{A33c}$$

where the Langevin noise terms are not included as their mean value is zero. Next, using Equation (A33) and the following expansions:

$$\sin(x+\Delta)\approx\sin(x)+\Delta\cos(x) \tag{A34a}$$

$$\cos(x+\Delta)\approx\cos(x)-\Delta\sin(x), \tag{A34b}$$

Equation (A31a) can be rewritten as:

$$i\Omega' S_\Delta \overset{(19)(A33)(20)}{=\,=\,\approx} \left(G_0+a_gN_\Delta-\left\{G_0+2[\kappa_2\cos(\phi_2)+\kappa_1\cos(\phi_1)]+\frac{R_{\text{sp}}}{S_0}\right\}\right)(S_0+S_\Delta)$$

$$+2\Big\{\kappa_2(S_0+S_\Delta)\left[\cos(\phi_2)-i\Omega'\phi_\Delta t_2\sin(\phi_2)\right]$$

$$-\kappa_2(S_0+S_\Delta)\frac{i\Omega' S_\Delta}{2S_0}t_2\left[\cos(\phi_2)-i\Omega'\phi_\Delta t_2\sin(\phi_2)\right]$$

$$+\kappa_1(S_0+S_\Delta)\left[\cos(\phi_1)-i\Omega'\phi_\Delta t_1\sin(\phi_1)\right]$$

$$+\kappa_1(S_0+S_\Delta)\frac{i\Omega' S_\Delta}{2S_0}t_1\left[\cos(\phi_1)-i\Omega'\phi_\Delta t_1\sin(\phi_1)\right]\Big\}+R_{\text{sp}}+F_S.$$

Simplifying this equation, and neglecting the quadratic terms yields:

$$i\Omega' S_\Delta \overset{(20)}{\approx} a_g N_\Delta S_0 - \zeta_s S_\Delta - i\Omega'\phi_\Delta 2\mathrm{K}_s S_0 + i\Omega' S_\Delta(\kappa_1^c - \kappa_2^c) + F_S. \tag{A35a}$$

In a similar way, Equation (A31b) can be rewritten using Equations (A33) and (20):

$$\begin{aligned} i\Omega'\phi_\Delta \approx\ & \frac{\alpha_H}{2}\left(G_0 + a_g N_\Delta - \tau_{ph}^{-1}\right) - \left[\alpha_H \frac{G_0 - \tau_{ph}^{-1}}{2} - \kappa_2 \sin(\phi_2) + \kappa_1 \sin(\phi_1)\right] \\ & - \kappa_2\left(1 - \frac{i\Omega' S_\Delta}{2S_0} t_2\right)\left[\sin(\phi_2) + i\Omega'\phi_\Delta t_2 \cos(\phi_2)\right] \\ & + \kappa_1\left(1 + \frac{i\Omega' S_\Delta}{2S_0} t_1\right)\left[\sin(\phi_1) + i\Omega'\phi_\Delta t_1 \cos(\phi_1)\right] + F_\phi \\ \Leftrightarrow 2i\Omega'\phi_\Delta \overset{(20a)(20b)}{=\!=}\ & \alpha_H a_g N_\Delta + 2(\kappa_1^c - \kappa_2^c) i\Omega'\phi_\Delta + \frac{i\Omega' S_\Delta}{S_0}\mathrm{K}_s + 2F_\phi. \end{aligned} \tag{A35b}$$

Finally, Equation (A31c) can be rewritten as:

$$\begin{aligned} i\Omega' N_\Delta \overset{(19)(A33)}{=\!=} \overset{(20)}{\approx}\ & G_0 S_0 - (G_0 + a_g N_\Delta) S_0 - (G_0 + a_g N_\Delta) S_\Delta - \tau_{sp}^{-1} N_\Delta + F_N \\ \Leftrightarrow i\Omega' N_\Delta \overset{(20h)}{=}\ & -\tau_e^{-1} N_\Delta - G_0 S_\Delta + F_N. \end{aligned} \tag{A35c}$$

The linearized rate equations of the laser under study are thus Equations (A35a)–(A35c), from which the power spectral density, and subsequently the intrinsic linewidth, can be computed.

Appendix E. Power Spectral Density

The next step is to find an expression for ϕ_{0p} from which the FN PSD, and thus the laser intrinsic linewidth, can be computed. The following definitions are convenient:

$$A_\phi \equiv \left(i\Omega + \tau_e^{-1}\right)\left(i\Omega\mathrm{K}_c + \frac{a_g G_0}{i\Omega + \tau_e^{-1}} S_0 + \zeta_s\right) \tag{A36a}$$

$$2A_S \equiv i\Omega\mathrm{K}_s \frac{i\Omega + \tau_e^{-1}}{S_0} - \alpha_H a_g G_0 \tag{A36b}$$

$$2A_N \equiv \frac{\alpha_H A_\phi + 2S_0 A_S}{i\Omega + \tau_e^{-1}} a_g \tag{A36c}$$

$$B_\phi \equiv \mathrm{K}_c A_\phi - \alpha_H a_g G_0 S_0 \mathrm{K}_s + i\Omega\mathrm{K}_s^2\left(i\Omega + \tau_e^{-1}\right). \tag{A36d}$$

First, N_{0p} is extracted from Equation (21c):

$$N_{0p} = \frac{\widehat{F}_N - G_0 S_{0p}}{i\Omega + \tau_e^{-1}}. \tag{A37}$$

Next, replacing Equation (A37) into Equation (21a) yields the expression for $S_{0\mathrm{p}}$:

$$\begin{aligned}
(i\Omega \mathrm{K}_c + \zeta_s) S_{0\mathrm{p}} &= a_g \frac{\widehat{F}_N - G_0 S_{0\mathrm{p}}}{i\Omega + \tau_e^{-1}} S_0 - 2i\Omega S_0 \mathrm{K}_s \phi_{0\mathrm{p}} + \widehat{F}_S \\
\Leftrightarrow \frac{A_\phi}{i\Omega + \tau_e^{-1}} S_{0\mathrm{p}} &= a_g S_0 \frac{\widehat{F}_N}{i\Omega + \tau_e^{-1}} - 2i\Omega S_0 \mathrm{K}_s \phi_{0\mathrm{p}} + \widehat{F}_S \\
\Leftrightarrow S_{0\mathrm{p}} &\overset{\text{(A36a)}}{=} \frac{a_g S_0 \widehat{F}_N - 2i S_0 \Omega \mathrm{K}_s (i\Omega + \tau_e^{-1}) \phi_{0\mathrm{p}} + (i\Omega + \tau_e^{-1}) \widehat{F}_S}{A_\phi}.
\end{aligned} \tag{A38}$$

Finally, inserting Equation (A37) and (A38) into Equation (21b) and grouping the terms with $\phi_{0\mathrm{p}}$, $\widehat{F}_S$, $\widehat{F}_\phi$ and $\widehat{F}_N$ yields:

$$\begin{aligned}
2i\Omega B_\phi \phi_{0\mathrm{p}} &\overset{\text{(A36)}}{=} \left[-\alpha_{\mathrm{H}} a_g G_0 S_0 + \alpha_{\mathrm{H}} A_\phi + i\Omega \mathrm{K}_s \left(i\Omega + \tau_e^{-1}\right)\right] \frac{a_g}{i\Omega + \tau_e^{-1}} \widehat{F}_N + 2A_S \widehat{F}_S + 2 A_\phi \widehat{F}_\phi \\
&\overset{\text{(A36b)}}{=} (\alpha_{\mathrm{H}} A_\phi + 2 S_0 A_S) \frac{a_g}{i\Omega + \tau_e^{-1}} \widehat{F}_N + A_S \widehat{F}_S + 2 A_\phi \widehat{F}_\phi \\
\Leftrightarrow \phi_{0\mathrm{p}} &\overset{\text{(A36c)}}{=} \frac{A_N \widehat{F}_N + A_S \widehat{F}_S + A_\phi \widehat{F}_\phi}{i\Omega B_\phi}.
\end{aligned} \tag{A39}$$

With Equation (A39) it is possible to calculate an expression for Equation (22):

$$\begin{aligned}
2\pi^2 |B_\phi|^2 S_f^{(1)} &\overset{(22)}{=} \Omega^2 |B_\phi|^2 \left\langle \phi_{0\mathrm{p}}(\Omega) \phi_{0\mathrm{p}}^*(\Omega) \right\rangle \\
&\overset{\text{(A39)}}{=} \Omega^2 |B_\phi|^2 \left\langle \frac{A_N \widehat{F}_N + A_S \widehat{F}_S + A_\phi \widehat{F}_\phi}{i\Omega B_\phi} \left(\frac{A_N \widehat{F}_N + A_S \widehat{F}_S + A_\phi \widehat{F}_\phi}{i\Omega B_\phi} \right)^* \right\rangle \\
&= \left\langle \left(A_N \widehat{F}_N + A_S \widehat{F}_S + A_\phi \widehat{F}_\phi\right) \left(A_N \widehat{F}_N + A_S \widehat{F}_S + A_\phi \widehat{F}_\phi\right)^* \right\rangle.
\end{aligned} \tag{A40}$$

It can be seen from Equation (A36) that the coefficients A_i are independent of time, and assuming an ergodic process they can be taken out of the average in Equation (A40), obtaining:

$$\pi^2 |B_\phi|^2 S_f^{(1)} \overset{\text{(17b)}}{=} |A_N|^2 D_{NN} + |A_S|^2 D_{SS} + |A_\phi|^2 D_{\phi\phi} + (A_S A_N^* + A_N A_S^*) D_{SN} \tag{A41}$$

$$\begin{aligned}
\Leftrightarrow 4\pi^2 |B_\phi|^2 S_f^{(1)} &\overset{\text{(A36)(23)}}{=} \left[(\zeta_s \alpha_{\mathrm{H}} a_g)^2 + \Omega^2 a_g^2 \mathrm{F}_2^2 \right] D_{NN} \\
&+ \left[\Omega^4 \mathrm{K}_c^2 + \Omega^2 \mathrm{F}_0 + \left(\tau_e^{-1} \zeta_s + a_g G_0 S_0 \right)^2 \right] 4 D_{\phi\phi} \\
&+ \left\{ \Omega^2 \left[\left(\tau_e^{-1} \frac{\mathrm{K}_s}{S_0} \right)^2 + 2\alpha_{\mathrm{H}} a_g G_0 \frac{\mathrm{K}_s}{S_0} \right] + \Omega^4 \left(\frac{\mathrm{K}_s}{S_0} \right)^2 + [\alpha_{\mathrm{H}} a_g G_0]^2 \right\} D_{SS} \\
&- 2 \left[-\alpha_{\mathrm{H}}^2 a_g^2 G_0 \zeta_s + \Omega^2 \frac{\mathrm{K}_s}{S_0} a_g \left(\tau_e^{-1} \mathrm{F}_2 - \zeta_s \alpha_{\mathrm{H}} \right) \right] D_{SS} \\
\Leftrightarrow 2\pi^2 S_f^{(1)} &\overset{(23)}{=} \frac{\Lambda_4 \Omega^4 + \Lambda_2 \Omega^2 + \Lambda_0}{2 |B_\phi|^2}.
\end{aligned} \tag{A42}$$

Furthermore, using the following definitions:

$$\Delta_4 \equiv \mathsf{F}_1^2 \tag{A43a}$$

$$\Delta_2 \equiv \mathsf{K}_c^2\zeta_s^2 + \tau_e^{-2}\mathsf{F}_1^2 - 2\mathsf{F}_1 a_g G_0 S_0 \mathsf{F}_3 \tag{A43b}$$

$$\Delta_0 \equiv \left(a_g G_0 S_0 \mathsf{F}_3 + \mathsf{K}_c\zeta_s\tau_e^{-1}\right)^2, \tag{A43c}$$

the expression for B_ϕ from Equation (A36d) can be rewritten as:

$$B_\phi = a_g G_0 S_0 \mathsf{F}_3 - \Omega^2\mathsf{F}_1 + i\Omega\left(\mathsf{K}_c\zeta_s + \tau_e^{-1}\mathsf{F}_1\right) + \mathsf{K}_c\zeta_s\tau_e^{-1} \tag{A44a}$$

$$\begin{aligned}\Leftrightarrow |B_\phi|^2 &= \Omega^2\left(\mathsf{K}_c\zeta_s + \tau_e^{-1}\mathsf{F}_1\right)^2 + \left(a_g G_0 S_0 \mathsf{F}_3 - \Omega^2\mathsf{F}_1 + \mathsf{K}_c\zeta_s\tau_e^{-1}\right)^2 \\ &\overset{(A43c)}{=} \Omega^4\Delta_4 + \Omega^2\left[\left(\mathsf{K}_c\zeta_s + \tau_e^{-1}\mathsf{F}_1\right)^2 - 2\mathsf{F}_1 a_g G_0 S_0 \mathsf{F}_3 - 2\mathsf{F}_1\mathsf{K}_c\zeta_s\tau_e^{-1}\right] + \Delta_0 \\ &\overset{(A43c)}{=} \Omega^4\Delta_4 + \Omega^2\Delta_2 + \Delta_0,\end{aligned} \tag{A44b}$$

with which the expression of the FN PSD becomes:

$$4\pi^2 S_f^{(1)} \overset{(23)}{=} \frac{\Lambda_4\Omega^4 + \Lambda_2\Omega^2 + \Lambda_0}{\Delta_4\Omega^4 + \Delta_2\Omega^2 + \Delta_0}. \tag{A45}$$

Appendix F. Expression for the Intrinsic Linewidth

The laser intrinsic linewidth can be found from Equation (25). Defining:

$$\beta_{Ag} \equiv \frac{\tau_e^{-1}\zeta_s}{a_g G_0 S_0} \quad ; \quad \delta_{Ag} \equiv \frac{\zeta_s}{G_0} \quad ; \quad \Delta f_0 = \frac{R_{sp}}{4\pi S_0}, \tag{A46}$$

where, following [3]:

$$\delta_{Ag} \simeq 0 \simeq \beta_{Ag}, \tag{A47}$$

as $\delta_{Ag} < 10^{-2}$, which accounts for shot noise in the generation and recombination of minority carriers, and β_{Ag} is inversely proportional to the laser power which above threshold becomes negligible. Starting from Equation (A45) and setting $\Omega = 0$ as required by Equation (25):

$$\begin{aligned} 4\pi\Delta f &= \Lambda_0/\Delta_0 \\ &\overset{(A43c)(23)}{=} \frac{(\zeta_s\alpha_H a_g)^2\left[R_{sp}S_0\left(1 + \frac{(G_0)^2}{\zeta_s^2} + \frac{2G_0}{\zeta_s}\right) + N\tau_{sp}^{-1}\right] + \left(\tau_e^{-1}\zeta_s + a_g G_0 S_0\right)^2\frac{R_{sp}}{S_0}}{\left(a_g G_0 S_0 \mathsf{F}_3 + \mathsf{K}_c\zeta_s\tau_e^{-1}\right)^2} \\ \Leftrightarrow \frac{\Delta f}{\Delta f_0} &= \frac{\left(\frac{\alpha_H}{G_0}\right)^2\left(\zeta_s^2 + G_0^2 + 2G_0\zeta_s + \frac{\zeta_s^2 N\tau_{sp}^{-1}}{R_{sp}S_0}\right) + \left(\beta_{Ag} + 1\right)^2}{\left(\mathsf{K}_c\beta_{Ag} + \mathsf{F}_3\right)^2} \\ &= \frac{\left(\beta_{Ag} + 1\right)^2 + \alpha_H^2\left[1 + 2\delta_{Ag} + \delta_{Ag}^2\left(1 + \frac{N\tau_{sp}^{-1}}{R_{sp}S_0}\right)\right]}{\left(\mathsf{K}_c\beta_{Ag} + \mathsf{F}_3\right)^2}\end{aligned}$$

$$\overset{(A47)}{\Leftrightarrow} \frac{\Delta f}{\Delta f_0 (1+\alpha_H^2)} \simeq \mathsf{F}_3^{-2}$$

$$\overset{(23)(20)}{=} \{1+\kappa_2 t_2[\cos(\phi_2) - \alpha_H \sin(\phi_2)] - \kappa_1 t_1[\cos(\phi_1) + \alpha_H \sin(\phi_1)]\}^{-2}$$

$$\overset{(A20a)(A20b)}{=} [1+\gamma_H \kappa_2 t_2 \cos(\phi_2+\theta_H) - \gamma_H \kappa_1 t_1 \cos(\phi_1 - \theta_H)]^{-2}. \quad \text{(A48)}$$

The presence of EOF from both sides of the laser cavity results in two terms in the linewidth expression, one for each side, as was seen in the threshold gain reduction and lasing frequency shift due to feedback. This result is discussed in Section 2.3.

Appendix G. Suplementary Images: Tolerances

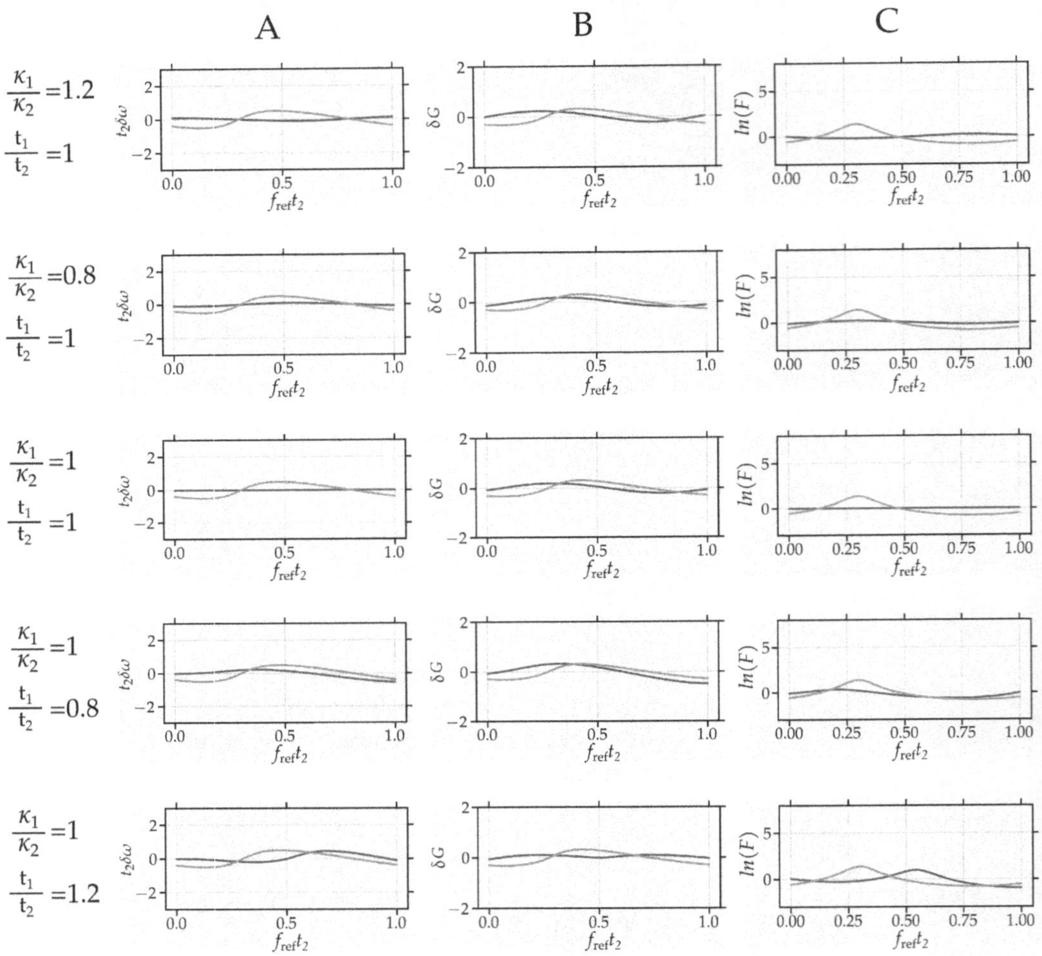

Figure A2. Numerical solutions for $C = 0.5$ under condition (12) for a $\pm 20\%$ variation of conditions (8) and (11). Double feedback case shown in blue, single feedback case shown in orange. Column (**A**) shows the lasing frequency shift results. Column (**B**) shows the threshold gain shift. Column (**C**) shows the intrinsic linewidth variations.

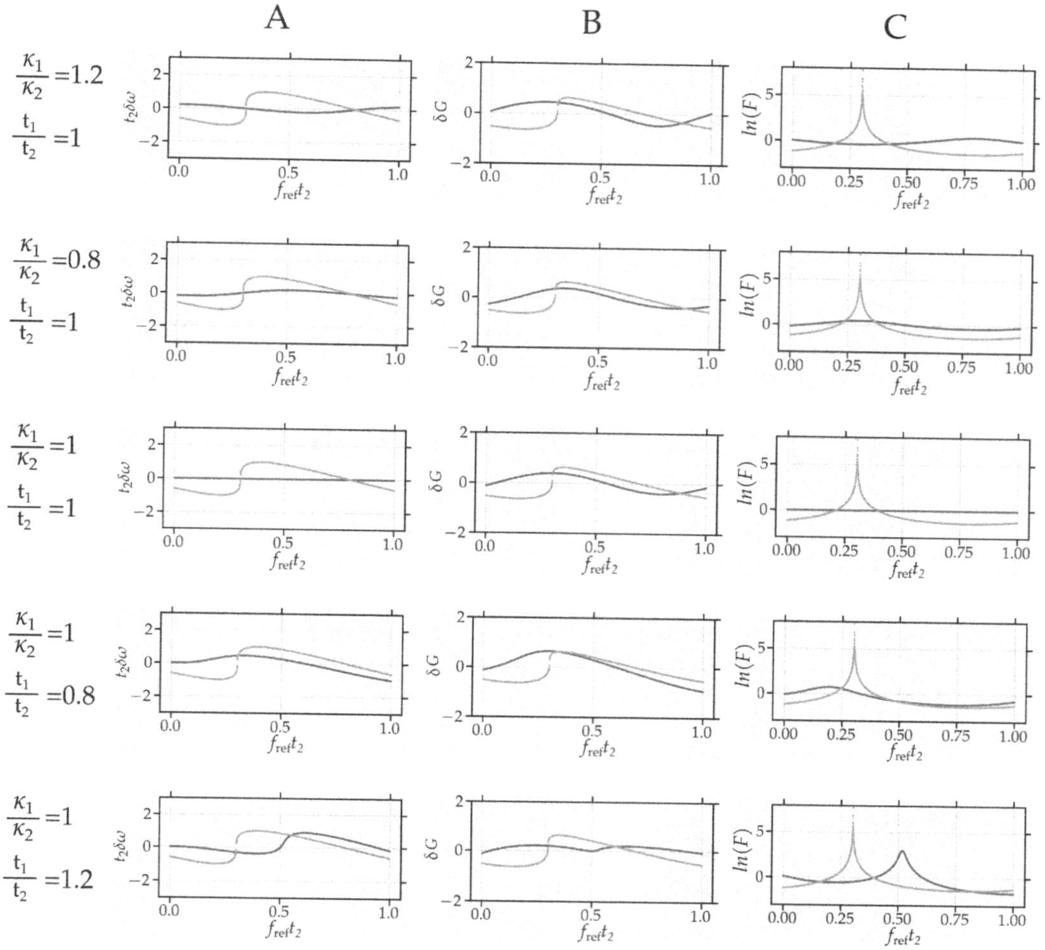

Figure A3. Numerical solutions for $C = 1$ under condition (12) for a ±20% variation of conditions (8) and (11). Double feedback case shown in blue, single feedback case shown in orange. Column (**A**) shows the lasing frequency shift results. Column (**B**) shows the threshold gain shift. Column (**C**) shows the intrinsic linewidth variations.

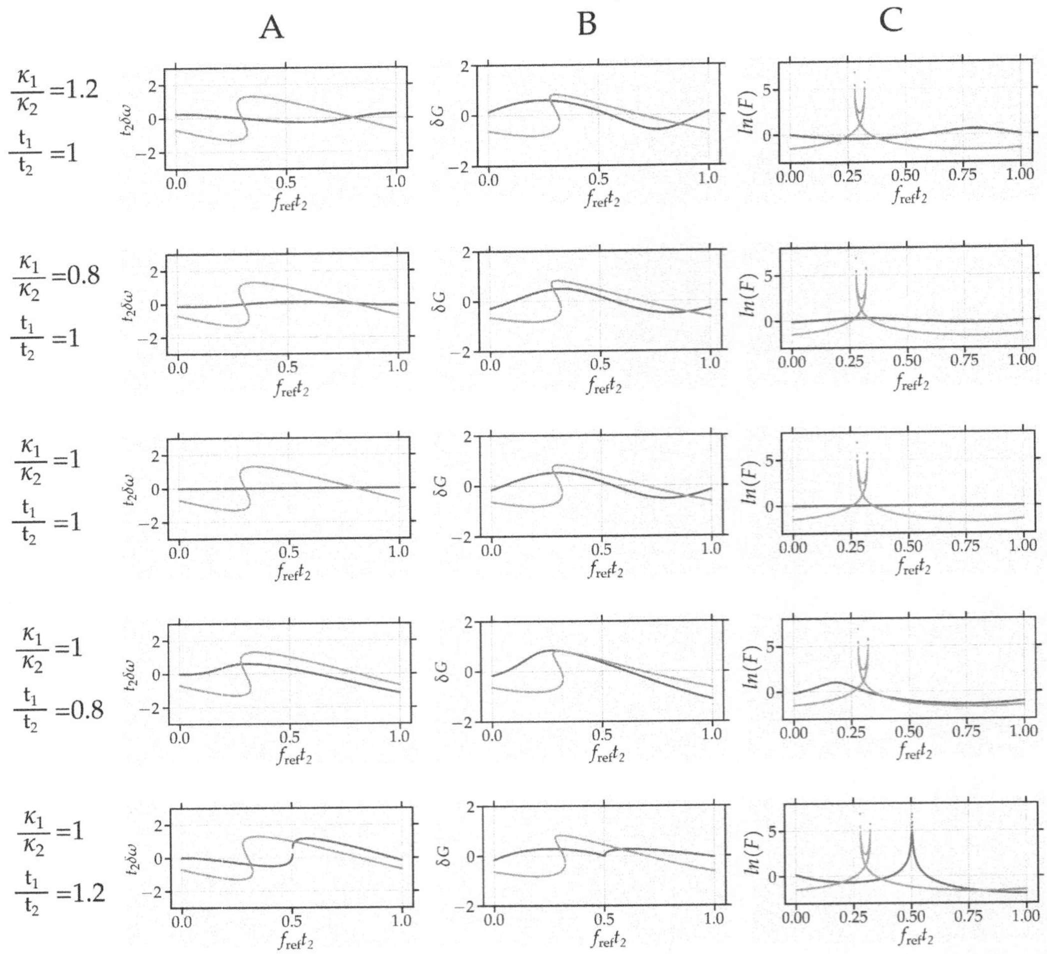

Figure A4. Numerical solutions for $C = 1.3$ under condition (12) for a $\pm 20\%$ variation of conditions (8) and (11). Double feedback case shown in blue, single feedback case shown in orange. Column (**A**) shows the lasing frequency shift results. Column (**B**) shows the threshold gain shift. Column (**C**) shows the intrinsic linewidth variations.

References

1. Sciamanna, M.; Shore, K.A. Physics and applications of laser diode chaos. *Nat. Photonics* **2015**, *9*, 151–162. [CrossRef]
2. Soriano, M.C.; García-Ojalvo, J.; Mirasso, C.R.; Fischer, I. Complex photonics: Dynamics and applications of delay-coupled semiconductors lasers. *Rev. Modern Phys.* **2013**, *85*, 421–470. [CrossRef]
3. Agrawal, G.P. Line Narrowing in a Single-Mode Injection Laser Due to External Optical Feedback. *IEEE J. Quantum Electron.* **1984**, *20*, 468–471. [CrossRef]
4. Olesen, H.; Osmundsen, J.H.; Tromborg, B. Nonlinear dynamics and spectral behaviour for an external cavity laser. *IEEE J. Quantum Electron.* **1986**, *22*, 762–773. [CrossRef]
5. Petermann, K. External optical feedback phenomena in semiconductor lasers. *IEEE J. Sel. Top. Quantum Electron.* **1995**, *1*, 480–489. [CrossRef]
6. Patzak, E.; Olesen, H.; Sugimura, A.; Saito, S.; Mukai, T. Spectral linewidth reduction in semiconductor lasers by an external cavity with weak optical feedback. *Electron. Lett.* **1983**, *19*, 938–940. [CrossRef]
7. Tkach, R.W.; Chraplyvy, A.R. Regimes of Feedback Effects in 1.5-μm Distributed Feedback Lasers. *J. Lightw. Technol.* **1986**, *4*, 1655–1661. [CrossRef]
8. Lenstra, D.; Verbeek, B.; Boef, A.D. Coherence collapse in single-mode semiconductor lasers due to optical feedback. *IEEE J. Quantum Electron.* **1985**, *21*, 674–679. [CrossRef]
9. Henry, C.; Kazarinov, R. Instability of semiconductor lasers due to optical feedback from distant reflectors. *IEEE J. Quantum Electron.* **1986**, *22*, 294–301. [CrossRef]
10. Ebisawa, S.; Komatsu, S. Orbital Instability of Chaotic Laser Diode with Optical Injection and Electronically Applied Chaotic Signal. *Photonics* **2020**, *7*, 25. [CrossRef]
11. Locquet, A. Routes to Chaos of a Semiconductor Laser Subjected to External Optical Feedback: A Review. *Photonics* **2020**, *7*, 22. [CrossRef]

12. Gomez, S.; Huang, H.; Duan, J.; Combrié, S.; Shen, A.; Baili, G.; de Rossi, A.; Grillot, F. High coherence collapse of a hybrid III–V/Si semiconductor laser with a large quality factor. *J. Phys. Photonics* **2020**, *2*, 025005. [CrossRef]
13. Zhao, D.; Andreou, S.; Yao, W.; Lenstra, D.; Williams, K.; Leijtens, X. Monolithically Integrated Multiwavelength Laser With Optical Feedback: Damped Relaxation Oscillation Dynamics and Narrowed Linewidth. *IEEE Photonics J.* **2018**, *10*, 6602108. [CrossRef]
14. Aoyama, K.; Yokota, N.; Yasaka, H. Strategy of optical negative feedback for narrow linewidth semiconductor lasers. *Opt. Express* **2018**, *26*, 21159–21169. [CrossRef]
15. Doerr, C.R.; Dupuis, N.; Zhang, L. Optical isolator using two tandem phase modulators. *Opt. Lett.* **2011**, *36*, 4293–4295. [CrossRef]
16. Van Schaijk, T.T.M.; Lenstra, D.; Williams, K.A.; Bente, E.A.J.M. Model and experimental validation of a unidirectional phase modulator. *Opt. Express* **2018**, *26*, 32388–32403. [CrossRef] [PubMed]
17. Septon, T.; Becker, A.; Gosh, S.; Shtendel, G.; Sichkovskyi, V.; Schnabel, F.; Sengül, A.; Bjelica, M.; Witzigmann, B.; Reithmaier, J.P.; et al. Large linewidth reduction in semiconductor lasers based on atom-like gain material. *Optica* **2019**, *6*, 1071–1077. [CrossRef]
18. Duan, J.; Huang, H.; Dong, B.; Jung, D.; Norman, J.C.; Bowers, J.E.; Grillot, F. 1.3-µm Reflection Insensitive InAs/GaAs Quantum Dot Lasers Directly Grown on Silicon. *IEEE Photonics Technol. Lett.* **2019**, *31*, 345–348. [CrossRef]
19. Yamasaki, K.; Kanno, K.; Matsumoto, A.; Akahane, K.; Yamamoto, N.; Naruse, M.; Uchida, A. Fast dynamics of low-frequency fluctuations in a quantum-dot laser with optical feedback. *Opt. Express* **2021**, *29*, 17962–17975. [CrossRef] [PubMed]
20. Huang, D.; Pintus, P.; Bowers, J.E. Towards heterogeneous integration of optical isolators and circulators with lasers on silicon. *Opt. Mat. Express* **2018**, *8*, 2471–2483. [CrossRef]
21. Yan, W.; Yang, Y.; Liu, S.; Zhang, Y.; Xia, S.; Kang, T.; Yang, W.; Qin, J.; Deng, L.; Bi, L. Waveguide-integrated high-performance magneto-optical isolators and circulators on silicon nitride platforms. *Optica* **2020**, *7*, 1555–1562. [CrossRef]
22. Lenstra, D.; van Schaijk, T.T.M.; Williams, K.A. Toward a Feedback-Insensitive Semiconductor Laser. *IEEE J. Sel. Top. Quantum Electron.* **2019**, *25*, 1–13. [CrossRef]
23. Khoder, M.; der Sande, G.V.; Danckaert, J.; Verschaffelt, G. Effect of External Optical Feedback on Tunable Micro-Ring Lasers Using On-Chip Filtered Feedback. *IEEE Photonics Technol. Lett.* **2016**, *28*, 959–962. [CrossRef]
24. Komljenovic, T.; Liang, L.; Chao, R.L.; Hulme, J.; Srinivasan, S.; Davenport, M.; Bowers, J.E. Widely-Tunable Ring-Resonator Semiconductor Lasers. *Appl. Sci.* **2017**, *7*, 732. [CrossRef]
25. Kasai, K.; Nakazawa, M.; Ishikawa, M.; Ishii, H. 8 kHz linewidth, 50 mW output, full C-band wavelength tunable DFB LD array with self-optical feedback. *Opt. Express* **2018**, *26*, 5675–5685. [CrossRef]
26. Morton, P.A.; Morton, M.J. High-Power, Ultra-Low Noise Hybrid Lasers for Microwave Photonics and Optical Sensing. *J. Lightw. Technol.* **2018**, *36*, 5048–5057. [CrossRef]
27. Xiang, C.; Morton, P.A.; Bowers, J.E. Ultra-narrow linewidth laser based on a semiconductor gain chip and extended Si_3N_4 Bragg grating. *Opt. Lett.* **2019**, *44*, 3825–3828. [CrossRef]
28. Liu, Y.; Ohtsubo, J. Dynamics and chaos stabilization of semiconductor lasers with optical feedback from an interferometer. *IEEE J. Quantum. Elect.* **1997**, *33*, 1163–1169. [CrossRef]
29. Mezzapesa, F.P.; Columbo, L.L.; Dabbicco, M.; Brambilla, M.; Scamarcio, G. QCL-based nonlinear sensing of independent targets dynamics. *Opt. Express* **2014**, *22*, 5867. [CrossRef] [PubMed]
30. Zhu, W.; Chen, Q.; Wang, Y.; Luo, H.; Wu, H.; Ma, B. Improvement on vibration measurement performance of laser self-mixing interference by using a pre-feedback mirror. *Opt. Laser Eng.* **2018**, *105*, 150–158. [CrossRef]
31. Ruan, Y.; Liu, B.; Yu, Y.; Xi, J.; Guo, Q.; Tong, J. High sensitive sensing by a laser diode with dual optical feedback operating at period-one oscillation. *Appl. Phys. Lett.* **2019**, *115*, 011102. [CrossRef]
32. Seimetz, M. *Laser Linewidth Limitations for Optical Systems with High-Order Modulation Employing Feed Forward Digital Carrier Phase Estimation*; Paper OTuM2; OFC: Auckland, New Zealand, 2008.
33. Liang, W.; Ilchenko, V.S.; Eliyahu, D.; Dale, E.; Savchenkov, A.A.; Seidel, D.; Matsko, A.B.; Maleki, L. Compact stabilized semiconductor laser for frequency metrology. *Appl. Opt.* **2015**, *54*, 3353. [CrossRef] [PubMed]
34. Barton, J.S.; Skogen, E.J.; Mašanović, M.L.; Denbaars, S.P.; Coldren, L.A. A widely tunable high-speed transmitter using an integrated SGDBR laser-semiconductor optical amplifier and Mach–Zehnder modulator. *IEEE J. Sel. Top. Quantum Electron.* **2003**, *9*, 1113–1117. [CrossRef]
35. Huang, D.; Tran, M.A.; Guo, J.; Peters, J.; Komljenovic, T.; Malik, A.; Morton, P.A.; Bowers, J.E. High-power sub-kHz linewidth lasers fully integrated on silicon. *Optica* **2019**, *5*, 745–752. [CrossRef]
36. Wang, H.; Kim, D.; Harfouche, M.; Santis, C.T.; Satyan, N.; Rakuljic, G.; Yariv, A. Narrow-Linewidth Oxide-Confined Heterogeneously Integrated Si III-V Semiconductor Lasers. *IEEE Photonics Technol. Lett.* **2017**, *29*, 2199–2202. [CrossRef]
37. Lin, Y.; Browning, C.; Timens, R.B.; Geuzebroek, D.H.; Roeloffzen, C.G.H.; Hoekman, M.; Geskus, D.; Oldenbeuving, R.M.; Heideman, R.G.; Fan, Y.; et al. Characterization of Hybrid InP-TriPleX Photonic Integrated Tunable Lasers Based on Silicon Nitride (Si_3N_4/SiO_2) Microring Resonators for Optical Coherent System. *IEEE Photonics J.* **2018**, *10*, 1400108. [CrossRef]
38. Li, B.; Jin, W.; Wu, L.; Chang, L.; Wang, H.; Shen, B.; Yuan, Z.; Feshali, A.; Paniccia, M.; Vahala, K.J.; et al. Reaching fiber-laser coherence in integrated photonics. *Opt. Lett.* **2021**, *46*, 5201–5204. [CrossRef]
39. Lang, R.; Kobayashi, K. External Optical Feedback Effects on Semiconductor Injection Laser Properties. *IEEE J. Quantum Electron.* **1980**, *16*, 347–355. [CrossRef]

40. Smit, M.; Leijtens, X.; Ambrosius, H.; Bente, E.; van der Tol, J.; Smalbrugge, B.; de Vries, T.; Geluk, E.J.; Bolk, J.; van Veldhoven, R.; et al. An introduction to InP-based generic integration technology. *Semicond. Sci. Technol.* **2014**, *29*, 083001. [CrossRef]
41. Augustin, L.M.; Santos, R.; den Haan, E.; Kleijn, S.; Thijs, P.J.A.; Latkowski, S.; Zhao, D.; Yao, W.; Bolk, J.; Ambrosius, H.; et al. InP-Based Generic Foundry Platform for Photonic Integrated Circuits. *IEEE J. Sel. Top. Quantum Electron.* **2018**, *24*, 6100210. [CrossRef]
42. Henry, C.H. Theory of the Linewidth of Semiconductor Lasers. *IEEE J. Quantum Electron.* **1982**, *18*, 259–264. [CrossRef]
43. Osinski, M.; Buus, J. Linewidth broadening factor in semiconductor lasers–An overview. *IEEE J. Quantum Electron.* **1987**, *23*, 9–29. [CrossRef]
44. Lax, M.; Noise, V.C. Noise in Self-Sustained Oscillators. *Phys. Rev.* **1967**, *160*, 290–307. [CrossRef]
45. Rowe, H.E. *Signals and Noise in Communication Systems*, 1st ed.; D. Van Nostrand Company. Inc.: Princeton, NJ, USA, 1965; Chapter 2.3.
46. Spano, P.; Piazzolla, S.; Tamburrini, M. Theory of noise in semiconductor lasers in the presence of optical feedback. *IEEE J. Quantum Electron.* **1984**, *20*, 350–357. [CrossRef]
47. Schawlow, A.L.; Townes, C.H. Infrared and Optical Masers. *Phys. Rev.* **1958**, *112*, 1940–1949. [CrossRef]
48. Yu, Y.; Giuliani, G.; Donati, S. Measurement of the Linewidth Enhancement Factor of Semiconductor Lasers Based on the Optical Feedback Self-Mixing Effect. *IEEE Photonics Technol. Lett.* **2004**, *16*, 990–992. [CrossRef]
49. Schunk, N.; Petermann, K. Numerical analysis of the feedback regimes for a single-mode semiconductor laser with external feedback. *IEEE J. Quantum Electron.* **1988**, *24*, 1242–1247. [CrossRef]
50. Panozzo, M.; Quintero-Quiroz, C.; Tiana-Alsina, J.; Torrent, M.C.; Masoller, C. Experimental characterization of the transition to coherence collapse in a semiconductor laser with optical feedback. *Chaos* **2017**, *27*, 114315. [CrossRef]
51. Happach, M.; Felipe, D.D.; Friedhoff, V.N.; Kresse, M.; Irmscher, G.; Kleinert, M.; Zawadzki, C.; Rehbein, W.; Brinker, W.; Mohrle, M.; et al. Influence of Integrated Optical Feedback on Tunable Lasers. *IEEE J. Quantum Electron.* **2020**, *56*. [CrossRef]
52. Petermann, K. *Laser Diode Modulation and Noise (Advances in Optoelectronics 3)*; Kluwer Academic Publishers: Dordrecht, The Netherlands, 1988; Chapter 2.

Nonlinear Dynamics of Two-State Quantum Dot Lasers under Optical Feedback

Abstract: A modified rate equation model was presented to theoretically investigate the nonlinear dynamics of solitary two-state quantum dot lasers (TSQDLs) under optical feedback. The simulated results showed that, for a TSQDL biased at a relatively high current, the ground-state (GS) and excited-state (ES) lasing of the TSQDL can be stimulated simultaneously. After introducing optical feedback, both GS lasing and ES lasing can exhibit rich nonlinear dynamic states including steady state (S), period one (P1), period two (P2), multi-period (MP), and chaotic (C) state under different feedback strength and phase offset, respectively, and the dynamic states for the two lasing types are always identical. Furthermore, the influences of the linewidth enhancement factor (LEF) on the nonlinear dynamical state distribution of TSQDLs in the parameter space of feedback strength and phase offset were also analyzed. For a TSQDL with a larger LEF, much more dynamical states can be observed, and the parameter regions for two lasing types operating at chaotic state are widened after introducing optical feedback.

Keywords: nonlinear dynamics; quantum dot lasers; optical feedback; chaotic; linewidth enhancement factor (LEF)

---

1. Introduction

After introducing external perturbations, semiconductor lasers (SLs) can exhibit rich nonlinear dynamics [1,2], which can be applied in many fields such as random number generation, secure communication, photonic microwave signal generation, all-optical logic gates, and reservoir computing [3–7].

Quantum dot (QD) lasers are self-assembled nanostructured SLs. Compared with traditional quantum well (QW) SLs, QD lasers have many advantages such as low threshold current density [8], high temperature stability [9], low chirp [10], and large modulation bandwidth [11]. Such unique characteristics make QD lasers become excellent candidate light sources in optical communication, optical interconnection, silicon photonic integrated circuits, and photonic microwave generation, etc. [12–16]. Due to strong three-dimension quantum confinement of the carriers, QD lasers have discrete energy levels and state densities, which lead to their unique emission performances. Related studies have shown that there exist two current thresholds in ordinary QD lasers. When the bias current is increased to the first threshold, QD lasers can emit on the ground-state (GS). Continuously increasing the bias current, the number of carriers at the excited-state (ES) increases rapidly. Once the bias current exceeds a certain value (the second threshold), QD lasers can simultaneously emit on GS and ES. Correspondingly, such QD lasers are named as two-state QD lasers (TSQDLs) [17,18]. Via some technologies, QD lasers can emit solely on GS or ES, and the corresponding QD lasers are named as GS-QD lasers and ES-QD lasers, respectively [19,20].

Previous studies have shown that different types of QD lasers can exhibit different performances. GS-QD lasers possess a low threshold current and low sensitivity to optical feedback owing to relatively low energy levels and strong damping of relaxation oscillation [21,22]. Compared with GS-QD lasers, ES-QD lasers possess larger modulation bandwidths and richer nonlinear dynamics under external perturbations owing to faster carrier capture rates [23–26]. Different from GS-QD lasers and ES-QD lasers, TSQDLs can lase at two wavelengths separated by several tens of nanometers [27] and exhibit lower intensity noise [28], which can be applied in many fields such as terahertz (THz) signal generation, two-color light sources, two color mode-locking, all-optical processing, and artificial optical neurons, etc. [29–32]. In recent years, the investigations on the nonlinear dynamics of TSQDLs under external perturbations have attracted special attention. Through introducing optical injection into GS, the ES emission in TSQDLs can be suppressed and the mode switching from ES to GS is triggered [33,34]. Through scanning the optical power of injection light along different varying routes, a bistable phenomenon can be observed [35,36]. After introducing optical feedback to TSQDLs, many interesting phenomena can be observed such as mode switching and mode competition between the GS and ES [37,38], energy exchanging among longitudinal modes [39], two-color oscillating [40], and anti-phase low frequency fluctuating [41]. However, to our knowledge, the nonlinear dynamical state evolution of TSQDLs under optical feedback has not been reported.

In this work, based on three-level model of QD lasers [42,43], a modified theoretical model for TSQDLs under optical feedback was presented to numerically investigate the nonlinear dynamical characteristics of TSQDLs under optical feedback. Moreover, the influences of the linewidth enhancement factor (LEF) on the nonlinear dynamical state distribution of TSQDLs in the parameter space of feedback strength and phase offset were also analyzed.

2. Rate Equation Model

The theoretical model in this work was based on the three-level model of QD lasers, which has been adopted to analyze the static and dynamic behaviors, noise characteristics of QD lasers operating at free-running [42,43], and the small-signal modulation response and relative intensity noise of QD lasers under optical injection-locking conditions [44]. Figure 1 shows the simplified schematic diagram of the carrier dynamics for two-state QD lasers (TSQDLs) based on the three-level model [45]. In this system, two relatively low energy levels involving ground state (GS) and the first excited state (ES) were taken into account. The electrons and holes were treated as neutral excitons (electron-hole pairs), and the stimulated emission can occur in GS and ES. It was assumed that all QDs had the same size and the active region consisted of only one QD ensemble. Therefore, the inhomogeneous broadening effect was ignored. As shown in the figure, the carriers were injected directly into the wetting layer (WL) from the electrodes. In the WL, owing to Auger recombination and phonon-assisted scattering processes [46,47], some carriers were captured into ES with a captured time τ_{ES}^{WL}. Some carriers relaxed directly into GS with a relaxation time τ_{GS}^{WL}. The rest of the carriers recombined spontaneously with a time τ_{WL}^{spon}. For the carriers in ES, some of them relaxed into GS with a relaxation time τ_{GS}^{ES} and the other carriers recombined spontaneously with an emission time τ_{ES}^{spon}. On the other hand, owing to the thermal excitation effect, some carriers were excited into WL with an escape time τ_{WL}^{ES}. Similarly, the carriers in GS were excited into ES with an escape time τ_{ES}^{GS}, and some carriers also recombined spontaneously with an emission time τ_{GS}^{spon}. Based on the three-level model, after referring to the optical feedback processing methods in Ref. [48], we propose modified rate equations for describing the nonlinear dynamics of TSQDLs under optical feedback as follows:

$$\frac{dN_{WL}}{dt} = \frac{\eta I}{q} + \frac{N_{ES}}{\tau_{WL}^{ES}} - \frac{N_{WL}}{\tau_{ES}^{WL}}(1-\rho_{ES}) - \frac{N_{WL}}{\tau_{GS}^{WL}}(1-\rho_{GS}) - \frac{N_{WL}}{\tau_{WL}^{spon}} \tag{1}$$

$$\frac{dN_{ES}}{dt} = \frac{N_{WL}}{\tau_{ES}^{WL}}(1-\rho_{ES}) + \frac{N_{GS}}{\tau_{ES}^{GS}}(1-\rho_{ES}) - \frac{N_{ES}}{\tau_{WL}^{ES}} - \frac{N_{ES}}{\tau_{GS}^{ES}}(1-\rho_{GS}) - \frac{N_{ES}}{\tau_{ES}^{spon}} - \Gamma_P v_g g_{ES} S_{ES} \tag{2}$$

$$\frac{dN_{GS}}{dt} = \frac{N_{WL}}{\tau_{GS}^{WL}}(1-\rho_{GS}) + \frac{N_{ES}}{\tau_{GS}^{ES}}(1-\rho_{GS}) - \frac{N_{GS}}{\tau_{ES}^{GS}}(1-\rho_{ES}) - \frac{N_{GS}}{\tau_{GS}^{spon}} - \Gamma_P v_g g_{GS} S_{GS} \tag{3}$$

$$\frac{dS_{GS}}{dt} = \left(\Gamma_P v_g g_{GS} - \frac{1}{\tau_p}\right) S_{GS} + \beta_{sp}\frac{N_{GS}}{\tau_{GS}^{spon}} + 2\frac{\text{k}}{\tau_{in}}\sqrt{S_{GS}(t)S_{GS}(t-\tau)}\cos(\Delta\phi_{GS}) \tag{4}$$

$$\frac{dS_{ES}}{dt} = \left(\Gamma_P v_g g_{ES} - \frac{1}{\tau_p}\right) S_{ES} + \beta_{sp}\frac{N_{ES}}{\tau_{ES}^{spon}} + 2\frac{k}{\tau_{in}}\sqrt{S_{ES}(t)S_{ES}(t-\tau)}\cos(\Delta\phi_{ES}) \tag{5}$$

$$\frac{d\phi_{GS}}{dt} = \frac{\alpha}{2}\left(\Gamma_P v_g g_{GS} - \frac{1}{\tau_p}\right) - \frac{k}{\tau_{in}}\sqrt{\frac{S_{GS}(t-\tau)}{S_{GS}(t)}}\sin(\Delta\phi_{GS}) \tag{6}$$

$$\frac{d\phi_{ES}}{dt} = \frac{\alpha}{2}\left(\Gamma_P v_g g_{ES} - \frac{1}{\tau_p}\right) - \frac{k}{\tau_{in}}\sqrt{\frac{S_{ES}(t-\tau)}{S_{ES}(t)}}\sin(\Delta\phi_{ES}) \tag{7}$$

where *WL*, *ES*, *GS* are the wetting layer, excited-state, and ground-state, respectively, and the superscript *spon* represents the spontaneous emission. N, S, ϕ are the carrier number, photon number, and phase, respectively. I is the injection current, η is the current injection efficiency, and q is the electron charge. Γ_p is the optical confinement factor, v_g (= c/n_r, where c is the light speed in vacuum and n_r the refractive index) is the group velocity. τ_p is the photon lifetime, τ_{in} is the round-trip time in the laser cavity, and τ (= $2\,l_{ex}/c$, where l_{ex} the external cavity length) is the round-trip time of external cavity. k is the feedback strength, and α is the linewidth enhancement factor. Considering that GS and ES have twofold degeneration and fourfold degeneration, respectively, the carrier occupation probabilities and the gains of GS and ES can be expressed as [42]:

$$\rho_{GS} = \frac{N_{GS}}{2N_B}; \rho_{ES} = \frac{N_{ES}}{4N_B} \tag{8}$$

$$g_{GS} = \frac{a_{GS}}{1+\xi_{GS}\frac{S_{GS}}{V_S}}\frac{N_B}{V_B}(2\rho_{GS}-1) \tag{9}$$

$$g_{ES} = \frac{a_{ES}}{1+\xi_{ES}\frac{S_{ES}}{V_S}}\frac{N_B}{V_B}(2\rho_{ES}-1) \tag{10}$$

where N_B is the number of quantum dots. a_{GS} and a_{ES} are the differential gain, ξ_{GS} and ξ_{ES} are the gain compression factor, vs. is the volume of the laser field inside the cavity, and V_B is the volume of the active region. The feedback phase variation can be described as:

$$\Delta\phi_{GS} = \phi_{GS}(t) - \phi_{GS}(t-\tau) + \omega_{GS}\tau \tag{11}$$

$$\Delta\phi_{ES} = \phi_{ES}(t) - \phi_{ES}(t-\tau) + \omega_{ES}\tau \tag{12}$$

where ω_{GS} and ω_{ES} are the angular frequencies for GS and ES lasing, respectively.

The rate equations can be numerically solved by the fourth-order Runge-Kutta method via MATLAB software. During the calculations, the used parameters and their values are given in Table 1 [42]:

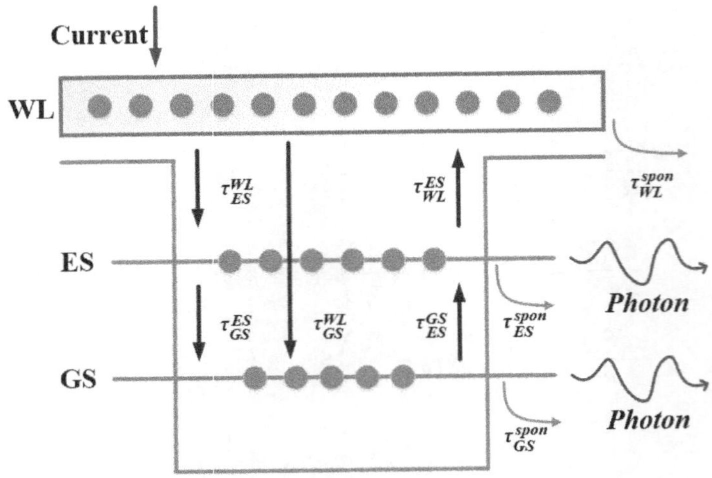

Figure 1. Schematic diagram of the carrier dynamics for QD lasers based on the three-level model. WL: wetting layer; GS: ground state; ES: excited state.

Table 1. Simulation parameters of the QD lasers.

Symbol	Parameter	Value
τ_{ES}^{WL}	Capture time from WL to ES	12.6 ps
τ_{GS}^{ES}	Capture time from ES to GS	8 ps
τ_{GS}^{WL}	Relaxation time from WL to GS	15 ps
τ_{ES}^{GS}	Escape time from GS to ES	10.4 ps
τ_{WL}^{ES}	Escape time from ES to WL	5.4 ns
τ_{WL}^{spon}	Spontaneous emission time from WL	0.5 ns
τ_{ES}^{spon}	Spontaneous emission time from ES	0.5 ns
τ_{GS}^{spon}	Spontaneous emission time from GS	1.2 ps
τ_P	Photon lifetime	4.1 ps
N_B	Total number of QD	1.0×10^7
Γ_p	Optical confinement factor	0.06
n_r	Refractive index	3.5
τ_{in}	Round-trip time	10 ps
a_{GS}	Differential gain from GS	5.0×10^{-15} cm^2
a_{ES}	Differential gain from ES	10.0×10^{-15} cm^2
ξ_{GS}	Gain compression factor from GS	1.0×10^{-16} cm^3
ξ_{ES}	Gain compression factor from ES	8.0×10^{-16} cm^3
β_{sp}	Spontaneous emission factor	5.0×10^{-6}
ω_{GS}	Angular frequency from GS	1.446×10^{15} rad/s
ω_{ES}	Angular frequency from ES	1.529×10^{15} rad/s
V_B	Active region volume	5.0×10^{-11} cm^3
V_S	Resonant cavity volume	0.833×10^{-15} cm^3
η	Injection efficiency	0.25
q	Elementary charge	1.6×10^{-19} C
τ	Feedback delay time	100 ps
α	Linewidth enhancement factor	3.5

3. Results and Discussion

Figure 2 shows the normalized output power of the GS and ES lasing as a function of the injection current for a TSQDL under free-running (solid lines) or optical feedback with a feedback strength of $k = 0.11$ (dotted lines). For the TSQDL operating at free-running, the threshold currents of the GS and ES lasing were 36 mA (I_{th}^{GS}) and 88 mA (I_{th}^{ES}), respectively. With the increase of the current from 36 mA to 88 mA, the power of GS lasing gradually increased while the ES lasing was always in a suppressed state. However, once the injection current was exceeded 88 mA, the ES lasing could be observed. Further increasing the current, the power of the ES lasing rapidly increased while the power of the GS lasing increased slowly. Above results are in agreement with those reported in Ref. [43]. After introducing an optical feedback of $k = 0.11$, the threshold current for GS slightly decreased, which is similar with that observed in a single-mode distributed feedback semiconductor laser under optical feedback. However, optical feedback raises the threshold of ES. The reason is that the predominant component in the feedback light is originating from GS lasing, and therefore the optical feedback enhances the competitiveness of the GS lasing. Correspondingly, a higher current is needed for ES to start oscillation. In the following, we fixed the current of the TSQDL at 120 mA, at which the power of GS lasing was more than that of ES lasing.

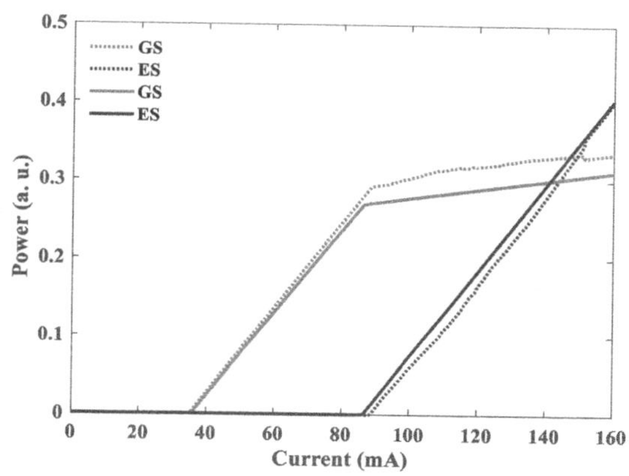

Figure 2. Normalized output power as a function of the injection current for a TSQDL under free-running (solid lines) or optical feedback with a feedback strength of $k = 0.11$ (dotted lines).

Figure 3 displays the time series, power spectra, and phase portraits of typical dynamic state output from GS lasing and ES lasing of a TSQDL biased at 120 mA under optical feedback with $\tau = 100$ ps and different k. For $k = 0.03$, the output intensity of GS lasing (Figure 3(a1)) was nearly a constant, the power spectrum was relatively smooth (Figure 3(a2)), and the phase portrait was a dot (Figure 3(a3)). Obviously, under this case, the dynamical state of GS lasing is a stable (S) state. For $k = 0.07$, the time series of GS lasing (Figure 3(b1)) exhibited a stable periodic oscillation with a fundamental frequency of about 6.3 GHz obtained from the power spectrum (Figure 3(b2)), and the phase portrait is a dense dot (Figure 3(b3)). Based on these characteristics, the dynamic state of GS lasing can be judged as a period-one (P1) state. For $k = 0.092$, the time series of GS lasing (Figure 3(c1)) behaves periodic oscillation with two peak intensities, both the subharmonic frequency (about 3.1 GHz) and the fundamental frequency (about 6.3 GHz) present clearly in the power spectrum (Figure 3(c2)), and the corresponding phase portrait (Figure 3(c3)) is two closed circles, which are typical characteristics of period-two (P2) state. For $k = 0.097$, the time series of GS lasing (Figure 3(d1)) exhibited multiple different peaks, a quarter-harmonic frequency component appeared in the power spectrum (Figure 3(d2)), and the phase portrait (Figure 3(d3)) showed multiple loops. These features mean that the dynamical state of GS lasing is a multi-period (MP) state. For $k = 0.154$, the time series of GS lasing (Figure 3(e1)) showed a disordered oscillation, and the power spectra were broadened (Figure 3(e2)). In addition, the corresponding phase portrait (Figure 3(e3)) showed a strange attractor. Therefore, the dynamic state of GS lasing can be determined to be the chaotic (C) state. Through comparing the characteristics of ES lasing with those of

GS lasing, it can be seen that the dynamical states of ES lasing are always the same as those of GS lasing.

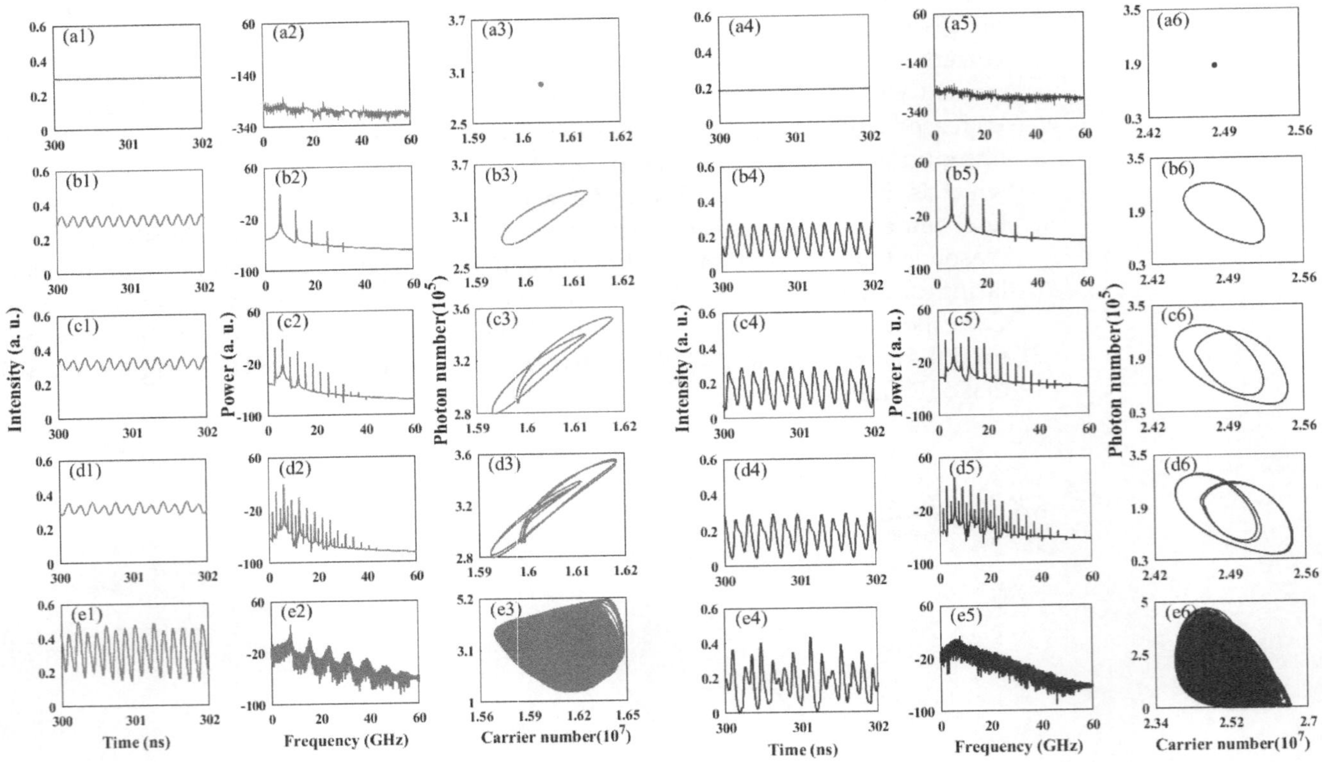

Figure 3. Time series, power spectra, and phase portraits output from GS lasing (red) and ES lasing (blue) in a TSQDL biased at 120 mA under optical feedback with τ = 100 ps and k = 0.03 (**a**), 0.07 (**b**), 0.092 (**c**), 0.097 (**d**), and 0.154 (**e**), respectively.

Above results show that, through setting feedback parameters at different values, some typical dynamical states can be observed for both ES and GS lasing. In order to inspect the evolution route of dynamical state with the feedback strength, Figure 4 presents the bifurcation diagrams of the power extreme and largest Lyapunov exponent (LLE) of the GS lasing and ES lasing as a function of feedback strength. LLE is an important indicator to measure the stability of a laser nonlinear dynamical system [49]. A positive LLE value means that the laser operates at a chaotic state while a negative LLE value corresponds to a steady state. For a laser operating at periodic states, the LLE value tends to approach zero. From this diagram, it can be seen that, with the increase of k from 0 to 0.043, the output of GS lasing and ES lasing remains in a stable state due to the relatively low feedback strength. Further increasing the feedback strength, the external cavity modes compete with the intrinsic oscillation frequency of the laser, and the dynamic states of GS lasing and ES lasing transform into periodic states including P1, P2, and MP. When the feedback strength exceeds 0.11, the TSQDL enters into the C state due to coherent collapse. As a result, the dynamics evolution routes of S-P1-P2-MP-C of the GS lasing and ES lasing are presented. Continuously increasing the feedback strength, the laser enters into the chaos state through period-doubling bifurcation, and such an evolution process repeats continuously.

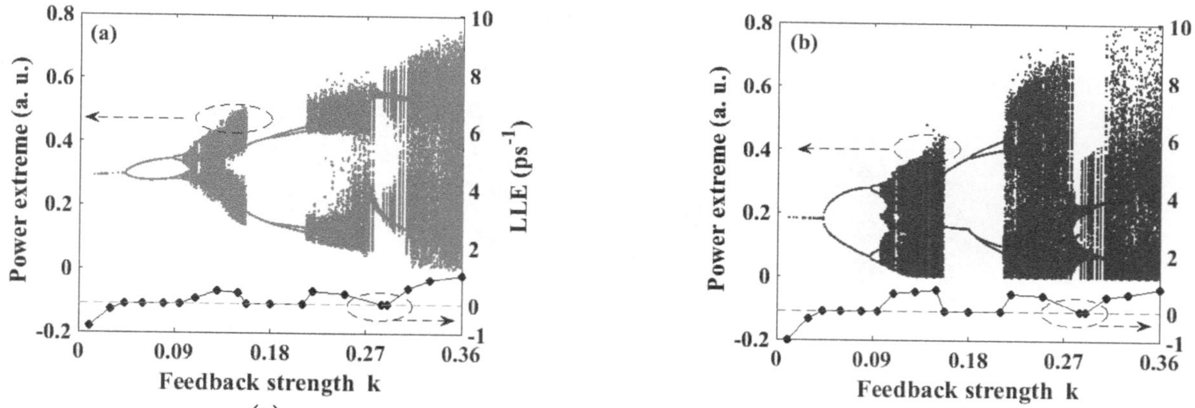

(a) (b)

Figure 4. Bifurcation diagrams of power extreme and largest Lyapunov exponent (LLE) as a function of feedback strength of the GS lasing (**a**) and ES lasing (**b**) in a TSQDL biased at 120 mA under optical feedback with τ = 100 ps.

Next, we discuss the influences of the round-trip time (τ) of the external cavity under a given feedback strength of k = 0.1. Here, we only consider the case that τ is varied around τ_0 = 100 ps within a very small range, in which the offset ($\Delta\tau$) of τ from τ_0 = 100 ps satisfies $-\pi/\omega_{GS} \leq \Delta\tau \leq \pi/\omega_{GS}$. Under this case, the phase offset $\varphi(=\Delta\tau\omega_{GS})$ of GS lasing is varied within $(-\pi, \pi)$, and the corresponding phase offset of ES lasing is varied within $(-1.06\pi, 1.06\pi)$. Figure 5 presents the bifurcation diagrams of the power extreme and LLE of the GS lasing and ES lasing as a function of phase offset under k = 0.1. With the increase of phase offset φ from $-\pi$ to π, the dynamics evolution routes are more diverse. There exist multiple chaotic evolution routes for GS lasing and ES lasing including P1-S-C, P2-P1-P2-C, and C -MP-P2-C.

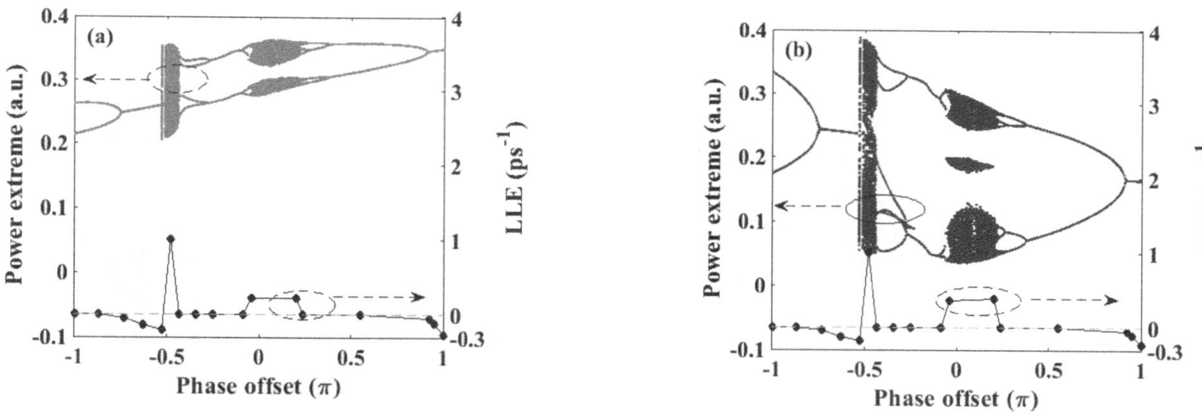

(a) (b)

Figure 5. Bifurcation diagrams of the power extreme; LLE as a function of phase offset of the GS lasing (**a**) and ES lasing (**b**) in a TSQDL under I = 120 mA and k = 0.1.

The above results demonstrate that the feedback strength and the round-trip time τ (equivalent to phase offset) of the external cavity are two crucial parameters affecting the nonlinear dynamics of TSQDLs. Therefore, it is essential to investigate the overall dynamical evolution in the parameter space of feedback strength and phase offset. Figure 6 presents the mapping of the dynamical states for GS lasing (a) and ES lasing (b) in the parameter space of feedback strength and phase offset. There are rich dynamic states including S, P1, P2, MP, and C in the parameter space. With the increase of feedback strength, the phase offset required for achieving a chaotic state is gradually widened. Although the dynamic state distributions of GS lasing and ES lasing are similar, there exist subtle differences at the boundary between two modes. Through observing this diagram carefully, it can be found that there are multiple evolution routes for driving the laser into the chaotic state such as S-P1-P2-MP-C, P1-P2-MP-C, and P1-MP-C.

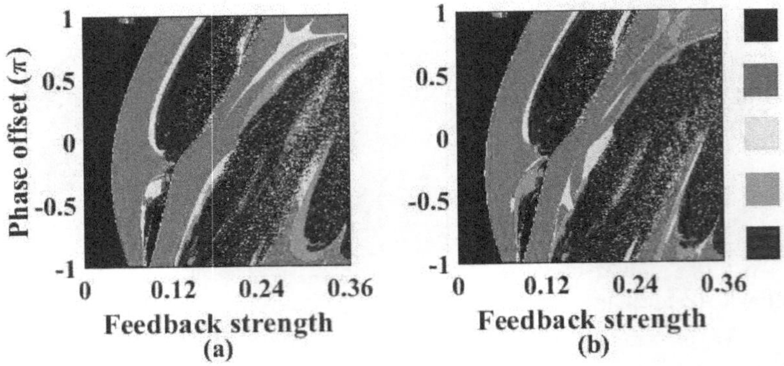

Figure 6. Mapping of the dynamical states for GS lasing (**a**) and ES lasing (**b**) of a TSQDL in the parameter space of feedback strength and phase offset. S: stable, P1: period-one, P2: period-two, MP: multi-period, and C: chaos.

Relevant research shows that the linewidth enhancement factor (LEF) α plays an important role for the nonlinear dynamics of SLs under external perturbations [50,51]. The above results were obtained under a fixed α taken as 3.5. Finally, we discuss the influences of LEF on the dynamical state distribution of a TSQDL under optical feedback. Figure 7 depicts mappings of dynamic states of GS lasing and ES lasing under different α. For $\alpha = 0.5$ (Figure 7(a1,a2), the dynamical states of GS and ES are relatively simple, which include S, P1, and C. In the whole parameter space, most of the region is in a stable state, and only a small region is in the chaotic state. For $\alpha = 2.5$, as shown in Figure 7(b1,b2), there are much richer dynamic states involving P2 and MP. For a larger α of 4.5 as shown in Figure 7(c1,c2), the chaotic state occupies a large area. Therefore, a large α is helpful for achieving chaotic state output.

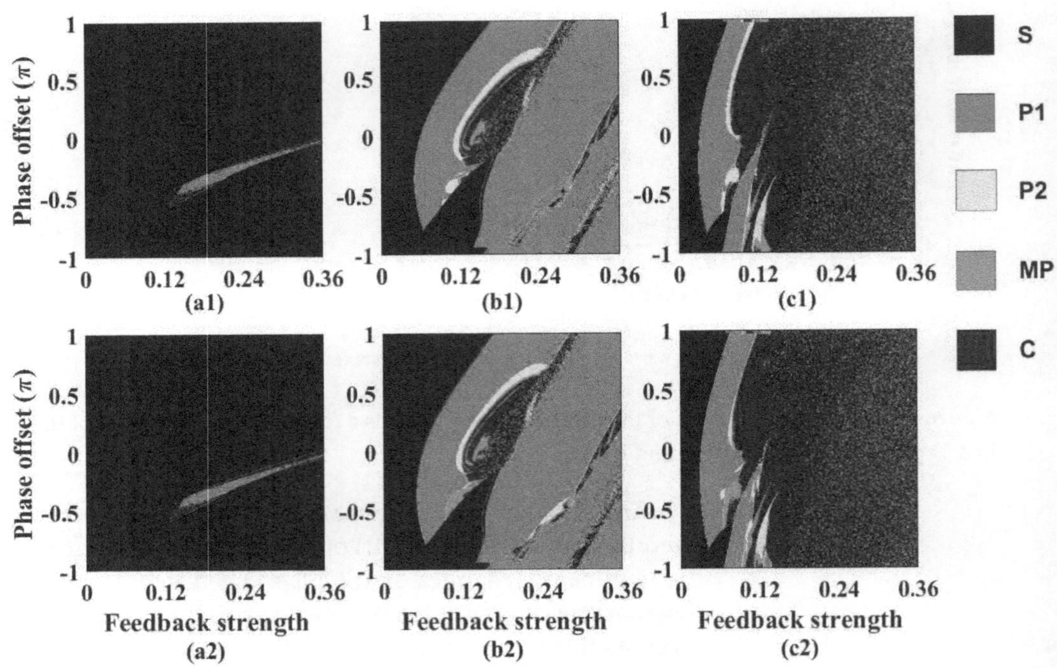

Figure 7. Mappings of the dynamical states of GS lasing (the first row) and ES lasing (the second row) in the parameter space of feedback strength and phase offset under different α, where (**a**) $\alpha = 0.5$, (**b**) $\alpha = 2.5$, (**c**) $\alpha = 4.5$. S: stable, P1: period-one, P2: period-two, MP: multi-period, and C: chaos.

Additionally, it should be pointed out that above results were obtained under the condition that the spontaneous emission noises were ignored. In fact, after considering the influence of spontaneous emission noise, the boundary of dynamical states may be changed slightly.

4. Conclusions

In summary, via a rate equation model used to characterize TSQDLs with optical feedback, the nonlinear dynamics of TSQDLs subject to optical feedback were investigated theoretically. For a TSQDL biased at 120 mA, both GS and ES lasing could be stimulated simultaneously, and the output power of GS emission was slightly larger than that of ES emission. After introducing optical feedback, multiple nonlinear dynamical states including S, P1, P2, MP, and C were observed for GS lasing and ES lasing under suitable feedback strengths and phase offset. Through mapping the evolution of dynamics state in the parameter space of feedback strength and phase offset, different evolution routes were revealed. In addition, the influences of the linewidth enhanced factor (LEF) on the dynamic state distribution of TSQDLs in the space parameter of feedback strength and phase shift were also presented. For a larger LEF, the parameter regions for GS lasing and ES lasing operating at chaotic state were wider. Although the dynamical behaviors of TSQDLs under optical feedback were similar to those observed in quantum well lasers under optical feedback, TSQDLs under optical feedback have the ability to provide two-channel chaotic signals with different lasing wavelengths, which are more promising for high-speed random number generation, wavelength-division multiplexing secure communication, and parallel-reservoir computing.

References

1. Lin, F.Y.; Tu, S.Y.; Huang, C.C.; Chang, S.M. Nonlinear dynamics of semiconductor lasers under repetitive optical pulse injection. *IEEE J. Sel. Top. Quantum Electron.* **2009**, *15*, 604–611. [CrossRef]
2. Yuan, G.H.; Yu, S.Y. Bistability and switching properties of semiconductor ring lasers with external optical injection. *IEEE J. Quantum Electron.* **2008**, *44*, 41–48. [CrossRef]
3. Kawaguchi, Y.; Okuma, T.; Kanno, K.; Uchida, A. Entropy rate of chaos in an optically injected semiconductor laser for physical random number generation. *Opt. Express* **2021**, *29*, 2442–2457. [CrossRef] [PubMed]
4. Jiang, N.; Xue, C.P.; Lv, Y.X.; Qiu, K. Physically enhanced secure wavelength division multiplexing chaos communication using multimode semiconductor lasers. *Nonlinear Dyn.* **2016**, *86*, 1937–1949. [CrossRef]
5. Li, N.Q.; Pan, W.; Xiang, S.Y.; Yan, L.S.; Luo, B.; Zou, X.H.; Zhang, L.Y.; Mu, P.H. Photonic generation of wideband time-delay-signature-eliminated chaotic signals utilizing an optically injected semiconductor laser. *IEEE J. Quantum Electron.* **2012**, *48*, 1339–1345. [CrossRef]
6. Salvide, M.F.; Masoller, C.; Torre, M.S. All-optical stochastic logic gate based on a VCSEL with tunable optical injection. *IEEE J. Quantum Electron.* **2013**, *49*, 886–893. [CrossRef]
7. Guo, X.X.; Xiang, S.Y.; Zhang, Y.H.; Lin, L.; Wen, A.J.; Hao, Y. Polarization multiplexing reservoir computing based on a VCSEL with polarized optical feedback. *IEEE J. Sel. Top. Quantum Electron.* **2020**, *26*, 1700109. [CrossRef]
8. Liu, G.T.; Stintz, A.; Li, H.; Malloy, K.J.; Lester, L.F. Extremely low room-temperature threshold current density diode lasers using InAs dots in $In_{0.15}Ga_{0.85}As$ quantum well. *Electron. Lett.* **1999**, *35*, 1163–1165. [CrossRef]
9. Mikhrin, S.S.; Kovsh, A.R.; Krestnikov, I.L.; Kozhukhov, A.V.; Livshits, D.A.; Ledentsov, N.N.; Shernyakov, Y.M.; Novikov, I.I.; Maximov, M.V.; Ustinov, V.M.; et al. High power temperature-insensitive 1.3 μm InAs/InGaAs/GaAs quantum dot lasers. *Semicond. Sci. Technol.* **2005**, *20*, 340–342. [CrossRef]

10. Saito, H.; Nishi, K.; Kamei, A.; Sugou, S. Low chirp observed in directly modulated quantum dot lasers. *IEEE Photonics Technol. Lett.* **2000**, *12*, 1298–1300. [CrossRef]
11. Kuntz, M.; Fiol, G.; Lammlin, M.; Schubert, C.; Kovsh, A.R.; Jacob, A.; Umbach, A.; Bimberg, D. 10Gb/s data modulation using 1.3 μm InGaAs quantum dot lasers. *Electron. Lett.* **2005**, *41*, 244–245. [CrossRef]
12. Norman, J.C.; Jung, D.; Zhang, Z.Y.; Wan, Y.T.; Liu, S.T.; Shang, C.; Herrick, R.W.; Chow, W.W.; Gossard, A.C.; Bowers, J.E. A review of high-performance quantum dot lasers on silicon. *IEEE J. Quantum Electron.* **2019**, *55*, 2000511. [CrossRef]
13. Norman, J.C.; Jung, D.; Wan, Y.T.; Bowers, J.E. Perspective: The future of quantum dot photonic integrated circuits. *APL Photonics* **2018**, *3*, 030901. [CrossRef]
14. Zhukov, A.E.; Kovsh, A.R. Quantum dot diode lasers for optical communication systems. *Quantum Electron.* **2008**, *38*, 409–423. [CrossRef]
15. Wang, C.; Raghunathan, R.; Schires, K.; Chan, S.C.; Lester, L.F.; Grillot, F. Optically injected InAs/GaAs quantum dot laser for tunable photonic microwave generation. *Opt. Lett.* **2016**, *41*, 1153–1156. [CrossRef]
16. Jiang, Z.F.; Wu, Z.M.; Yang, W.Y.; Hu, C.X.; Lin, X.D.; Jin, Y.H.; Dai, M.; Cui, B.; Yue, D.Z.; Xia, G.Q. Numerical simulations on narrow-linewidth photonic microwave generation based on a QD laser simultaneously subject to optical injection and optical feedback. *Appl. Opt.* **2020**, *59*, 2935–2941. [CrossRef]
17. Viktorov, E.A.; Mandel, P.; Tanguy, Y.; Houlihan, J.; Huyet, G. Electron-hole asymmetry and two-state lasing in quantum dot lasers. *Appl. Phys. Lett.* **2005**, *87*, 053113. [CrossRef]
18. Meinecke, S.; Lingnau, B.; Rohm, A.; Ludge, K. Stability of optically injected two-state quantum-dot lasers. *Ann. Phys.* **2017**, *529*, 1600279. [CrossRef]
19. Li, Q.Z.; Wang, X.; Zhang, Z.Y.; Chen, H.M.; Huang, Y.Q.; Hou, C.C.; Wang, J.; Zhang, R.Y.; Ning, J.Q.; Min, J.H.; et al. Development of modulation p-doped 1310 nm InAs/GaAs quantum dot laser materials and ultrashort cavity Fabry-Perot and distributed-feedback laser diodes. *ACS Photonics* **2018**, *5*, 1084–1093. [CrossRef]
20. Grillot, F.; Norman, J.C.; Duan, J.; Zhang, Z.; Dong, B.; Huang, H.; Chow, W.W.; Bowers, J.E. Physics and applications of quantum dot lasers for silicon photonics. *J. Nanophotonics* **2020**, *9*, 1271–1286. [CrossRef]
21. O'Brien, D.; Hegarty, S.P.; Huyet, G.; McInerney, J.G.; Kettler, T.; Laemmlin, M.; Bimberg, D.; Ustinov, V.M.; Zhukov, A.E.; Mikhrin, S.S.; et al. Feedback sensitivity of 1.3 mu m InAs/GaAs quantum dot lasers. *Electron. Lett.* **2003**, *39*, 1819–1820. [CrossRef]
22. Huyet, G.; O'Brien, D.; Hegarty, S.P.; McInerney, J.G.; Uskov, A.V.; Bimberg, D.; Ribbat, C.; Ustinov, V.M.; Zhukov, A.E.; Mikhrin, S.S.; et al. Quantum dot semiconductor lasers with optical feedback. *Phys. Status Solidi A* **2003**, *201*, 345–352. [CrossRef]
23. Stevens, B.J.; Childs, D.T.D.; Shahid, H.; Hogg, R.A. Direct modulation of excited state quantum dot lasers. *Appl. Phys. Lett.* **2009**, *95*, 061101. [CrossRef]
24. Lee, C.S.; Bhattacharya, P.; Frost, T.; Guo, W. Characteristics of a high speed 1.22μm tunnel injection p-doped quantum dot excited state laser. *Appl. Phys. Lett.* **2011**, *98*, 011103. [CrossRef]
25. Lin, L.C.; Chen, C.Y.; Huang, H.M.; Arsenijevic, D.; Bimberg, D.; Grillot, F.; Lin, F.Y. Comparison of optical feedback dynamics of InAs/GaAs quantum-dot lasers emitting solely on ground or excited states. *Opt. Lett.* **2018**, *43*, 210–213. [CrossRef] [PubMed]
26. Jiang, Z.F.; Wu, Z.M.; Jayaprasath, E.; Yang, W.Y.; Hu, C.X.; Xia, G.Q. Nonlinear dynamics of exclusive excited-state emission quantum dot lasers under optical injection. *Photonics* **2019**, *6*, 58. [CrossRef]
27. Xu, P.F.; Yang, T.; Ji, H.M.; Cao, Y.L.; Gu, Y.X.; Liu, Y.; Ma, W.Q.; Wang, Z.G. Temperature-dependent modulation characteristics for 1.3 μm InAs/GaAs quantum dot lasers. *J. Appl. Phys.* **2010**, *107*, 013102.
28. Pawlus, R.; Columbo, L.L.; Bardella, P.; Breuer, S.; Gioannini, M. Intensity noise behavior of an InAs/InGaAs quantum dot laser emitting on ground states and excited states. *Opt. Lett.* **2018**, *43*, 867–870. [CrossRef] [PubMed]
29. Naderi, N.A.; Grillot, F.; Yang, K.; Wright, J.B.; Gin, A.; Lester, L.F. Two-color multi-section quantum dot distributed feedback laser. *Opt. Express* **2010**, *18*, 27028–27035. [CrossRef] [PubMed]
30. Cataluna, M.A.; Nikitichev, D.I.; Mikroulis, S.; Simos, H.; Simos, C.; Mesaritakis, C.; Syvridis, D.; Krestnikov, I.; Livshits, D.; Rafailov, E.U. Dual-wavelength mode-locked quantum-dot laser, via ground and excited state transitions: Experimental and theoretical investigation. *Opt. Express* **2012**, *18*, 12832–12838. [CrossRef]
31. Breuer, S.; Rossetti, M.; Drzewietzki, L.; Bardella, P.; Montrosset, I.; Elsasser, W. Joint experimental and theoretical investigations of two state mode locking in a strongly chirped reverse biased monolithic quantum dot laser. *IEEE J. Quantum Electron.* **2011**, *47*, 1320–1329. [CrossRef]
32. Mesaritakis, C.; Kapsalis, A.; Bogris, A.; Syvridis, D. Artificial neuron based on integrated semiconductor quantum dot mode-locked lasers. *Sci. Rep.* **2016**, *6*, 39317. [CrossRef] [PubMed]
33. Tykalewicz, B.; Goulding, D.; Hegarty, S.P.; Huyet, G.; Byrne, D.; Phelan, R.; Kelleher, B. All-optical switching with a dual-state, single-section quantum dot laser via optical injection. *Opt. Lett.* **2014**, *39*, 4607–4610. [CrossRef]
34. Viktorov, E.A.; Dubinkin, I.; Fedorov, N.; Erneux, T.; Tykalewicz, B.; Hegarty, S.P.; Huyet, G.; Goulding, D.; Kelleher, B. Injection-induced, tunable, all-optical gating in a two-state quantum dot laser. *Opt. Lett.* **2016**, *41*, 3555–3558. [CrossRef]
35. Tykalewicz, B.; Goulding, D.; Hegarty, S.P.; Huyet, G.; Dubinkin, I.; Fedorov, N.; Erneux, T.; Viktorov, E.A.; Kelleher, B. Optically induced hysteresis in a two-state quantum dot laser. *Opt. Lett.* **2016**, *41*, 1034–1037. [CrossRef] [PubMed]
36. Jiang, Z.F.; Wu, Z.M.; Jayaprasath, E.; Yang, W.Y.; Hu, C.X.; Cui, B.; Xia, G.Q. Power-induced lasing state switching and bistability in a two-state quantum dot laser subject to optical injection. *Opt. Appl.* **2020**, *50*, 257–269.

37. Virte, M.; Breuer, S.; Sciamanna, M.; Panajotov, K. Switching between ground and excited states by optical feedback in a quantum dot laser diode. *Appl. Phys. Lett.* **2014**, *105*, 121109. [CrossRef]
38. Virte, M.; Panajotov, K.; Sciamanna, M. Mode competition induced by optical feedback in two-color quantum dot lasers. *IEEE J. Quantum Electron.* **2013**, *49*, 578–585. [CrossRef]
39. Virte, M.; Pawlus, R.; Sciamanna, M.; Panajotov, K.; Breuer, S. Energy exchange between modes in a multimode two-color quantum dot laser with optical feedback. *Opt. Lett.* **2016**, *41*, 3205–3208. [CrossRef]
40. Meinecke, S.; Kluge, L.; Hausen, J.; Lingnau, B.; Ludge, K. Optical feedback induced oscillation bursts in two-state quantum-dot lasers. *Opt. Express* **2020**, *28*, 3361–3377. [CrossRef]
41. Viktorov, E.A.; Mandel, P.; O'Driscoll, I.; Carroll, O.; Huyet, G.; Houlihan, J.; Tanguy, Y. Low-frequency fluctuations in two-state quantum dot lasers. *Opt. Lett.* **2006**, *31*, 2302–2304. [CrossRef]
42. Zhou, Y.G.; Duan, J.N.; Grillot, F.; Wang, C. Optical noise of dual-state lasing quantum dot lasers. *IEEE J. Quantum Electron.* **2020**, *56*, 2001207. [CrossRef]
43. Veselinov, K.; Grillot, F.; Cornet, C.; Even, J.; Bekiarski, A.; Gioannini, M.; Loualiche, S. Analysis of the double laser emission occurring in 1.55-mu m InAs-InP (113) B quantum-dot lasers. *IEEE J. Quantum Electron.* **2007**, *43*, 810–816. [CrossRef]
44. Dehghaninejad, A.; Sheikhey, M.M.; Baghban, H. Dynamic behavior of injection-locked two-state quantum dot lasers. *J. Opt. Soc. Am. B* **2019**, *36*, 1518–1524. [CrossRef]
45. Wang, C.; Grillot, F.; Even, J. Impacts of wetting layer and excited state on the modulation response of quantum-dot lasers. *IEEE J. Quantum Electron.* **2012**, *48*, 1144–1150. [CrossRef]
46. Ohnesorge, B.; Albrecht, M.; Oshinowo, J.; Forchel, A.; Arakawa, Y. Rapid carrier relaxation in self-assembled InxGa1-xAs/GaAs quantum dots. *Phys. Rev. B* **1996**, *54*, 11532–11538. [CrossRef]
47. Ignatiev, I.V.; Kozin, I.E.; Nair, S.V.; Ren, H.W.; Sugou, S.; Masumoto, Y. Carrier relaxation dynamics in InP quantum dots studied by artificial control of nonradiative losses. *Phys. Rev. B* **2000**, *61*, 15633–15636. [CrossRef]
48. Grillot, F.; Wang, C.; Nederi, N.A.; Even, J. Modulation Properties of Self-Injected Quantum-Dot Semiconductor Diode Lasers. *IEEE J. Sel. Top. Quantum Electron.* **2013**, *19*, 1900812. [CrossRef]
49. Chen, H.; Lei, T.F.; Lu, S.; Dai, W.P.; Qiu, L.J.; Zhong, L. Dynamics and Complexity Analysis of Fractional-Order Chaotic Systems with Line Equilibrium Based on Adomian Decomposition. *Complexity* **2020**, *2020*, 5710765. [CrossRef]
50. AL-Hosiny, N.M. Effect of linewidth enhancement factor on the stability map of optically injected distributed feedback laser. *Opt. Rev.* **2014**, *21*, 261–264. [CrossRef]
51. Abdulrhmann, S.; Yamada, M.; Ahmed, M. Numerical modeling of the output and operations of semiconductor lasers subject to strong optical feedback and its dependence on the linewidth-enhancement factor. *Int. J. Numer. Model. Electron. Netw. Devices Fields* **2011**, *24*, 218–229. [CrossRef]

Experimental and Numerical Study of Locking of Low-Frequency Fluctuations of a Semiconductor Laser with Optical Feedback

Abstract: We study the output of a semiconductor laser with optical feedback operated in the low-frequency fluctuations (LFFs) regime and subject to weak sinusoidal current modulation. In the LFF regime, the laser intensity exhibits abrupt drops, after which it recovers gradually. Without modulation, the drops occur at irregular times, while, with weak modulation, they can lock to the external modulation and they can occur, depending on the parameters, every two or every three modulation cycles. Here, we characterize experimentally the locking regions and use the well-known Lang–Kobayashi model to simulate the intensity dynamics. We analyze the effects of several parameters and find that the simulations are in good qualitative agreement with the experimental observations.

Keywords: semiconductor lasers; optical feedback; modulation; locking; low-frequency fluctuations

---

1. Introduction

Locking is a phenomenon that ubiquitously occurs in oscillators that are subject to an external periodic forcing, and refers to the synchronization, or to the adjustment, of the oscillator's rhythm, to that of the external forcing. Locking has many applications, for example, for cardiac re-synchronization after arrhythmia, for deep brain stimulation, jet lag re-adjustment, etc. [1–3].

A semiconductor laser whose pump current is periodically modulated is a stochastic nonlinear oscillator that can shows bistability and a chaotic output [4,5], and that allows controlled experiments in order to understand how locking emerges and how it depends on the parameters of the laser and of the external signal. With weak optical feedback, the laser intensity shows feedback-induced fluctuations that, under appropriate conditions, can be controlled by periodic current modulation. In particular, a weak modulation of the laser current can control the low-frequency fluctuations (LFFs) that occur when the laser operates near the threshold [4]. Without current modulation, the laser intensity shows irregular and abrupt drops (that in the following we will refer to as *spikes*), while with current modulation, under appropriate conditions, the spikes lock to the modulation (see Figure 1), and they occur with a rhythm that depends on the frequency of the modulation [6–17].

In recent years, we have performed detailed experiments on the modulated LFF dynamics, characterizing the temporal correlations in the spike times [18,19], the role of the modulation waveform [20], and the degree of locking [21] and we compared the spiking LFF dynamics with simulations of a weakly forced neuron [22]. We have also discovered that weak sinusoidal modulation can generate time-crystal-like behavior [23,24] because it can produce highly regular subharmonic locking, but not harmonic locking [25,26]. The lack of harmonic locking under weak sinusoidal modulation could be understood when using the well-known Lang–Kobayashi (LK) model [27] to simulate the intensity dynamics [28].

The goal of this work is to perform an in-depth comparison of experimental observations and the predictions of the LK model. This paper is organized as follows. In Section 2, we present the LK model, in Section 3 we describe the experimental setup and datasets, in Section 4 we present the comparison of observations and simulations, and in Section 5, we present the discussion and the conclusions.

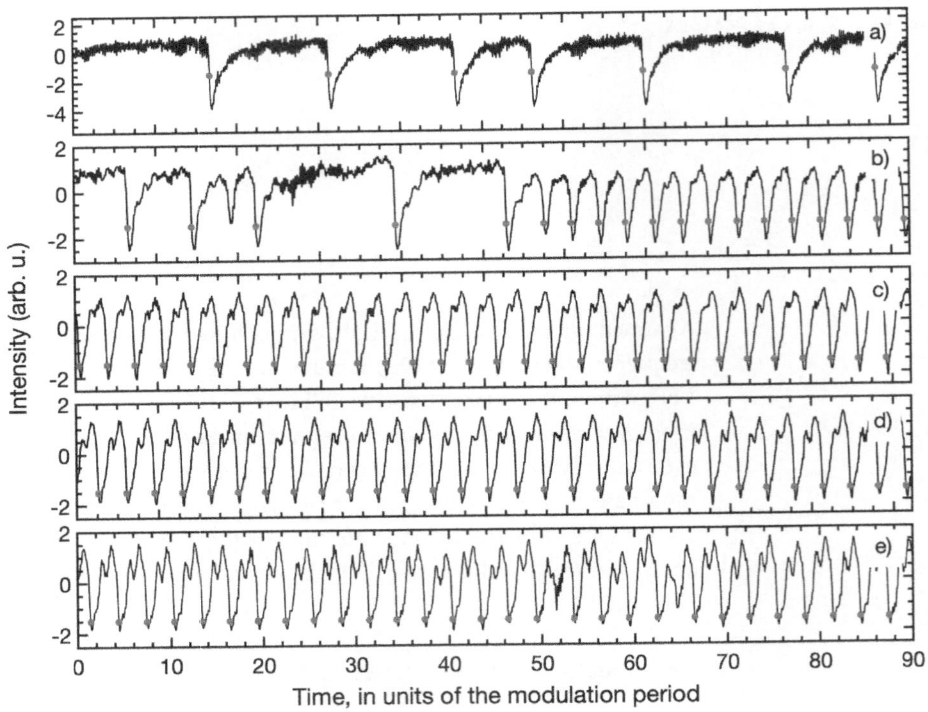

Figure 1. Available online: examples (accessed on) of intensity time series recorded with different modulation amplitudes. (**a**) $A_{mod} = 0\%$, (**b**) 0.73% (**c**) 1.22%, (**d**) 1.70%, and (**e**) 2.43% of the dc level $I_{dc} = 26$ mA. The modulation frequency is $f_{mod} = 44$ MHz. The red dots indicate the detected spike times.

2. Model

The Lang–Kobayashi rate equations describing a single-mode semiconductor laser with weak optical feedback and sinusoidal pump current modulation are [26–28]:

$$\dot{E} = k(1+i\alpha)(G-1)E + \eta E(t-\tau)e^{-i\omega_0\tau} + \sqrt{D}\xi, \quad (1)$$

$$\dot{N} = \gamma_N(\mu_{dc} + a_{mod}\sin(2\pi f_{mod}t) - N - G|E|^2). \quad (2)$$

Here, E represents the slowly varying complex optical field ($|E|^2$ is proportional to the laser intensity) and N, the carrier density. η, τ, and $\omega_0\tau$ are the feedback strength, the delay time, and the feedback phase, respectively; $k = 1/(2\tau_p)$ where τ_p is the photon lifetime, $\gamma_N = 1/\tau_N$ where τ_N is the carrier lifetime, $G = N/(1+\epsilon|E|^2)$ is the gain and ϵ is the gain saturation coefficient, α is the linewidth enhancement factor. ξ is a complex Gaussian white noise that takes into account spontaneous emission and D is the strength of the noise. μ_{dc} is the dc value of the pump current parameter, which is proportional to $I_{dc}/I_{th,sol}$ [29], with I_{dc} being the dc value of the pump current and $I_{th,sol}$ the threshold current without feedback. a_{mod} and f_{mod} are the modulation amplitude and frequency, respectively.

The model equations were integrated with the same procedure and parameters as in [26,28] that fit the experimental conditions: $k = 300$ ns^{-1}, $\gamma_N = 1$ ns^{-1}, $\alpha = 4$, $\epsilon = 0.01$, η = 30 ns^{-1}, $\tau = 5$ ns, $\mu_{dc} = 0.99$, and $D = 10^{-5}$ ns^{-2}. To detect the spike times, the intensity time series, $|E(t)|^2$, was band-pass filtered to simulate the finite bandwidth of the experimental detection system, and was then normalized to zero mean and unit variance. Then, a spike was detected whenever the intensity dropped below a threshold, $Th = -1.1$.

To analyze the statistical characteristics of the spikes, for each set of parameters five intensity time series were simulated, starting from random initial conditions and using different noise seeds, and from them, after disregarding a transient time, the average number of spikes, the average inter-spike interval (IS), and the average standard deviation of the distribution of ISIs were calculated.

3. Experimental Setup and Datasets

The experimental setup and datasets were described in [20]. A diode laser with center wavelength of 685 nm (Thorlabs HL6750MG, Newton, NJ, USA) was used, whose temperature and current were stabilized with an accuracy of 0.01 C and 0.01 mA, respectively. A beam splitter sent 90% of the light reflected by a mirror back to the laser, and the other 10% to the detection system: a high-speed photo-detector (Thorlabs Det10A/M) connected to an amplifier (Femto HSA-Y-2-40, Berlin, Germany) whose output was recorded with a digital oscilloscope (Agilent Technologies Infiniium DSO9104A, 1 GHz bandwidth). A 500 MHz Bias-T in the laser mount was used to modulate the laser current with a periodic signal generated by an arbitrary waveform generator (Agilent 81150A, Santa Clara, CA, USA). The length of the external cavity was 70 cm, which gave a feedback delay time of 4.7 ns. The threshold current of the free-running laser was $I_{th,sol} = 26.62$ mA, and with optical feedback, it was reduced to $I_{th} = 24.70$ mA (7.2% reduction). In the experiments, three modulation parameters were varied, the dc value of the laser current, I_{dc}, the amplitude, A_{mod}, and frequency, f_{mod}, of the driving signal, and for each set of parameters, three modulation waveforms were used (pulse-down, pulse-up, and sinusoidal). Here, we analyze the data recorded with sinusoidal modulation. Specifically, we analyze the ISIs recorded with different I_{dc}, A_{mod}, and f_{mod}.

4. Results

Figure 1 presents experimental intensity time series recorded without current modulation (top panel) and with current modulation of increasing amplitude (panels b–e). We see that for intermediate modulation amplitudes the spikes become periodic and a spike occurs every three modulation cycles (locking 3:1). Model simulations that show good agreement with these observations were presented in [26] (Figure 4).

To perform a systematic comparison of experiments and model simulations, we analyze how the statistics of the ISIs depend on the amplitude and on the frequency of the modulation. Specifically, we analyze the number of spikes, the mean ISI normalized to the modulation period, T_{mod}, and the coefficient of variation, C_v, that measures the relative width of the ISI distribution ($C_v = \sigma_{ISI}/\langle ISI \rangle$ where $\langle ISI \rangle$ and σ_{ISI} are the mean and the standard deviation of the ISI distribution). The results are presented in Figures 2 and 3, for the experimental and the simulated data, respectively. We see a good qualitative agreement: the number of spikes increases in the regions of locking, which are seen as a well-defined cyan region (locking 2:1, $\langle ISI \rangle = 2T_{mod}$) and a yellow region (locking 3:1, $\langle ISI \rangle = 3T_{mod}$), and in these regions C_v is small, revealing a narrow ISI distribution.

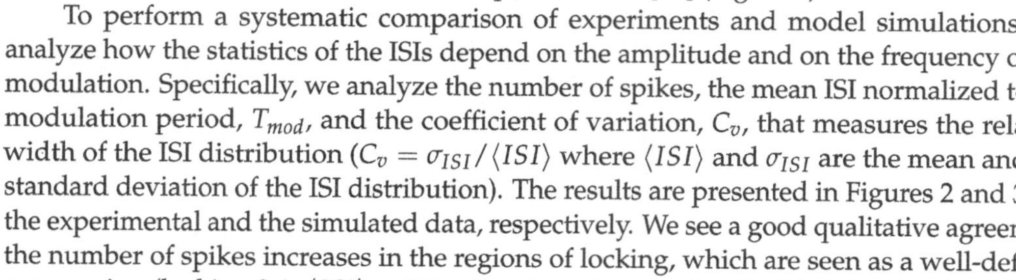

Figure 2. Analysis of the experimental inter-spike intervals (ISIs). (**a**): number of spikes (in color code) vs. the modulation amplitude and frequency; (**b**): mean ISI normalized to the modulation period (the red color represents $\langle ISI \rangle / T_{mod} \geq 5$); (**c**) coefficient of variation. The dc value of the pump current is as in Figure 1.

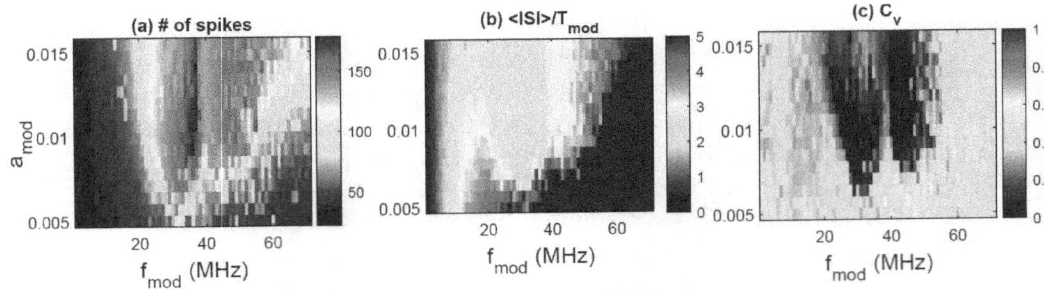

Figure 3. As in Figure 2, but obtained from model simulations with $\mu_{dc} = 0.99$, other parameters are as indicated in the text.

Let us next compare the combined effect of varying the dc value of pump current and the modulation amplitude, keeping the modulation frequency fixed. In Figures 4 and 5, we present the analysis of experimental and simulated ISIs, respectively. In both figures, from top to bottom, $f_{mod} = 26$ MHz, 44 MHz, and 55 MHz. We again observe a very good qualitative agreement between experiments and simulations. As I_{dc} or μ_{dc} increase, we see that the number of spikes increases (left column) and the mean ISI decreases (in the blue regions, the mean ISI becomes equal to or smaller than the modulation period). However, we see in the right column that the coefficient of variation increases with I_{dc} or μ_{dc}, which indicates that 1:1 locking is not achieved.

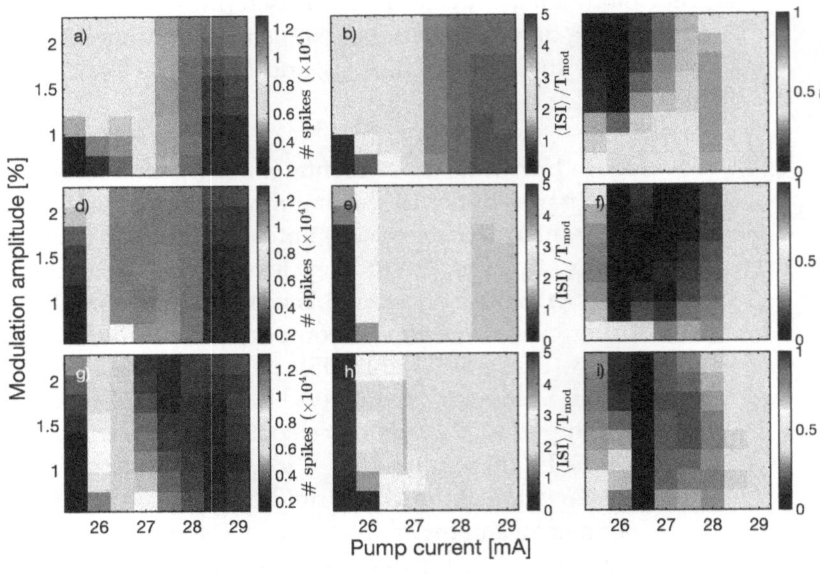

Figure 4. Experimental characterization of the locking region as a function of the modulation amplitude and the dc value of the pump current, for different modulation frequencies. **(a,d,g)** number of spikes; **(b,e,h)** $\langle ISI \rangle / T_{mod}$; **(c,f,h)** coefficient of variation. **(a–c)** $f_{mod} = 26$ MHz; **(d–f)** f_{mod} =44 MHz; **(g–i)** f_{mod} =55 MHz.

In Figure 4b, we note that for large enough I_{dc} the mean ISI is approximately equal to the modulation period, but there is no 1:1 locking because the ISI distribution is quite broad (the coefficient of variation is ≈ 0.5). One could wonder if for other modulation frequencies, harmonic locking could be obtained. To address this point, we examine the statistics of the ISIs as a function of the modulation amplitude and frequency. The results are presented in Figure 6 (experimental data recorded with $I_{dc} = 27$ mA) and in Figure 7 (simulated data with $\mu_{dc} = 1.01$). In Figure 6b, we see, for low modulation frequencies, a blue region that indicates $\langle ISI \rangle / T_{mod} \sim 1$, but in this region C_v is large ($C_v \sim 0.5$). In Figure 7b, we also see a blue region with similar characteristics. In contrast with the experiments, in the simulation, 3:1 locking is not seen because the yellow region in Figure 7b is quite narrow, and in this region, C_v is large.

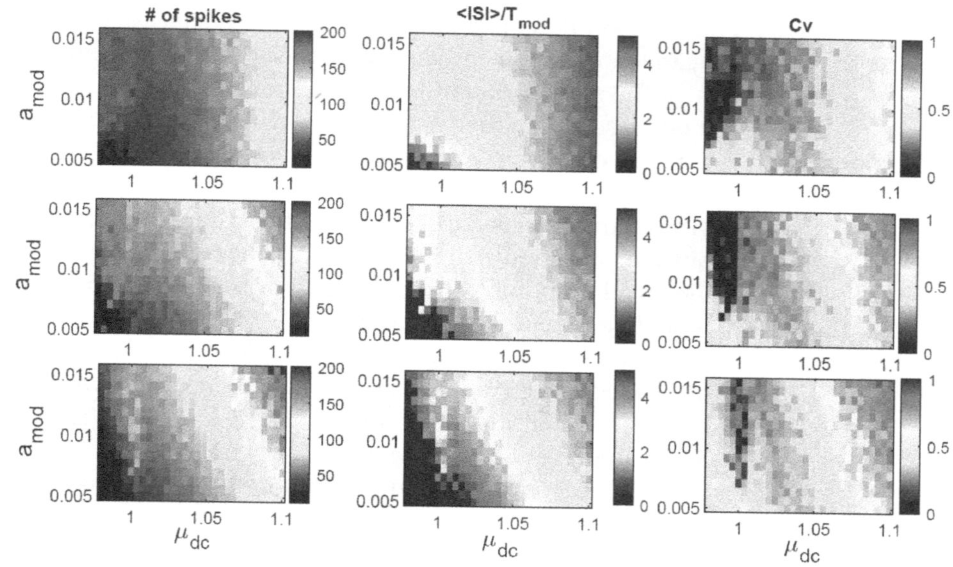

Figure 5. As Figure 4, but obtained from model simulations. First row, $f_{mod} = 26$ MHz; second row, $f_{mod} = 44$ MHz; and third row, $f_{mod} = 55$ MHz. We again see a tendency of the mean ISI to decrease as the dc value of the pump current increases, but no 1:1 locking is obtained because in the region where $\langle ISI \rangle / T_{mod} \sim 1$, the width of the distribution of ISIs, measured by the coefficient of variation, is quite large.

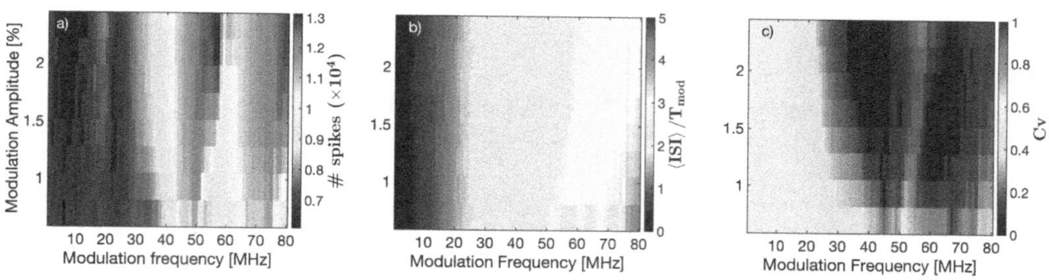

Figure 6. As Figure 2, but the dc value of the pump current is $I_{dc} = 27$ mA. (**a**) number of spikes; (**b**) $\langle ISI \rangle / T_{mod}$; (**c**) coefficient of variation.

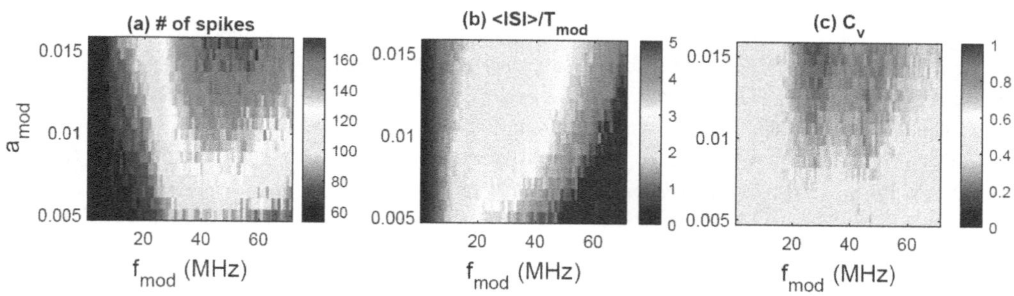

Figure 7. As Figure 3, but $\mu_{dc} = 1.01$, other parameters are as indicated in the text.

5. Conclusions

We have studied the dynamics of a semiconductor laser with optical feedback and current modulation, which operates in the LFF regime. We have analyzed how the number of spikes and how the mean and the standard deviation of the ISI distribution vary with the modulation parameters: the dc value, the amplitude, and the frequency. We have found a very good agreement between experimental observations and the simulations of the LK model. With increasing I_{dc}, $\langle ISI \rangle$ tends to decrease, but, at least in the range of modulation amplitudes studied here, no 1:1 locking was found. Harmonic locking can be observed with larger modulation amplitudes [28], but in that case the intensity dynamics are dominated by the modulation and the feedback-induced spikes are rather small.

The main motivation of our study was to understand the locking phenomena of a diode laser, from the point of view of nonlinear dynamics. In that sense, model simulations have been performed to further understand why small-amplitude sinusoidal current modulation does not produce harmonic locking. Since we have found well-defined regions of subharmonic locking (exploring the parameters space pump current, modulation amplitude, and modulation frequency) our results may be of interest for applications that use small-amplitude electric modulation to generate highly regular optical pulsing with a repetition rate in the MHz range.

It will be interesting for future work to understand how the locking behavior depends on the feedback strength, i.e., to characterize the transition from locked sinusoidal-like oscillations (without optical feedback) to locked feedback-induced spikes. It will also be interesting to analyze if the interplay of noise, delayed feedback, and current modulation can produce locking regions where the spikes are emitted with a very regular timing.

References

1. Winfree, A.T. *The Geometry of Biological Time*; Springer: London, UK, 2001.
2. Pikovsky, A.; Rosenblum, M.; Kurths, J. *Synchronization: A Universal Concept in Nonlinear Sciences*; Cambridge University Press: Cambridge, UK, 2001.
3. Glass, L. Synchronization and rhythmic processes in physiology. *Nature* **2001**, *410*, 277. [CrossRef] [PubMed]
4. Ohtsubo, J. *Semiconductor Lasers: Stability, Instability and Chaos*, 4th ed.; Springer: Berlin, Germany, 2017.
5. Erneux, T.; Glorieux, P. *Laser Dynamics*; Cambridge University Press: Cambridge, UK, 2010.
6. Baums, D.; Elsasser, W.; Gobel, E.O. Farey tree and Devil's staircase of a modulated external-cavity semiconductor laser. *Phys. Rev. Lett.* **1989**, *63*, 155. [CrossRef]
7. Sacher, S.; Baums, D.; Panknin, P.; Elsasser, W.; Gobel, E.O. Intensity instabilities of semiconductor-lasers under current modulation, external light injection, and delayed feedback'. *Phys. Rev. A* **1992**, *45*, 1893. [CrossRef]
8. Takiguchi, Y.; Liu, Y.; Ohtsubo, J. Low-frequency fluctuation induced by injection-current modulation in semiconductor lasers with optical feedback. *Opt. Lett.* **1998**, *23*, 1369. [CrossRef] [PubMed]
9. Sukow, D.W.; Gauthier, D.J. Entraining power-dropout events in an external-cavity semiconductor laser using weak modulation of the injection current. *IEEE J. Quantum Electron.* **2000**, *36*, 175. [CrossRef]
10. Mendez, J.M.; Laje, R.; Giudici, M.; Aliaga, J.; Mindlin, G.B. Dynamics of periodically forced semiconductor laser with optical feedback. *Phys. Rev. E* **2001**, *63*, 066218. [CrossRef] [PubMed]
11. Lawrence, J.S.; Kane, D.M. Nonlinear dynamics of a laser diode with optical feedback systems subject to modulation. *IEEE J. Quantum Electron.* **2002**, *38*, 185. [CrossRef]
12. Marino, F.; Giudici, M.; Barland, S.; Balle, S. Experimental evidence of stochastic resonance in an excitable optical system. *Phys. Rev. Lett.* **2002**, *88*, 040601. [CrossRef] [PubMed]
13. Lam, W.-S.; Guzdar, P.N.; Roy, R. Effect of spontaneous emission noise and modulation on semiconductor lasers near threshold with optical feedback. *Int. J. Mod. Phys. B* **2003**, *17*, 4123. [CrossRef]
14. Toomey, J.P.; Kane, D.M.; Lee, M.W.; Shore, K.A. Nonlinear dynamics of semiconductor lasers with feedback and modulation. *Opt. Express* **2010**, *18*, 16955. [CrossRef]
15. Ahmed, M.; El-Sayed, N.Z.; Ibrahim, H. Chaos and noise control by current modulation in semiconductor lasers subject to optical feedback. *Eur. Phys. J. D* **2012**, *66*, 141. [CrossRef]
16. Spitz, O.; Wu, J.G.; Carras, M.; Wong, C.W.; Grillot, F. Chaotic optical power dropouts driven by low frequency bias forcing in a mid-infrared quantum cascade laser. *Sci. Rep.* **2019**, *9*, 4451. [CrossRef] [PubMed]
17. Spitz, O.; Wu, J.G.; Herdt, A.; Carras, M.; Elsasser, W.; Wong, C.W.; Grillot, F. Investigation of chaotic and spiking dynamics in mid-infrared quantum cascade lasers operating continuous-waves and under current modulation. *IEEE J. Sel. Top. Quantum Electron.* **2019**, *25*, 1200311. [CrossRef]

18. Aragoneses, A.; Sorrentino, T.; Perrone, S.; Gauthier, D. J.; Torrent, M.C.; Masoller, C. Experimental and numerical study of the symbolic dynamics of a modulated external-cavity semiconductor laser. *Opt. Express* **2014**, *22*, 4705. [CrossRef] [PubMed]
19. Sorrentino, T.; Quintero-Quiroz, C.; Aragoneses, A.; Torrent, M.C.; Masoller, C. Effects of periodic forcing on the temporally correlated spikes of a semiconductor laser with feedback. *Opt. Express* **2015**, *23*, 5571. [CrossRef]
20. Tiana-Alsina, J.; Quintero-Quiroz, C.; Panozzo, M.; Torrent, M.C.; Masoller, C. Experimental study of modulation waveforms for entraining the spikes emitted by a semiconductor laser with optical feedback. *Opt. Express* **2018**, *26*, 9298. [CrossRef] [PubMed]
21. Tiana-Alsina, J.; Quintero-Quiroz, C.; Torrent, M.C.; Masoller, C. Quantifying the degree of locking in weakly forced stochastic systems. *Phys. Rev. E* **2019**, *99*, 022207. [CrossRef]
22. Tiana-Alsina, J.; Quintero-Quiroz, C.; Masoller, C. Comparing the dynamics of periodically forced lasers and neurons. *New J. Phys* **2019**, *21*, 103039. [CrossRef]
23. Yao, N.Y.; Nayak, C. Time crystals in periodically driven systems. *Phys. Today* **2018**, *71*, 40. [CrossRef]
24. Sacha K.; Zakrzewski, J. Time crystals: A review. *Rep. Prog. Phys.* **2018**, *81*, 016401. [CrossRef]
25. Tiana-Alsina, J.; Masoller, C. Time crystal dynamics in a weakly modulated stochastic time delayed system. *Bull. Am. Phys. Soc.* 2022, *Submitted.*
26. Tiana-Alsina, J.; Masoller, C. Dynamics of a semiconductor laser with feedback and modulation: Experiments and model comparison. *Opt. Express* 2022, *in press.* [CrossRef]
27. Lang, R.; Kobayashi, K. External optical feedback effects on semiconductor injection-laser properties. *IEEE J. Quantum Electron.* **1980**, *16*, 347. [CrossRef]
28. Tiana-Alsina, J.; Masoller, C. Locking phenomena in semiconductor lasers near threshold with optical feedback and sinusoidal current modulation. *Appl. Sci.* **2021**, *11*, 7871. [CrossRef]
29. Barland, S.; Spinicelli, P.; Giacomelli, G.; Marin, F. Measurement of the working parameters of an air-post vertical-cavity surface-emitting laser. *IEEE J. Quantum Electron.* **2005**, *41*, 1235. [CrossRef]

Effects of Asymmetric Coupling Strengthon Nonlinear Dynamics of Two Mutually Long-Delay-Coupled Semiconductor Lasers

Abstract: This study investigates the effects of asymmetric coupling strength on nonlinear dynamics of two mutually long-delay-coupled semiconductor lasers through both experimental and numerical efforts. Dynamical maps and spectral features of dynamical states are analyzed as a function of the coupling strength and detuning frequency for a fixed coupling delay time. Symmetry in the coupling strength of the two lasers, in general, symmetrizes their dynamical behaviors and the corresponding spectral features. Slight to moderate asymmetry in the coupling strength moderately changes their dynamical behaviors from the ones when the coupling strength is symmetric, but does not break the symmetry of their dynamical behaviors and the corresponding spectral features. High asymmetry in the coupling strength not only strongly changes their dynamical behaviors from the ones when the coupling strength is symmetric, but also breaks the symmetry of their dynamical behaviors and the corresponding spectral features. Evolution of the dynamical behaviors from symmetry to asymmetry between the two lasers is identified. Experimental observations and numerical predictions agree not only qualitatively to a high extent but also quantitatively to a moderate extent.

Keywords: semiconductor lasers; nonlinear dynamics; mutual coupling; asymmetric coupling strength; symmetry breaking

1. Introduction

Nonlinear dynamics of two mutually delay-coupled semiconductor lasers has attracted much research interest due to its profound physics and promising applications. By simply adjusting the operating conditions of the two lasers, including bias current, coupling strength, and detuning frequency, various dynamical behaviors can be induced, such as mutual injection locking, period-one (P1) dynamics, period-two (P2) dynamics, quasi-periodic dynamics, and chaos. The unique temporal and spectral features found in these dynamical behaviors have been proposed, respectively, to improve performance characteristics of existing technologies, such as enhancing the bandwidth of direct modulation [1–5] and suppressing nonlinear distortion due to direct modulation [6–8], or to provide alternatives for novel applications, such as tunable microwave generation [9–12], chaotic synchronization [13–16], reservoir computing [17–19], and decision making [20]. For these technological applications, the bias currents of the two lasers are, in general, adjusted independently and differently so that specific characteristics or functionalities are achieved. This inevitably leads to a difference in the coupling strength between the two lasers, i.e., the coupling strength is asymmetric.

Prior studies [21–25] that investigate nonlinear dynamical behaviors and their features in mutually coupled lasers mainly considered symmetric coupling strength only. The dynamical behaviors of the two lasers are mainly identical, i.e., symmetric, even though symmetry breaking in their behaviors does happen over a limited range of operating condition. An interesting yet fundamental question to ask is whether the dynamical behaviors of the two lasers with asymmetric coupling strength are kept symmetric. For example, the result of a recent study [26], of which purpose focuses on showing that coupling strength asymmetry makes the mutually coupled laser system behave like a unidirectionally coupled laser system, indicates that their dynamical behaviors are still identical even when their coupling strength becomes slightly or moderately asymmetric. Would the dynamical behavior symmetry still hold if the extent of the coupling strength asymmetry enhances? The answer to this question is important not only for fundamental understandings about how and to what extent such a laser system responds to asymmetric coupling strength, but also for technological applications where such a laser system is expected to operate at a specific dynamical behavior all the time even when the coupling strength becomes asymmetric. However, this issue has not been much emphasized yet, and is thus numerically and experimentally investigated in this study using two mutually coupled lasers with a delay time longer than the relaxation resonance period of the lasers at free running. As shown in the following analyses, slight to moderate asymmetry in the coupling strength does not break the symmetry between the dynamical behaviors of the two lasers. Symmetry breaking of the dynamical behaviors happens when the coupling strength is highly asymmetric. Evolution of the dynamical behaviors from symmetry to asymmetry between the two lasers is observed, where numerical predictions and experimental observations agree not only qualitatively to a high extent but also quantitatively to a moderate extent.

The remainder of this paper is outlined as follows. In Section 2, the numerical model for two mutually delay-coupled semiconductor lasers, which is derived from the well-known Lang–Kobayashi equations, used in this study is described, and numerical predictions are demonstrated. In Section 3, the experimental setup of the laser system used in this study is introduced, and experimental observations are presented and compared with the numerical predictions shown in Section 2. Finally, a summary is given in Section 4.

2. Numerical Prediction

The dynamical behaviors of two mutually delay-coupled semiconductor lasers are numerically investigated in this section to obtain a picture of when, how, and to what extent changes in their dynamical behaviors happen if their coupling strength varies from symmetry to asymmetry. The numerical results would serve as a proper guidance for an experiment study demonstrated in Section 3 to verify the numerical predictions. Optical and microwave spectra presented here are obtained by considering the spontaneous emission noise of both lasers in the numerical calculation so that a fair comparison can be made with those obtained in the experimental study. Temporal evolutions shown here are calculated without taking into account the spontaneous emission noise of both lasers so that an easy comparison can be made between the outputs of the two lasers.

2.1. Numerical Model

Two mutually delay-coupled semiconductor lasers under study can be mathematically described by the following Lang–Kobayashi equations [27–30]:

Laser Diode 1 (LD1):

$$\frac{dA_1}{dt} = -\frac{\gamma_{c1}}{2}A_1 + i(\omega_{01}-\omega_{c1})A_1 + \frac{\Gamma_1}{2}g_1(1-ib_1)A_1 + \eta_{21}A_2(t-\tau_2)e^{i\omega_{02}\tau_2} - i\Omega A_1 + F_{sp1} \quad (1)$$

$$\frac{dN_1}{dt} = \frac{J_1}{ed_1} - \gamma_{s1}N_1 - g_1S_1. \quad (2)$$

Laser Diode 2 (LD2):

$$\begin{aligned}\frac{dA_2}{dt} &= -\frac{\gamma_{c2}}{2}A_2 + i(\omega_{02}-\omega_{c2})A_2 + \frac{\Gamma_2}{2}g_2(1-ib_2)A_2 \\ &\quad +\eta_{12}A_1(t-\tau_1)e^{i\omega_{02}\tau_1} + F_{sp2}\end{aligned} \tag{3}$$

$$\frac{dN_2}{dt} = \frac{J_2}{ed_2} - \gamma_{s2}N_2 - g_2S_2. \tag{4}$$

Here, A_j is the total complex intracavity field amplitude of LDj, where j = 1 or 2, γ_{cj} is the cavity decay rate, ω_{0j} is the free-running oscillation frequency, ω_{cj} is the angular frequency of the cold cavity, Γ_j is the confinement factor describing the spatial overlap between the gain medium and the optical mode, b_j is the linewidth enhancement factor relating the dependence of the refractive index on changes in the optical gain, g_j is the optical gain parameter which is a function of the charge carrier density N_j and the intracavity photon density S_j, $F_{spj} = F_{rj} + iF_{ij}$ is the complex field noise, η_{12} and η_{21} are the injection coupling rates from LD1 to LD2 and from LD2 to LD1, respectively, τ_1 and τ_2 are the coupling delay times from LD1 to LD2 and from LD2 to LD1, respectively, $f_i = \Omega/2\pi = (\omega_{01}-\omega_{02})/2\pi$ is the detuning frequency between LD1 and LD2 at free running, J_j is the bias current density, e is the electron charge, d_j is the active layer thickness, and γ_{sj} is the spontaneous carrier decay rate. The photon density is related to the intracavity field by:

$$S_j = \frac{2\epsilon_0 n_j^2}{\hbar\omega_{0j}}|A_j|^2 \tag{5}$$

where ϵ_0 is the free-space permittivity, n_j is the refractive index, and $\hbar$ is the reduced Plank's constant. The gain coefficient g_j is a function of the photon density and carrier density described as:

$$g_j = \frac{\gamma_{cj}}{\Gamma_j} + \gamma_{nj}\frac{N_j - N_{0j}}{S_{0j}} - \gamma_{pj}\frac{S_j - S_{0j}}{\Gamma_j S_{0j}} \tag{6}$$

where γ_{nj} represents the differential carrier relaxation rate, γ_{pj} describes the nonlinear carrier relaxation rate, N_{0j} indicates the free-running carrier density, and S_{0j} expresses the free-running photon density, respectively.

For the purpose of numerical calculation, Equations (1)–(4) are recast about the steady-state, free-running operating point of each laser, where $A_j = |A_{0j}|(a_{rj} + ia_{ij})$ and $N_j = N_{0j}(1+\tilde{n}_j)$ are used, and A_{0j} is the free-running field amplitude.

Laser Diode 1 (LD1):

$$\begin{aligned}\frac{da_{r1}}{dt} &= \frac{1}{2}\left[\frac{\gamma_{c1}\gamma_{n1}}{\gamma_{s1}\tilde{J}_1}\tilde{n}_1 - \gamma_{p1}(a_{r1}^2 + a_{i1}^2 - 1)\right](a_{r1} + b_1a_{i1}) \\ &\quad +\Omega a_{i1} + \xi_{21}^s\gamma_{c1}[a_{r2}(t-\tau_2)\cos\omega_{02}\tau_2 - a_{i2}(t-\tau_2)\sin\omega_{02}\tau_2] + F_{a_{r1}}\end{aligned} \tag{7}$$

$$\begin{aligned}\frac{da_{i1}}{dt} &= \frac{1}{2}\left[\frac{\gamma_{c1}\gamma_{n1}}{\gamma_{s1}\tilde{J}_1}\tilde{n}_1 - \gamma_{p1}(a_{r1}^2 + a_{i1}^2 - 1)\right](-b_1a_{r1} + a_{i1}) \\ &\quad -\Omega a_{r1} + \xi_{21}^s\gamma_{c1}[a_{i2}(t-\tau_2)\cos\omega_{02}\tau_2 + a_{r2}(t-\tau_2)\sin\omega_{02}\tau_2] + F_{a_{i1}}\end{aligned} \tag{8}$$

$$\begin{aligned}\frac{d\tilde{n}_1}{dt} &= -\left[\gamma_{s1} + \gamma_{n1}(a_{r1}^2 + a_{i1}^2)\right]\tilde{n}_1 - \gamma_{s1}\tilde{J}_1(a_{r1}^2 + a_{i1}^2 - 1) \\ &\quad +\frac{\gamma_{s1}\gamma_{p1}}{\gamma_{c1}}\tilde{J}_1(a_{r1}^2 + a_{i1}^2)(a_{r1}^2 + a_{i1}^2 - 1).\end{aligned} \tag{9}$$

Laser Diode 2 (LD2):

$$\frac{da_{r2}}{dt} = \frac{1}{2}\left[\frac{\gamma_{c2}\gamma_{n2}}{\gamma_{s2}\tilde{J}_2}\tilde{n}_2 - \gamma_{p2}(a_{r2}^2 + a_{i2}^2 - 1)\right](a_{r2} + b_2 a_{i2}) + \xi_{12}^{s}\gamma_{c2}[a_{r1}(t-\tau_1)\cos\omega_{02}\tau_1 - a_{i1}(t-\tau_1)\sin\omega_{02}\tau_1] + F_{a_{r2}} \tag{10}$$

$$\frac{da_{i2}}{dt} = \frac{1}{2}\left[\frac{\gamma_{c2}\gamma_{n2}}{\gamma_{s2}\tilde{J}_2}\tilde{n}_2 - \gamma_{p2}(a_{r2}^2 + a_{i2}^2 - 1)\right](-b_2 a_{r2} + a_{i2}) + \xi_{12}^{s}\gamma_{c2}[a_{i1}(t-\tau_1)\cos\omega_{02}\tau_1 + a_{r1}(t-\tau_1)\sin\omega_{02}\tau_1] + F_{a_{i2}} \tag{11}$$

$$\frac{d\tilde{n}_2}{dt} = -\left[\gamma_{s2} + \gamma_{n2}(a_{r2}^2 + a_{i2}^2)\right]\tilde{n}_2 - \gamma_{s2}\tilde{J}_2(a_{r2}^2 + a_{i2}^2 - 1) + \frac{\gamma_{s2}\gamma_{p2}}{\gamma_{c2}}\tilde{J}_2(a_{r2}^2 + a_{i2}^2)(a_{r2}^2 + a_{i2}^2 - 1). \tag{12}$$

Here $\xi_{12}^{s} = \eta_{12}|A_{01}|/\gamma_{c2}|A_{02}|$ and $\xi_{21}^{s} = \eta_{21}|A_{02}|/\gamma_{c1}|A_{01}|$ represent the strength of coupling from LD1 to LD2 and from LD2 to LD1, respectively. A superscript s is used for both symbols to distinguish the coupling strength defined here from the one defined in the experimental study presented in Section 3. The normalized bias level is described by $\tilde{J}_j = (J_j/ed_j - \gamma_{sj}N_j)/\gamma_{sj}N_j$. The phase factor $\omega_{02}\tau_j$ is set to zero throughout the numerical calculation in order to simplify the study. The normalized Langevin noise-source parameters $F_{a_{rj}} = F_{rj}/|A_{0j}|$ and $F_{a_{ij}} = F_{ij}/|A_{0j}|$ describe the real and imaginary parts of the normalized spontaneous emission parameters, respectively, and are characterized by a spontaneous emission rate as [31]:

$$\left\langle F_{a_{rj}}(t)F_{a_{rj}}(t')\right\rangle = \left\langle F_{a_{ij}}(t)F_{a_{ij}}(t')\right\rangle = \frac{R_{spj}}{2|A_{0j}|^2}\delta(t-t') \tag{13}$$

$$\left\langle F_{a_{rj}}(t)F_{a_{ij}}(t')\right\rangle = \left\langle F_{a_{ij}}(t)F_{a_{rj}}(t')\right\rangle = 0 \tag{14}$$

where R_{spj} represents the fraction of the spontaneous emission noise into the laser mode.

The values of the intrinsic laser parameters used for the numerical calculation here, which are experimentally measured using the four-wave mixing method [32], are shown in Table 1. Throughout the numerical calculation, the intrinsic laser parameters of LD1 and LD2 are set identical in order to simplify the study. Under this condition, the relaxation resonance frequency of either free-running laser is 10.25 GHz. A second-order Runge–Kutta method with the measured laser parameters is used to solve Equations (7)–(12). Throughout the numerical study, a time duration of about 0.47 ps is used for one integration step, and a time duration of 1 μs is adopted for complete integration.

2.2. *Dynamics Behaviors under Symmetric Coupling Strength*

For the purpose of comparison, the dynamical behavior of the mutually delay-coupled laser system is first investigated when the coupling strength is symmetric, i.e., $\xi_{12}^{s} = \xi_{21}^{s}$, in this subsection. To obtain a global understanding of how the laser system behaves at a fixed coupling delay time of 40.15 ns under study, maps of dynamical states as a function of ξ_{12}^{s} and f_i for LD1 and LD2 are presented in Figure 1a and 1b, respectively. Regions of mutual injection locking, P1 dynamics, P2 dynamics, and chaos are marked by red, yellow, blue, and black, respectively. Periodic dynamics with periods higher than two are included in the regions of chaos. Comparing Figure 1a with Figure 1b demonstrates that the dynamical behaviors of both lasers are generally identical over the range of ξ_{12}^{s} and f_i under study when the coupling strength is symmetric. In addition, each different dynamical state generally appears symmetrically with respect to $f_i = 0$. The mutual injection locking states emerge at weak coupling strength and small frequency detuning. The P1 dynamical states appear when ξ_{12}^{s} is smaller than 0.044 over the range of f_i under study. The chaotic states start to emerge when ξ_{12}^{s} is greater than 0.007. Note that the coupling delay time, 40.15 ns, is chosen here according to the one used in the experimental setup described in

Section 3 so that fair comparisons can be made between numerical and experimental results demonstrated in Section 3.

Table 1. The values of laser parameters used in the numerical calculation.

Parameter	Symbol	Value
Linewidth enhancement factor	b_1, b_2	3
Normalized bias level	$\tilde{J}_1, \tilde{J}_2$	1.222
Coupling delay time	τ_1, τ_2	40.15 ns
Cavity decay rate	γ_{c1}, γ_{c2}	5.36×10^{11} s^{-1}
Spontaneous carrier relaxation rate	γ_{s1}, γ_{s2}	5.96×10^{9} s^{-1}
Differential carrier relaxation rate	γ_{n1}, γ_{n2}	7.53×10^{9} s^{-1}
Nonlinear carrier relaxation rate	γ_{p1}, γ_{p2}	1.91×10^{11} s^{-1}
Spontaneous emission rate	R_{sp1}, R_{sp2}	4.7×10^{18} V^{2}m^{-1}s^{-1} [31]

Figure 2 shows the typical optical spectrum, microwave spectrum, and temporal evolution for each different dynamical state of LD1 (red curve) and LD2 (black curve) presented in Figure 1. Note that the frequency axes of all the optical spectra shown in this study are relative to the free-running oscillation frequency of LD2. As Figure 2(a-i) shows, where (ξ^s_{12}, f_i) = (0.009, −2.9 GHz), both LD1 and LD2 oscillate at the same offset frequency of −1.52 GHz, indicating that mutual injection locking is established between the two lasers. Two relaxation resonance sidebands appear 10 GHz away from the principal oscillation with the lower one being slightly stronger due to the positive value of b. As Figure 2(b-i) presents, photodetection of the optical signal generates a spectral component at 10 GHz due to the relaxation resonance and a small bump around 0 GHz. The bump actually consists of several spectral components that are equally separated by 12.45 MHz, as the inset shows. The frequency separation corresponds to the loop frequency of the round-trip delay coupling between the two lasers, i.e., the reciprocal of the summation of the two coupling delay times. The appearance of such loop modes is a typical feature of a delay-coupled system because an additional resonance condition given by the round-trip delay coupling is required for the system to satisfy. As Figure 2(c-i) shows, the intensity of both lasers is constant over time yet with an extremely weak modulation at the loop frequency, 12.45 MHz, which can be hardly observed with bare eyes. Note that the intensity value shown in the figures of this section is calculated by removing the direct-current component of each signal. As Figure 2(c-i) also presents, the temporal evolution of the intensity is identical between the two lasers. In fact, LD1 leads LD2 by about 40.15 ns (i.e., the coupling delay time) in Figure 2(c-i) where the temporal evolution of the LD1 intensity is shifted by about 40.15 ns for easy comparison.

As Figure 2(a-ii) shows, where (ξ^s_{12}, f_i) = (0.011, −20 GHz), either LD1 or LD2 oscillates at a frequency that is slightly red-shifted from its free-running oscillation frequency due to the injection pushing effect [33]. Moreover, oscillation sidebands appear around the principal oscillation of each laser, which are equally separated by an oscillation frequency of f_0 = 20.06 GHz. Such a spectral feature is a typical signature of the P1 dynamics. This generates a microwave at f_0 = 20.06 GHz and its harmonics after photodetection, as illustrated in Figure 2(b-ii), which is highly advantageous for high-frequency microwave generation [34–38]. Due to the round-trip delay coupling, there also appears a small bump around 0 GHz, which consists of several spectral components equally separated by 12.45 MHz, as those shown in the inset of Figure 2(b-i). Similar closely-spaced spectral components also appear on top of each P1 spectral component shown in Figure 2(b-ii). As Figure 2(c-ii) shows, the intensity of either laser oscillates sinusoidally with a single period equal to the reciprocal of f_0 = 20.06 GHz. The sinusoidal intensity oscillation of either laser is, in fact, extremely weakly modulated at the loop frequency, 12.45 MHz, which can be hardly observed with bare eyes. As Figure 2(c-ii) also presents, the temporal evolution of the intensity oscillation is identical between the two lasers yet with LD1 leading

LD2 by about 40.15 ns, corresponding to the coupling delay time. For easy comparison, the temporal evolution of the LD1 intensity is shifted by about 40.15 ns in Figure 2(c-ii).

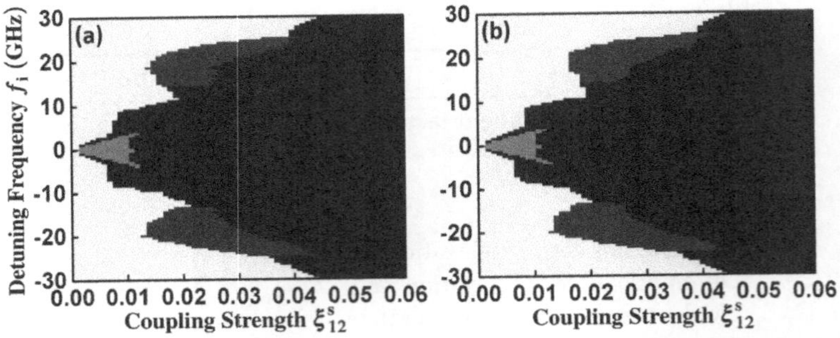

Figure 1. Maps of dynamical states for (**a**) LD1 and (**b**) LD2, respectively, in the mutually-coupled laser system when $\xi^s_{12} = \xi^s_{21}$. Regions of mutual injection locking, P1 dynamics, P2 dynamics, and chaos are marked by red, yellow, blue, and black, respectively.

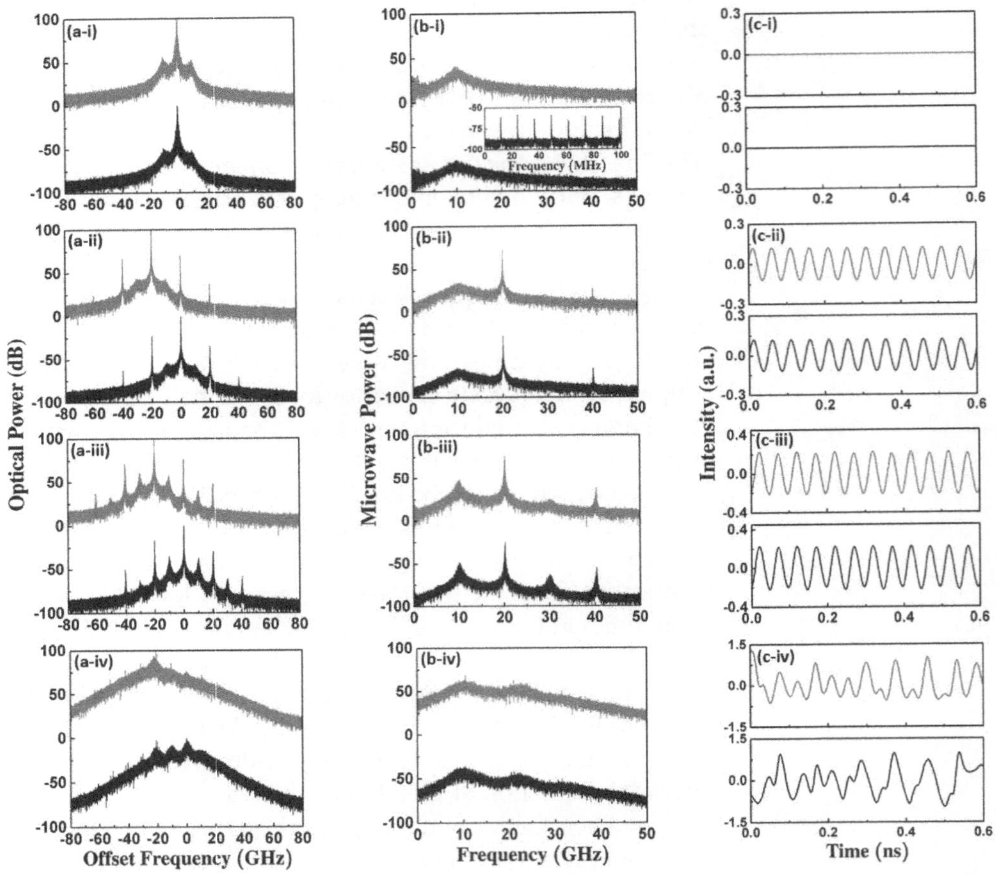

Figure 2. (**a**) Optical spectra, (**b**) microwave spectra, and (**c**) temporal evolutions of LD1 (red curve) and LD2 (black curve) for (**i**) mutual injection locking at (ξ^s_{12}, f_i) = (0.009, −2.9 GHz), (**ii**) P1 dynamics at (ξ^s_{12}, f_i) = (0.011, −20 GHz), (**iii**) P2 dynamics at (ξ^s_{12}, f_i) = (0.02, −20 GHz), and (**iv**) chaos at (ξ^s_{12}, f_i) = (0.05, −20 GHz). The inset of (**b-i**) shows the enlargement of the microwave spectrum for LD2 around 0 GHz. The x-axes in (**a**) are relative to the free-running oscillation frequency of LD2. The red curves in (**a**,**b**) are up-shifted by 100 dB for clear visibility.

By increasing the coupling strength so that (ξ^s_{12}, f_i) = (0.02, −20 GHz), as Figure 2(a-iii) shows, while the spectral components observed in Figure 2(a-ii) for either laser are similarly kept with a slight increase in their frequency separation, leading to f_0 = 20.21 GHz, subharmonics emerge in the midway between the spectral components. Such a spectral feature is a typical signature of the P2 dynamics. The beating between the spectral components at the photodetector not only gives rise to a microwave at f_0 = 20.21 GHz and its harmonics, but also leads to subharmonics at the midway between the spectral components, as Figure 2(b-iii) shows. Due to the round-trip delay coupling, there also appears a small bump around 0 GHz, which consists of several spectral components equally separated by 12.45 MHz, as those shown in the inset of Figure 2(b-i). Similar closely-spaced spectral components also appear on top of each P2 spectral component shown in Figure 2(b-iii). As Figure 2(c-iii) shows, not only does the intensity of either laser oscillate sinusoidally with a period equal to the reciprocal of f_0 = 20.21 GHz, but the intensity oscillation is also moderately modulated with a period equal to two times the reciprocal of f_0 = 20.21 GHz. Such a moderately modulated intensity oscillation is also extremely weakly modulated at the loop frequency, 12.45 MHz, which can be hardly observed with bare eyes. As Figure 2(c-iii) also presents, the temporal evolution of the moderately modulated intensity oscillation is almost identical between the two lasers. In fact, LD1 leads LD2 by about 40.15 ns (i.e., the coupling delay time) in Figure 2(c-iii) where the temporal evolution of the LD1 intensity is shifted by about 40.15 ns for easy comparison.

By continuing to increase the coupling strength so that (ξ^s_{12}, f_i) = (0.05, −20 GHz), as Figure 2(a-iv) shows, a broad and continuous spectral distribution appears for either laser, which is a typical signature of chaos. After photodetection, as Figure 2(b-iv) presents, such a spectral feature generates a broadband chaotic microwave with a spectral distribution of more than 40 GHz, which is highly advantageous for chaos-based applications, such as high-resolution chaotic radars [39–42], high-speed chaotic communication [43–46], and high-entropy random number generation [47–50]. Owing to the round-trip delay coupling, spectral components that are equally separated by 12.45 MHz, as those shown in the inset of Figure 2(b-i) yet with much weaker intensity, also emerge on top of the spectral distribution in Figure 2(b-iv). As Figure 2(c-iv) shows, the intensity of both lasers oscillates irregularly, and is extremely weakly modulated at the loop frequency, 12.45 MHz, which can be hardly be observed with bare eyes. The temporal evolution of the intensity oscillation is similar between the two lasers with LD1 leading LD2 by about 40.15 ns, corresponding to the coupling delay time. For easy comparison, the temporal evolution of the LD1 intensity is shifted by about 40.15 ns in Figure 2(c-iv).

As observed from Figure 2(a-ii) to Figure 2(a-iv), the laser system follows a period-doubling route to chaos as ξ^s_{12} increases at $f_i = -20$ GHz. A similar route is also found when f_i falls between −24 GHz and −13 GHz and between 15 GHz and 26 GHz, as Figure 1 presents. The results obtained in either Figure 1 or Figure 2 conclude that the dynamical behaviors of both lasers are, in general, symmetric when the coupling strength is symmetric, which agrees with the observations in prior studies [23–25].

2.3. *Dynamics Behaviors under Asymmetric Coupling Strength*

In the following analyses, to investigate how the two lasers react when the coupling strength becomes asymmetric, the strength of the coupling from LD2 to LD1 is fixed at ξ^s_{21} = 0.01 and 0.001, respectively, while the strength of the coupling from LD1 to LD2 is varied from ξ^s_{12} = 0 to 0.06. As noted, while ξ^s_{21} = 0.01 is about the same order of magnitude as ξ^s_{12}, ξ^s_{21} = 0.001 is about an order of magnitude smaller than ξ^s_{12}. Maps of dynamical states as a function of ξ^s_{12} and f_i for LD1 and LD2 when ξ^s_{21} = 0.01 are presented in Figure 3(a-i) and 3(b-i), respectively, at a fixed coupling delay time of 40.15 ns. Note that periodic dynamics with periods higher than two are included in the regions of chaos. Comparing Figure 3(a-i) and Figure 3(b-i) with Figure 1 shows that, while the regions of chaos suppress moderately, the regions of mutual injection locking and P1 dynamics expand moderately. The spectral features of different nonlinear dynamical states in Figure 3(a-i) and Figure 3(b-i) are closely similar to those presented in Figure 2. Comparing Figure 3(a-i) with Figure 3(b-i) demonstrates that the dynamical behaviors of the two lasers are generally symmetric at the extent of the coupling strength asymmetry under study here.

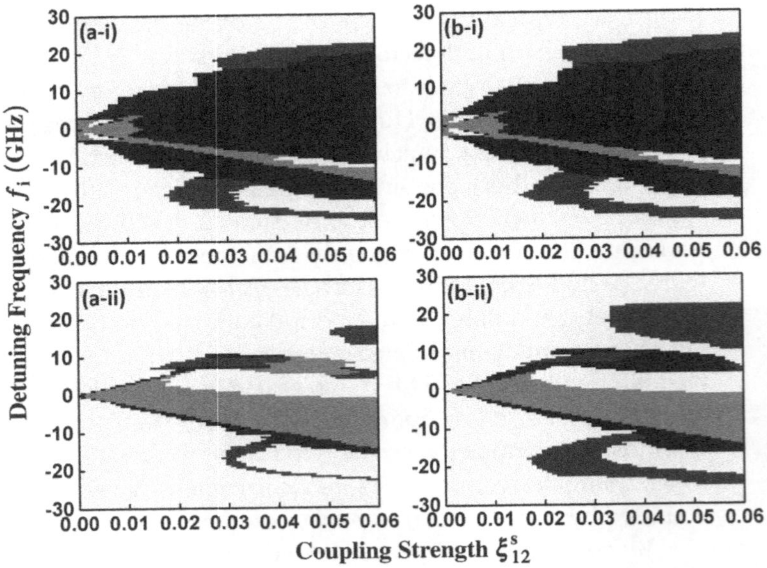

Figure 3. Maps of dynamical states for **(a)** LD1 and **(b)** LD2, respectively, in the mutually coupled laser system when **(i)** $\xi^s_{21} = 0.01$ and **(ii)** $\xi^s_{21} = 0.001$. Regions of mutual injection locking, P1 dynamics, P2 dynamics, and chaos are marked by red, yellow, blue, and black, respectively.

Such dynamical behavior symmetry is, however, not guaranteed if the extent of the coupling strength asymmetry increases, as Figure 3(a-ii) and Figure 3(b-ii) demonstrate, where ξ^s_{21} is reduced from 0.01 to 0.001. While the dynamical behaviors of the two lasers are symmetric over most of the operating conditions considered here, asymmetry happens over a region where ξ^s_{12} falls between 0.034 and 0.06 and f_i is between 11 GHz and 22 GHz, a region where ξ^s_{12} falls between 0.03 and 0.055 and f_i is between 7 GHz and 10 GHz, and a region where ξ^s_{12} falls between 0.018 and 0.06 and f_i is between −25 GHz and −12 GHz. Comparing Figure 3(a-ii) and Figure 3(b-ii) with Figure 3(a-i) and Figure 3(b-i) shows that, as ξ^s_{21} is reduced, the regions of mutual injection locking and P1 dynamics continue to expand and thus become dominant, while the regions of chaos continues to suppress.

To investigate how the dynamical behaviors of both lasers evolve from symmetry to asymmetry when $\xi^s_{21} = 0.001$, a development of optical spectra, microwave spectra, and temporal evolutions for LD1 (red curve) and LD2 (black curve) is presented in Figure 4 when ξ^s_{12} is adjusted and f_i is fixed at 9 GHz. At $\xi^s_{12} = 0.0019$, either LD1 or LD2 behaves as a P1 dynamical state with an oscillation frequency of about 9 GHz, as either Figure 4(a-i), Figure 4(b-i), or Figure 4(c-i) presents. The temporal evolution of the intensity oscillation is identical between the two lasers yet with LD1 lagging LD2 by about 40.15 ns, as Figure 4(c-i) shows where the temporal evolution of the LD2 intensity is shifted by about 40.15 ns.

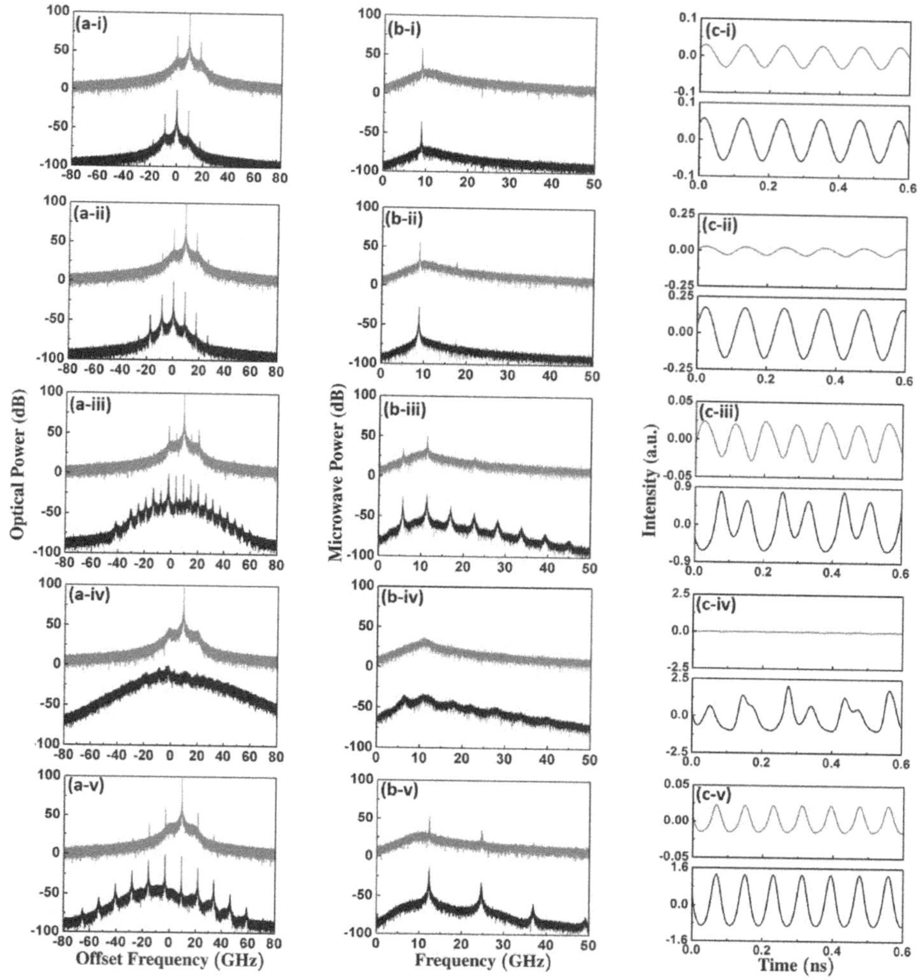

Figure 4. (**a**) Optical spectra, (**b**) microwave spectra, and (**c**) temporal evolutions of LD1 (red curve) and LD2 (black curve) for (**i**) $\xi^{s}_{12} = 0.0019$, (**ii**) $\xi^{s}_{12} = 0.006$, (**iii**) $\xi^{s}_{12} = 0.028$, (**iv**) $\xi^{s}_{12} = 0.048$, and (**v**) $\xi^{s}_{12} = 0.055$, respectively, when $\xi^{s}_{21} = 0.001$ and $f_i = 9$ GHz. The *x*-axes in (**a**) are relative to the free-running oscillation frequency of LD2. The red curves in (**a**,**b**) are up-shifted by 100 dB for clear visibility.

As ξ^{s}_{12} is increased to 0.006, either LD1 or LD2 still behaves as a P1 dynamical state yet with an oscillation frequency of about 8.87 GHz, as either Figure 4(a-ii), Figure 4(b-ii), or Figure 4(c-ii) presents. The temporal evolution of the intensity oscillation is identical between the two lasers yet with LD1 lagging LD2 by about 40.15 ns, as Figure 4(c-ii) presents where the temporal evolution of the LD2 intensity is shifted by about 40.15 ns. While the microwave spectral features and temporal evolutions between the two lasers look highly similar in Figure 4(b-ii) and Figure 4(c-ii), respectively, a slight deviation exists in their optical spectral features in Figure 4(a-ii). Not only a few more spectral components appear in LD2, but also the principal oscillation becomes less dominant, making the optical spectral profile of LD2 more widely distributed. This implies that the two lasers start to behave differently in a subtle manner even though they both behave as a P1 dynamical state. Such a deviation becomes more significant when ξ^{s}_{12} is further increased to 0.028, as Figure 4(a-iii) shows. Both lasers now evolve into a P2 dynamical state, as more evidently observed in Figure 4(b-iii) where subharmonics emerge in the midway of spectral components at the integral multiples of 11.29 GHz, and also in Figure 4(c-iii) where an intensity oscillation with a period equal to the reciprocal of 11.29 GHz is moderately modulated with a period equal to two times the reciprocal of 11.29 GHz. As noted, the temporal evolution of the moderately modulated intensity oscillation becomes moderately dissimilar between the two lasers with LD1 leading LD2 by about 40.15 ns, as Figure 4(c-iii) presents where the temporal evolution of the LD1 intensity is shifted by about 40.15 ns.

As ξ_{12}^{s} continues to increase to 0.048, the optical spectra, microwave spectra, and temporal evolutions of both lasers shown in Figure 4(a-iv), Figure 4(b-iv), and Figure 4(c-iv), respectively, exhibit completely different features and profiles. On one hand, as Figure 4(a-iv) shows, LD1 oscillates at an offset frequency of 9 GHz that is surrounded by two low-intensity spectral components about 11 GHz away. The two low-intensity components result from the modified relaxation resonance of LD1 due to the optical injection from LD2. Such a modification leads to the enhancement of the relaxation resonance frequency, which is more clearly identified in Figure 4(b-iv) where the microwave spectrum peaks at around 11 GHz. This indicates that LD1 now emits a continuous-wave optical output with a slightly higher relaxation resonance frequency as compared with its free-running condition, which is verified by Figure 4(c-iv) where the LD1 intensity remains constant over time. On the other hand, as either Figure 4(a-iv) or Figure 4(b-iv) shows, a broad and continuous spectral distribution is observed for LD2, indicating that LD2 now behaves as a chaotic state, which is verified by Figure 4(c-iv) where the LD2 intensity oscillates irregularly.

By further increasing ξ_{12}^{s} to 0.055, either LD1 or LD2 behaves as a P1 dynamical state with an oscillation frequency of about 12.28 GHz, as either Figure 4(a-v), Figure 4(b-v), or Figure 4(c-v) presents. While the microwave spectral features and temporal evolutions of both lasers look highly similar in Figure 4(b-v) and Figure 4(c-v), a distinct deviation exists in their optical spectral features in Figure 4(a-v). Not only do a few more spectral components emerge in LD2, but the principal oscillation also becomes less dominant, making the optical spectral profile of LD2 more widely distributed. This implies that the two lasers behave differently in a subtle manner even though they both behave as a P1 dynamical state.

The extremest case for the dynamical behavior asymmetry happens when $\xi_{21}^{s} = 0$. This indicates that no optical injection is introduced from LD2 to LD1 and the laser system therefore work as a unidirectional optical injection system. Under such an operating condition, the distribution of dynamical states as a function of ξ_{12}^{s} and f_i for LD2 is greatly similar to the one presented in Figure 3(b-ii), while LD1 is kept at its free-running operation and thus emits a continuous-wave optical output no matter how ξ_{12}^{s} and f_i are adjusted.

The results shown in Figures 1–4 indicate that the dynamical behavior of the laser system could change when the coupling strength becomes asymmetric. This suggests that if a specific dynamical behavior is used for applications where a difference in the coupling strength between the two lasers is likely to happen in order to achieve certain features or functionalities, care must be taken so that the laser system is operated at the same dynamical behavior even when the coupling strength becomes asymmetric during operation. In addition, the results also demonstrate that the dynamical behaviors of the two lasers could become asymmetric when the coupling strength becomes highly asymmetric. This suggests that if both lasers are expected to simultaneously operate at a specific dynamical behavior all the time for applications, care must be taken either to avoid the operation of the laser system with highly asymmetric coupling strength, or to avoid the operation of the laser system over regions where symmetry breaking in the dynamical behavior happens.

3. Experimental Observation

In the previous section, the dynamical behaviors of two mutually delay-coupled semiconductor lasers are numerically investigated when their coupling strength becomes asymmetric. Slight to moderate asymmetry in the coupling strength moderately changes their dynamical behaviors from the ones when the coupling strength is symmetric, but does not break the symmetry of their dynamical behaviors and spectral features. High asymmetry in the coupling strength, however, not only strongly changes their dynamical behaviors from the ones when the coupling strength is symmetric, but also breaks the symmetry of their dynamical behaviors and spectral features. In this section, an experimental study is carried out to verify the numerical predictions.

3.1. Experimental Setup

A schematic diagram of a mutually long-delay-coupled laser system consisting of two single-mode distributed feedback semiconductor lasers, LD1 (Furukawa FRL15DCW5-A81) and LD2 (Furukawa FRL15DCW5-A81), is presented in Figure 5a. The two lasers are mutually coupled by optical injection from one to the other through an optical circulator in each optical injection route, as the blue or red path indicates. For LD2, its bias current and temperature are fixed at 70 mA and 18.9 °C, respectively, throughout the study. This results in a free-running oscillation frequency of 193.28 THz, an output power of 15.48 mW, and a relaxation resonance frequency of 10 GHz. For LD1, while its bias current is fixed at 70 mA throughout the study, its temperature is slightly adjusted around 25.57 °C in order to detune its free-running oscillation frequency away from 193.28 THz (i.e., the free-running oscillation frequency of LD2) by f_i for the excitation of possible dynamical behaviors. The free-running LD1 therefore emits an output power varying slightly around 13.43 mW, depending on the temperature adjustment, with a relaxation resonance frequency of 10 GHz. A variable optical attenuator in each optical injection route adjusts the power of the optical injection (i.e., the coupling strength) from one laser to the other. For the experimental analysis, the coupling strength received by LD2, ξ_{12}^{e}, is defined as the square root of the power ratio between the optical injection from LD1 and the free-running LD2. Similarly, the coupling strength received by LD1, ξ_{21}^{e}, is defined as the square root of the power ratio between the optical injection from LD2 and the free-running LD1. Note that a superscript e is used for both symbols to distinguish the coupling strength defined here from the one defined in the numerical investigation presented in Section 2. These definitions differ by a factor of η_{12}/γ_{c2} for coupling from LD1 to LD2 and η_{21}/γ_{c1} for coupling from LD2 to LD1, respectively. According to the values of η_{12}, η_{21}, γ_{c1}, and γ_{c2} used in the simulation of this study, as previously indicated, the coupling strength defined in the experimental study is about an order of magnitude larger than that defined in the numerical investigation for a given ratio between the fields of the optical injection and the injected laser. Polarization maintaining fibers are used for all the optical devices and components to keep the polarization states of both lasers unchanged. Both optical injection routes have approximately the same fixed length, which corresponds to a coupling delay time of about 40.15 ns from one laser to the other. Such a delay time is longer than the relaxation resonance period of either laser used here. To investigate the spectral features of LD1 and LD2 outputs, respectively, one output port of each fiber coupler in Figure 5a is connected to a detection system consisting of an optical spectrum analyzer (Advantest Q8384) and a microwave spectrum analyzer (Keysight PXAN9030A) following a 50-GHz photodetector (u2t Photonics XPDV2120R), as shown in Figure 5b.

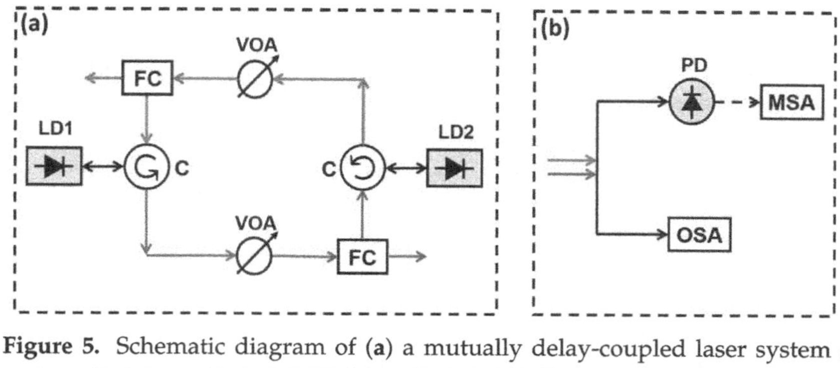

Figure 5. Schematic diagram of (**a**) a mutually delay-coupled laser system and (**b**) a detection system. LD1, laser diode 1; LD2, laser diode 2; FC, fiber coupler; C, circulator; VOA, variable optical attenuator; PD, photodetector; OSA, optical spectrum analyzer; and MSA, microwave spectrum analyzer.

3.2. Dynamical Behaviors under Symmetric Coupling Strength

The dynamical behavior of the mutually delay-coupled laser system is first studied when the coupling strength is symmetric, i.e., $\xi^e_{12} = \xi^e_{21}$, in this subsection. To obtain a global understanding of how the laser system behaves at a fixed coupling delay time of 40.15 ns under consideration, maps of dynamical states as a function of ξ^e_{12} and f_i for LD1 and LD2 are presented in Figure 6a and 6b, respectively. Regions of mutual injection locking, P1 dynamics, P2 dynamics, and chaos are marked by red, yellow, blue, and black, respectively. Periodic dynamics with periods higher than two are included in the regions of chaos. Comparing Figure 6a with Figure 6b demonstrates that the dynamical behaviors of both lasers are generally identical over the range of ξ^e_{12} and f_i under study when the coupling strength is symmetric. This experimental observation is consistent with the numerical prediction presented in Figure 1. Except for the P2 dynamical states that only appear over a small region where f_i falls between −26 GHz and −14 GHz, other nonlinear dynamical states generally appear symmetrically with respect to $f_i = 0$. The mutual injection locking states emerge at weak coupling strength and small frequency detuning. The P1 dynamical states appear when ξ^e_{12} is smaller than 0.21 over the range of f_i under study. The chaotic states start to emerge when ξ^e_{12} is greater than 0.1. Comparing Figure 6 with Figure 1 shows that, except for the P2 dynamical states that appear over a region where f_i falls between 15 GHz and 26 GHz only in the numerical result, the distribution of different dynamical states as a function of ξ^s_{12} and f_i is highly similar. For example, the mutual injection locking states appear when ξ^e_{12} is smaller than 0.125 in Figure 6 and when ξ^s_{12} is smaller than 0.012 in Figure 1, while the chaotic states start to emerge when ξ^e_{12} is greater than 0.1 in Figure 6 and when ξ^s_{12} is greater than 0.007 in Figure 1. Considering that ξ^e_{12} is by definition about an order of magnitude larger than ξ^s_{12}, these results demonstrate that the numerical model used here reproduces the experimental observations not only qualitatively to a high extent but also quantitatively to a moderate extent.

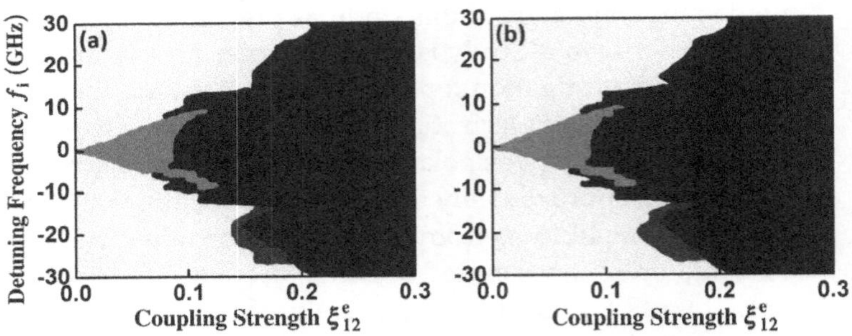

Figure 6. Maps of dynamical states for (**a**) LD1 and (**b**) LD2, respectively, in the mutually coupled laser system when $\xi^e_{12} = \xi^e_{21}$. Regions of mutual injection locking, P1 dynamics, P2 dynamics, and chaos are marked by red, yellow, blue, and black, respectively.

Figure 7 shows the typical optical and microwave spectra for each different dynamical state of LD1 (red curve) and LD2 (black curve) presented in Figure 6. Note that the frequency axes of all the optical spectra shown in this study are relative to the free-running oscillation frequency of LD2. As Figure 7(a-i) shows, where (ξ^e_{12}, f_i) = (0.06, −5 GHz), both LD1 and LD2 oscillate at the same offset frequency of −2.84 GHz, indicating that mutual injection locking is established between the two lasers. Photodetection of such an optical signal only generates a small bump around 0 GHz, as Figure 7(b-i) presents. The bump actually consists of several spectral components that are equally separated by 12.45 MHz, as the inset shows, which corresponds to the loop frequency of the round-trip delay coupling between the two lasers. The loop modes are not observed in Figure 7(a-i) due to the limited resolution, about 0.01 nm at the wavelength of 1550 nm, of the optical spectrum analyzer used in this study.

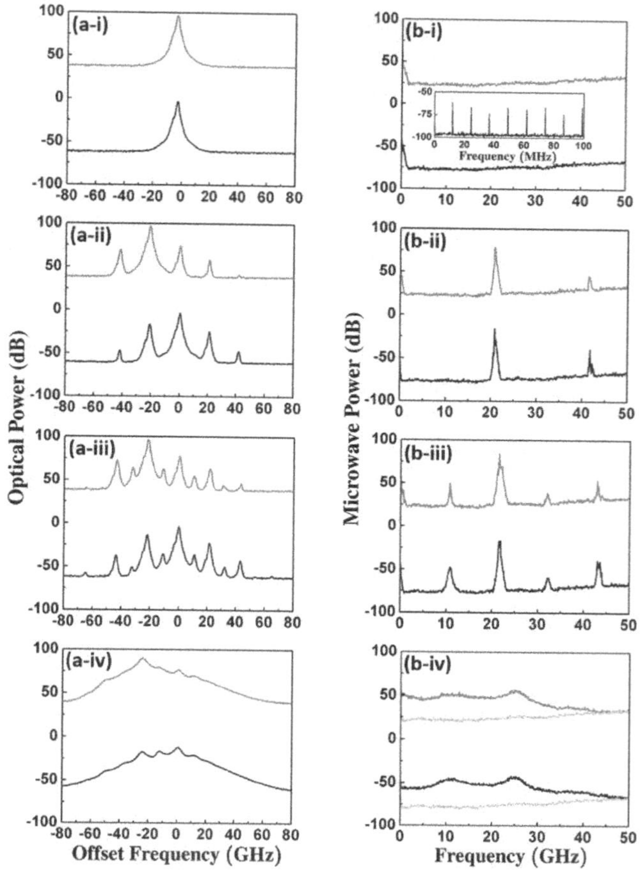

Figure 7. (**a**) Optical spectra and (**b**) microwave spectra of LD1 (red curve) and LD2 (black curve) for (**i**) mutual injection locking at (ξ_{12}^{e}, f_i) = (0.06, −5 GHz), (**ii**) P1 dynamics at (ξ_{12}^{e}, f_i) = (0.1, −20 GHz), (**iii**) P2 dynamics at (ξ_{12}^{e}, f_i) = (0.15, −20 GHz), and (**iv**) chaos at (ξ_{12}^{e}, f_i) = (0.28, −20 GHz). The inset of (**b-i**) shows the enlargement of the microwave spectrum for LD2 around 0 GHz. The *x*-axes in (**a**) are relative to the free-running oscillation frequency of LD2. The red curves in (**a**,**b**) are up-shifted by 100 dB for clear visibility. The gray curves in (**b-iv**) show the noise floor of the laser system.

As Figure 7(a-ii) shows, where (ξ_{12}^{e}, f_i) = (0.1, −20 GHz), either LD1 or LD2 oscillates at a frequency that is slightly red-shifted from its free-running oscillation frequency. In addition, oscillation sidebands emerge around the principal oscillation of each laser, which are equally separated by an oscillation frequency of f_0 = 20.8 GHz. Such a spectral feature is a typical signature of the P1 dynamics. Photodetection of the optical signal generates a microwave at f_0 = 20.8 GHz and its harmonics, as illustrated in Figure 7(b-ii). Due to the round-trip delay coupling, there also appears a small bump around 0 GHz, which consists of several spectral components equally separated by 12.45 MHz, as those shown in the inset of Figure 7(b-i). Similar closely-spaced spectral components also appear on top of each P1 spectral component shown in Figure 7(b-ii).

By increasing the coupling strength so that (ξ_{12}^{e}, f_i) = (0.15, −20 GHz), as Figure 7(a-iii) shows, while the spectral components observed in Figure 7(a-ii) for either laser are similarly kept with a slight increase in their frequency separation, leading to f_0 = 21.5 GHz, subharmonics emerge in the midway between the spectral components. Such a spectral feature is a typical signature of the P2 dynamics. The beating between the spectral components at the photodetector not only gives rise to a microwave at f_0 = 21.5 GHz and its harmonics, but also leads to subharmonics at the midway between the spectral components, as Figure 7(b-iii) shows. As observed, due to the round-trip delay coupling, there also appears a small bump around 0 GHz, which consists of several spectral components equally separated by 12.45 MHz, as those shown in the inset of Figure 7(b-i). Similar closely-spaced spectral components also appear on top of each P2 spectral component shown in Figure 7(b-iii).

By continuing to increase the coupling strength so that (ξ^e_{12}, f_i) = (0.28, -20 GHz), as Figure 7(a-iv) shows, a broad and continuous spectral distribution appears for either laser, which is a typical signature of chaos. After photodetection, as Figure 7(b-iv) presents, such a spectral feature generates a broadband chaotic microwave with a spectral distribution of more than 40 GHz. Due to the round-trip delay coupling, spectral components that are equally separated by 12.45 MHz, as those shown in the inset of Figure 7(b-i) yet with much weaker intensity, also emerge on top of the spectral distribution in Figure 7(b-iv).

As noted from Figure 7(a-ii) to Figure 7(a-iv), the laser system follows a period-doubling route to chaos as ξ^e_{12} increases at $f_i = -20$ GHz, which agrees with the numerical prediction shown in Figure 2. A similar route is also found when f_i falls between -26 GHz and -14 GHz, as demonstrated in Figure 7. The observations found in either Figure 6 or Figure 7 conclude that the dynamical behaviors of both lasers are, in general, symmetric when the coupling strength is symmetric, which verifies the numerical predictions demonstrated in either Figure 1 or Figure 2.

3.3. *Dynamics Behaviors under Asymmetric Coupling Strength*

To study how the two lasers respond when the coupling strength becomes asymmetric, the strength of the coupling from LD2 to LD1 is fixed at ξ^e_{21} = 0.01, while the strength of the coupling from LD1 to LD2 is varied from ξ^e_{12} = 0 to 0.3. Note that ξ^e_{21} = 0.01 is an order of magnitude smaller than ξ^e_{12}, and is so chosen that the dynamical behavior asymmetry could happen based on the numerical prediction found in Section 2.3. Maps of dynamical states as a function of ξ^e_{12} and f_i for LD1 and LD2 are presented in Figure 8a and 8b, respectively, at a fixed coupling delay time of 40.15 ns. Note that periodic dynamics with periods higher than two are included in the regions of chaos. Comparing Figure 8 with Figure 6 shows that, while the regions of chaos shrink dramatically, the regions of mutual injection locking and P1 dynamics largely expand and become dominant. Comparing Figure 8a with Figure 8b demonstrates that, while the dynamical behaviors of the two lasers are symmetric over most of the operating conditions considered here, asymmetry breaking happens mainly over a region where ξ^e_{12} falls between 0.17 and 0.24 and f_i is around 10 GHz and mildly over a region where ξ^e_{12} is around 0.16 and f_i is around -20 GHz. Compared with Figure 3(a-ii) and Figure 3(b-ii), the experimental observations on the distribution of different dynamical states greatly agree with the numerical predictions, except for the P2 dynamical states appearing on the right-upper corner of Figure 3.

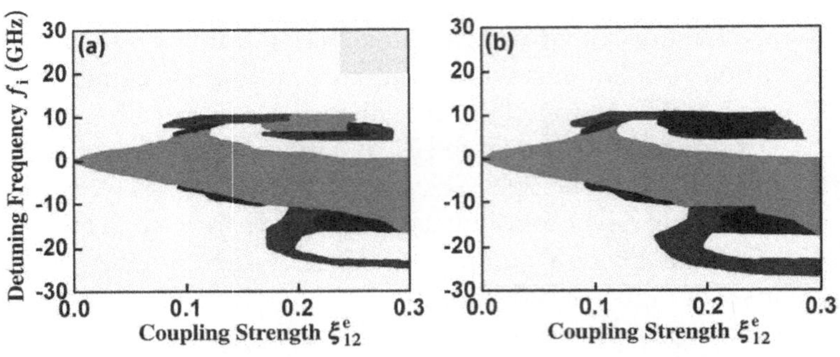

Figure 8. Maps of dynamical states for (**a**) LD1 and (**b**) LD2, respectively, in the mutually coupled laser system when ξ^e_{21} = 0.01. Regions of mutual injection locking, P1 dynamics, P2 dynamics, and chaos are marked by red, yellow, blue, and black, respectively.

To study how the dynamical behaviors of both lasers develop from symmetry to asymmetry when ξ^e_{21} = 0.01, a progression of optical and microwave spectra for LD1 (red curve) and LD2 (black curve) is presented in Figure 9a and 9b, respectively, when ξ^e_{12} is adjusted and f_i is fixed at 10 GHz. At ξ^e_{12} = 0.019, either LD1 or LD2 behaves as a P1 dynamical state with an oscillation frequency of about 10 GHz, as either Figure 9(a-i) or Figure 9(b-i) demonstrates. As ξ^e_{12} is increased to 0.064, either LD1 or LD2 still behaves as a P1 dynamical state yet with an oscillation frequency of about 9.4 GHz, as either Figure 9(a-

ii) or Figure 9(b-ii) presents. While the microwave spectral features of both lasers look highly similar in Figure 9(b-ii), a slight deviation exists in their optical spectral features in Figure 9(a-ii). Not only do a few more spectral components emerge in LD2, but the principal oscillation also becomes less dominant, making the optical spectral profile of LD2 more widely distributed. This implies that the two lasers start to behave differently in a subtle manner even though they both behave as a P1 dynamical state. Such a deviation becomes more significant when ξ_{12}^{e} is further increased to 0.151, as Figure 9(a-iii) shows. Both lasers now evolve into a P2 dynamical state, as more evidently observed in Figure 9(b-iii), where subharmonics emerge in the midway of spectral components at the integral multiples of 11.9 GHz.

As ξ_{12}^{e} is continued to increase to 0.213, both optical and microwave spectra of the two lasers shown in Figure 9(a-iv,b-iv) exhibit completely different spectral features. On one hand, as Figure 9(a-iv) presents, LD1 oscillates at an offset frequency of 9.47 GHz that is surrounded by two low-intensity spectral components about 11.7 GHz away. The two low-intensity components result from the modified relaxation resonance of LD1 due to the optical injection from LD2, which is more clearly identified in Figure 9(b-iv) where a small bump appears at around 11.7 GHz. This indicates that LD1 now emits a continuous-wave optical output with a slightly higher relaxation resonance frequency as compared with its free-running condition. On the other hand, as either Figure 9(a-iv) or Figure 9(b-iv) shows, a broad and continuous spectral distribution is observed for LD2, indicating that LD2 now behaves as a chaotic state. By further increasing ξ_{12}^{e} to 0.3, either LD1 or LD2 behaves as a P1 dynamical state with an oscillation frequency of about 16 GHz, as either Figure 9(a-v) or Figure 9(b-v) demonstrates. While the microwave spectral features of both lasers look similar in Figure 9(b-v), a slight deviation exists in their optical spectral features in Figure 9(a-v). Not only do a few more spectral components emerge in LD2, but the principal oscillation also becomes less dominant, making the optical spectral profile of LD2 more widely distributed. This implies that the two lasers behave differently in a subtle manner even though they both behave as a P1 dynamical state. Comparing Figure 9 with Figure 4 demonstrates that the experimental observations on the evolution of the dynamical behaviors from symmetry to asymmetry agree well with the numerical predictions.

The extremest case for the dynamical behavior asymmetry happens when no optical injection is introduced from LD2 to LD1, i.e., $\xi_{21}^{e} = 0$. Under such an operating condition, the distribution of dynamical states as a function of ξ_{12}^{e} and f_{i} for LD2 is greatly similar to the one presented in Figure 8b, while LD1 is kept at its free-running operation and thus emits a continuous-wave optical output no matter how ξ_{12}^{e} and f_{i} are adjusted.

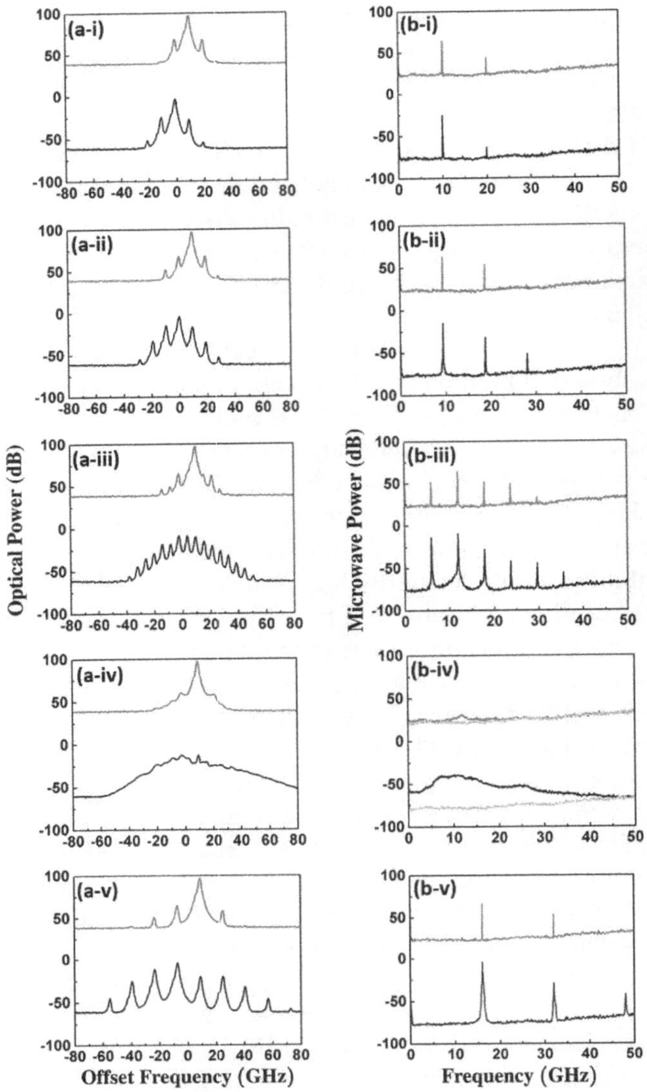

Figure 9. (**a**) Optical spectra and (**b**) microwave spectra of LD1 (red curve) and LD2 (black curve) for (**i**) $\xi^e_{12} = 0.019$, (**ii**) $\xi^e_{12} = 0.064$, (**iii**) $\xi^e_{12} = 0.151$, (**iv**) $\xi^e_{12} = 0.213$, and (**v**) $\xi^e_{12} = 0.3$, respectively, when $\xi^e_{21} = 0.01$ and $f_i = 10$ GHz. The *x*-axes in (**a**) are relative to the free-running oscillation frequency of LD2. The red curves in (**a,b**) are up-shifted by 100 dB for clear visibility. The gray curves in (**b-iv**) show the noise floor of the laser system.

4. Conclusions

This study experimentally and numerically investigates the effects of asymmetric coupling strength on nonlinear dynamics of two mutually coupled semiconductor lasers with a delay time longer than the relaxation resonance period of either laser at free running. Symmetry in the coupling strength of the two lasers, in general, symmetrizes their dynamical behaviors and the corresponding spectral features. Slight to moderate asymmetry in the coupling strength moderately changes their dynamical behaviors from the ones when the coupling strength is symmetric, but does not break the symmetry of their dynamical behaviors and the corresponding spectral features. The former suggests that if a specific dynamical behavior is used for applications where a difference in the coupling strength between the two lasers is likely to happen in order to achieve certain features or functionalities, care must be taken so that the laser system is operated at the same dynamical behavior even when the coupling strength becomes asymmetric during operation. High asymmetry in the coupling strength not only strongly changes their dynamical behaviors from the ones when the coupling strength is symmetric, but also breaks the symmetry of their dynamical behaviors and the corresponding spectral features. This suggests that if both lasers are expected to simultaneously operate at a specific dynamical behavior all the time for applications, care must be taken either to avoid the operation of the laser system with highly asymmetric coupling strength, or to avoid the operation of the laser system over regions where symmetry breaking in the dynamical behavior happens. Evolution of the dynamical behaviors from symmetry to asymmetry between the two lasers is observed. The numerical model used here reproduces the experimental observations not only qualitatively to a high extent but also quantitatively to a moderate extent.

References

1. Chrostowski, L.; Shi, W. Monolithic injection-locked high-speed semiconductor ring lasers. *J. Light. Technol.* **2008**, *26*, 3355–3362.
2. Chow, W.W.; Yang, Z.S.; Vawter, G.A.; Skogen, E.J. Modulation response improvement with isolator-free injection-locking. *IEEE Photonics Technol. Lett.* **2009**, *21*, 839–841.
3. Tauke-Pedretti, A.; Vawter, G.A.; Skogen, E.J.; Peake, G.; Overberg, M.; Alford, C.; Chow, W.W.; Yang, Z.S.; Torres, D.; Cajas, F. Mutual injection locking of monolithically integrated coupled-cavity DBR lasers. *IEEE Photonics Technol. Lett.* **2011**, *23*, 908–910.
4. Yang, Z.; Tauke-Pedretti, A.; Vawter, G.A.; Chow, W.W. Mechanism for modulation response improvement in mutually injection-locked semiconductor lasers. *IEEE J. Quantum Electron.* **2011**, *47*, 300–305.
5. Xiao, Z.X.; Huang, Y.Z.; Yang, Y.D.; Tang, M.; Xiao, J.L. Modulation bandwidth enhancement for coupled twin-square microcavity lasers. *Opt. Lett.* **2017**, *42*, 3173–3176.
6. Sun, C.; Liu, D.; Xiong, B.; Luo, Y.; Wang, J.; Hao, Z.; Han, Y.; Wang, L.; Li, H. Modulation characteristics enhancement of monolithically integrated laser diodes under mutual injection locking. *IEEE J. Sel. Top. Quantum Electron.* **2015**, *21*, 628–635.
7. Zhang, Y.; Li, L.; Zhou, Y.; Zhao, G.; Shi, Y.; Zheng, J.; Zhang, Z.; Liu, Y.; Zou, L.; Zhou, Y.; et al. Modulation properties enhancement in a monolithic integrated two-section DFB laser utilizing side-mode injection locking method. *Opt. Express* **2017**, *25*, 27595–27608.
8. Zheng, J.; Zhao, G.; Zhou, Y.; Zhang, Z.; Pu, T.; Shi, Y.; Zhang, Y.; Liu, Y.; Li, L.; Lu, J.; et al. Experimental demonstration of amplified feedback DFB Laser with modulation bandwidth enhancement based on the reconstruction equivalent chirp technique. *IEEE Photonics J.* **2017**, *9*, 1–8.
9. Chien, C.-Y.; Lo, Y.-H.; Wu, Y.-C.; Hsu, S.-C.; Tseng, H.-R.; Lin, C.-C. Compact photonic integrated chip for tunable microwave generation. *IEEE Photonics Technol. Lett.* **2014**, *26*, 490–493.
10. Lo, Y.H.; Wu, Y.C.; Hsu, S.C.; Hwang, Y.C.; Chen, B.C.; Lin, C.C. Tunable microwave generation of a monolithic dual-wavelength distributed feedback laser. *Opt. Express* **2014**, *22*, 13125–13137.
11. Zhang, X.; Zheng, J.; Pu, T.; Zhang, Y.; Shi, Y.; Li, J.; Li, Y.; Zhu, H.; Chen, X. Simple frequency-tunable optoelectronic oscillator using integrated multi-section distributed feedback semiconductor laser. *Opt. Express* **2019**, *27*, 7036–7046.
12. Zhou, Y.; Lu, Z.; Li, L.; Zhang, Y.; Zheng, J.; Du, Y.; Zou, L.; Shi, Y.; Zhang, X.; Chen, Y.; et al. Tunable microwave generation utilizing monolithic integrated two-section DFB laser. *Laser Phys.* **2019**, *29*, 046201.
13. Fujino, H.; Ohtsubo, J. Synchronization of chaotic oscillations in mutually coupled semiconductor lasers. *Opt. Rev.* **2001**, *8*, 351–357.
14. Gross, N.; Kinzel, W.; Kanter, I.; Rosenbluh, M.; Khaykovich, L. Synchronization of mutually versus unidirectionally coupled chaotic semiconductor lasers. *Opt. Commun.* **2006**, *267*, 464–468.
15. Mengue, A.D.; Essimbi, B.Z. Secure communication using chaotic synchronization in mutually coupled semiconductor lasers. *Nonlinear Dyn.* **2012**, *70*, 1241–1253.
16. Yan, S.L. Chaotic synchronization of two mutually coupled semiconductor lasers for optoelectronic logic gates. *Commun. Nonlinear Sci. Numer. Simul.* **2012**, *17*, 2896–2904.
17. Hou, Y.S.; Xia, G.Q.; Jayaprasath, E.; Yue, D.Z.; Yang, W.Y.; Wu, Z.M. Prediction and classification performance of reservoir computing system using mutually delay-coupled semiconductor lasers. *Opt. Commun.* **2019**, *433*, 215–220.
18. Hou, Y.S.; Xia, G.Q.; Jayaprasath, E.; Yue, D.Z.; Wu, Z.M. Parallel information processing using a reservoir computing system based on mutually coupled semiconductor lasers. *Appl. Phys. B-Lasers Opt.* **2020**, *126*, 40.

19. Liang, W.Y.; Xu, S.R.; Jiang, L.; Jia, X.H.; Lin, J.B.; Yang, Y.L.; Liu, L.M.; Zhang, X. Design of parallel reservoir computing by mutually-coupled semiconductor lasers with optoelectronic feedback. *Opt. Commun.* **2021**, *495*, 127120.
20. Mihana, T.; Mitsui, Y.; Takabayashi, M.; Kanno, K.; Sunada, S.; Naruse, M.; Uchida, A. Decision making for the multi-armed bandit problem using lag synchronization of chaos in mutually coupled semiconductor lasers. *Opt. Express* **2019**, *27*, 26989–27008.
21. Heil, T.; Fischer, I.; Elsasser, W.; Mulet, J.; Mirasso, C.R. Chaos synchronization and spontaneous symmetry-breaking in symmetrically delay-coupled semiconductor lasers. *Phys. Rev. Lett.* **2001**, *86*, 795–798.
22. Rogister, F.; Garcia-Ojalvo, J. Symmetry breaking and high-frequency periodic oscillations in mutually coupled laser diodes. *Opt. Lett.* **2003**, *28*, 1176–1178.
23. Mulet, J.; Mirasso, C.; Heil, T.; Fischer, I. Synchronization scenario of two distant mutually coupled semiconductor lasers. *J. Opt. B Quantum Semiclass. Opt.* **2004**, *6*, 97–105.
24. Junges, L.; Gavrielides, A.; Gallas, J.A.C. Synchronization properties of two mutually delay-coupled semiconductor lasers. *J. Opt. Soc. Am. B-Opt. Phys.* **2016**, *33*, C65–C71.
25. Seifikar, M.; Amann, A.; Peters, F.H. Dynamics of two identical mutually delay-coupled semiconductor lasers in photonic integrated circuits. *Appl. Opt.* **2018**, *57*, E37–E44.
26. Lingnau, B.; Perrott, A.H.; Dernaika, M.; Caro, L.; Peters, F.H.; Kelleher, B. Dynamics of on-chip asymmetrically coupled semiconductor lasers. *Opt. Lett.* **2020**, *45*, 2223–2226.
27. Simpson, T.B.; Liu, J.M. Phase and amplitude characteristics of nearly degenerate four-wave mixing in Fabry-Perot semiconductor lasers. *J. Appl. Phys.* **1993**, *73*, 2587–2589.
28. Liu, J.M.; Simpson, T.B. Four-wave mixing and optical modulation in a semiconductor laser. *IEEE J. Quantum Electron.* **1994**, *30*, 957–965.
29. Simpson, T.B.; Liu, J.M.; Huang, K.F.; Tai, K. Nonlinear dynamics induced by external optical injection in semiconductor lasers. *Quantum Semiclass. Opt.* **1997**, *9*, 765–784.
30. Hwang, S.K.; Liu, J.M. Dynamical characteristics of an optically injected semiconductor laser. Optics Communications. *Opt. Commun.* **2000**, *183*, 195–205.
31. Simpson, T.B.; Liu, J.M. Spontaneous emission, nonlinear optical coupling, and noise in laser diodes. *Opt. Commun.* **1994**, *112*, 43–47.
32. Hwang, S.K.; Liu, J.M.; White, J.K. 35-GHz intrinsic bandwidth for direct modulation in 1.3-μm semiconductor lasers subject to strong injection locking. *IEEE Photonics Technol. Lett.* **2004**, *16*, 972–974.
33. Chan, S.C. Analysis of an optically injected semiconductor laser for microwave generation. *IEEE J. Quantum Electron.* **2010**, *46*, 421–428.
34. Simpson, T.B.; Liu, J.M.; AlMulla, M.; Usechak, N.G.; Kovanis, V. Linewidth sharpening via polarization-rotated feedback in optically injected semiconductor laser oscillators. *IEEE J. Sel. Top. Quantum Electron.* **2013**, *19*, 1500807.
35. Lo, K.H.; Hwang, S.K.; Donati, S. Optical feedback stabilization of photonic microwave generation using period-one nonlinear dynamics of semiconductor lasers. *Opt. Express* **2014**, *22*, 18648–18661.
36. Lo, K.H.; Hwang, S.K.; Donati, S. Numerical study of ultrashort-optical-feedback-enhanced photonic microwave generation using optically injected semiconductor lasers at period-one nonlinear dynamics. *Opt. Express* **2017**, *25*, 31595–31611.
37. Zhang, L.; Chan, S.C. Cascaded injection of semiconductor lasers in period-one oscillations for millimeter-wave generation. *Opt. Lett.* **2019**, *44*, 4905–4908.
38. Tseng, C.H.; Lin, C.T.; Hwang, S.K. V- and W-band microwave generation and modulation using semiconductor lasers at period-one nonlinear dynamics. *Opt. Lett.* **2020**, *45*, 6819–6822.
39. Lin, F.Y.; Liu, J.M. Chaotic radar using nonlinear laser dynamics. *IEEE J. Quantum Electron.* **2004**, *40*, 815–820.
40. Xu, H.; Wang, B.J.; Han, H.; Liu, L.; Li, J.X.; Wang, Y.C.; Wang, A.B. Remote imaging radar with ultra-wideband chaotic signals over fiber links. *Int. J. Bifurc. Chaos* **2015**, *25*, 1530029.
41. Wang, L.S.; Guo, Y.Y.; Li, P.; Zhao, T.; Wang, Y.C.; Wang, A.B. White-chaos radar with enhanced range resolution and anti-jamming capability. *IEEE Photon. Technol. Lett.* **2017**, *29*, 1723–1726.
42. Tseng, C.H.; Hwang, S.K. Broadband chaotic microwave generation through destabilization of period-one nonlinear dynamics in semiconductor lasers for radar applications. *Opt. Lett.* **2020**, *45*, 3777–3780.
43. Pecora, L.M.; Carroll, T.L.; Johnson, G.A.; Mar, D.J.; Heagy, J.F. Fundamentals of synchronization in chaotic systems, concepts, and applications. *Chaos* **1997**, *7*, 520–543.
44. VanWiggeren, G.D.; Roy, R. Communication with chaotic lasers. *Science* **1998**, *279*, 1198–1200.
45. Argyris, A.; Syvridis, D.; Larger, L.; Annovazzi-Lodi, V.; Colet, P.; Fischer, I.; Garcia-Ojalvo, J.; Mirasso, C.R.; Pesquera, L.; Shore, K.A. Chaos-based communications at high bit rates using commercial fibre-optic links. *Nature* **2005**, *438*, 343–346.
46. Uchida, A.; Rogister, F.; Garcia-Ojalvo, J.; Roy, R. Synchronization and communication with chaotic laser systems. *Prog. Opt.* **2005**, *48*, 203–341.
47. Uchida, A.; Amano, K.; Inoue, M.; Hirano, K.; Naito, S.; Someya, H.; Oowada, I.; Kurashige, T.; Shiki, M.; Yoshimori, S.; et al. Fast physical random bit generation with chaotic semiconductor lasers. *Nat. Photonics* **2008**, *2*, 728–732.
48. Hart, J.D.; Terashima, Y.; Uchida, A.; Baumgartner, G.B.; Murphy, T.E.; Roy, R. Recommendations and illustrations for the evaluation of photonic random number generators. *APL Photonics* **2017**, *2*, 090901-1–090901-22.

49. Wang, A.; Wang, L.; Li, P.; Wang, Y. Minimal-post-processing 320-Gbps true random bit generation using physical white chaos. *Opt. Express* **2017**, *25*, 3153–3164.
50. Tseng, C.H.; Funabashi, R.; Kanno, K.; Uchida, A.; Wei, C.C.; Hwang, S.K. High-entropy chaos generation using semiconductor lasers subject to intensity-modulated optical injection for certified physical random number generation. *Opt. Lett.* **2021**, *46*, 3384–3387.

Mapping the Stability and Dynamics of Optically Injected Dual State Quantum Dot Lasers

Abstract: Optical injection is a key nonlinear laser configuration both for applications and fundamental studies. An important figure for understanding the optically injected laser system is the two parameter stability mapping of the dynamics found by examining the output of the injected laser under different combinations of the injection strength and detuning. We experimentally and theoretically generate this map for an optically injected quantum dot laser, biased to emit from the first excited state and optically injected near the ground state. Regions of different dynamical behaviours including phase-locking, excitability, and bursting regimes are identified. At the negatively detuned locking boundary, ground state dropouts and excited state pulses are observed near a hysteresis cycle for low injection strengths. Higher injection strengths reveal µs duration square wave trains where the intensities of the ground state and excited state operate in antiphase. A narrow region of extremely slow oscillations with periods of several tens of milliseconds is observed at the positively detuned boundary. Two competing optothermal couplings are introduced and are shown to reproduce the experimental results extremely well. In fact, the dynamics of the system are dominated by these optothermal effects and their interplay is central to reproducing detailed features of the stability map.

Keywords: optical injection; quantum dot lasers; semiconductor lasers; excitability; nonlinear dynamics; neuromorphic dynamics; chaos

---

1. Introduction

Optical injection is a key technique in many modern photonic systems and particularly unidirectional optical injection, where light from a primary laser is injected into the cavity of a secondary laser. It is used in high sensitivity signal sensing, to improve coherence of high power lasers, for spectral density enhancement in optical communications, and for arbitrary pulse shaping, among many other applications. Mapping the behaviour of an optically injected laser over control parameter space is thus of central importance for many applications. A stability map where one varies the injection level and the detuning (the frequency of the primary laser minus that of the secondary laser) and generates a diagram characterising the output of the injected device is one of the most effective tools in this regard [1]. Stability maps can take several forms and can be as simple as displaying the region where phase locking is obtained. Indeed, for many applications this is the most important regime, and this simple map is all that is needed. On the other hand, optically injected semiconductor lasers are also rich sources of fundamental non-linear dynamics, and the stability map can also be used to classify regions of different behaviour including, but not limited to, excitability, multistability, oscillatory behaviour, and chaos [1–3]. Such regions can be of interest both for fundamental studies and for applications. For example, period 1 dynamics have been used for tunable microwave generation [4,5], chaotic dynamics have attracted attention for random number generation [6], and the potential of excitability for use in neuromorphic photonics has been highlighted in [7] and elsewhere. Thus, a stability map detailing the different dynamical regimes is also of great importance. In [2] an automated, efficient, and quick technique for creating such an experimental map is presented. In [2] the device studied is a two frequency quantum well (QW) based device, but the technique is applicable to any device type. In short, the injection power is swept up and down at fixed values of the detuning. The output power is measured at regular intervals using fast detectors and a fast, real-time oscilloscope. Calculating the mean and standard deviation of the power then allows for easy discrimination of different regions in parameter space and simultaneously allows for identification of hysteretic regions.

Rate equation modelling of the system has proven to be extremely accurate with quantitative agreement between the maps produced experimentally and those arising from the model. A particularly striking example of this agreement is shown in [1] for a single frequency QW based laser. In this example, the map takes the form of a two parameter (injection level and detuning) bifurcation diagram, identifying where qualitative changes in the laser's behaviour occur. Analytic studies can also be performed, such as those in [3,8–10], and in particular, analytic expressions can be found for the two most important bifurcations for the generation of phase locking, namely the saddle node bifurcation and the Hopf bifurcation. In [11–13] numerical analyses of the Lyapunov exponents were used to produce the stability maps. One can also perform numerical studies analogous to the experimental technique to produce the mapping as shown in [2]. The superb agreement between experiment and theory allows great trust to be placed in simulations and predictions that arise from theory.

Optically injected InAs/GaAs based quantum dot (QD) lasers have attracted substantial attention in recent years. They display many different dynamical regimes, including several novel regimes heretofore unobserved with other devices, such as optothermal excitability [14], and mixed mode oscillations and canard explosions [15,16]. In [3,9], the stability map for a single mode QD laser is analysed both experimentally and analytically with excellent qualitative agreement between the experimental and analytical figures. The striking similarity to the Class A stability diagram [10] is also discussed. Microscopic rate equation analysis has also been performed [17], again agreeing well with the experiment. This model allows for a more accurate analysis of phase-amplitude coupling going beyond the simplified constant α factor typically assumed. In [18], excellent agreement between experimentally and numerically generated maps for multimode optically injected QD lasers is demonstrated with the importance of including spatial hole-burning emphasised.

A unique characteristic of InAs/GaAs QD lasers is their ability to lase from multiple different energy levels [19,20]. In particular they typically display ground state (GS) emission close to 1300 nm and first excited state (ES) emission at approximately 1215 nm. Each of these states has its own threshold and which state lases depends on many factors including the optical losses and the pump current. A typical evolution as the pump current is increased is presented in [21]. First, the GS threshold is reached, at which GS only light is emitted. This is followed by the ES threshold, where the GS and ES emit simultaneously. Further increasing the pump current typically results in a quenching of the GS emission after which only the ES lases. However, in other cases—such as with short device lengths—there may not be sufficient gain available in the GS to overcome the losses below the ES threshold. Then, as the current is increased, the ES threshold can be reached without any lasing ever being achieved from the GS. In this case, no emission from the GS is found at any pump current. This is the case for the device investigated in this work.

This multi-state structure allows for unique dual state injection scenarios. In particular, one can bias a QD device to emit from the ES only and then inject the device with light at the GS frequency from an external laser. In this configuration, it has been shown that the ES can be made to turn off and the GS to turn on, with the output phase locked to that of the injecting primary laser [21]. This behaviour can be obtained regardless of whether the QD device can be made to emit from the GS while free-running. Several dynamical regimes have been reported and analysed in this configuration, such as fast state switching [22], all-optical gating [23], dual state excitability [24], and neuromorphic bursting [15], with excellent agreement between experiment and theory.

We present a comprehensive combined experimental and theoretical analysis of the stability map for an optically injected dual state QD device. As well as mapping out previously discovered dynamical regions, we uncover new features such as slow oscillations with a period of 10s of milliseconds near the positively unlocked boundary. Optothermal effects are shown to play an important role in the system. Superb agreement is obtained between experiment and theory. We also note a marked absence of chaos in the system.

2. Results

2.1. Experiment

The device under investigation is a 300 μm QD laser composed of InAs quantum dots on a GaAs. It has the same epitaxial structure as the device used in [14,21] but is significantly shorter (300 μm here compared to 600–900 μm in [14,21]). As a result, the device under test here never lases from the GS when free-running, and instead, lases only from the ES. It is pumped at 75 mA (1.3 times threshold at 20.5 °C). A schematic of the experimental setup is shown in Figure 1. The device is mounted and placed on an xyz stage. The primary laser (PL) is a commercial tunable laser source with minimum step size of 0.1 pm (~0.0178 GHz). Light from the PL is injected into the secondary laser (SL) — the QD laser — via an optical circulator with an isolation greater than 30 dB. A polarisation controller is used to set the polarisation of the injected light and maximise coupling. The light from the SL enters the second port of the circulator and is directed to a filter where the ES and GS are separated. The ES emission is sent directly to a 12 GHz detector connected to a high speed, real-time oscilloscope. 10% of the GS power is sent to a power meter (PM) to monitor alignment and 90% goes directly to another 12 GHz detector, again connected to the oscilloscope.

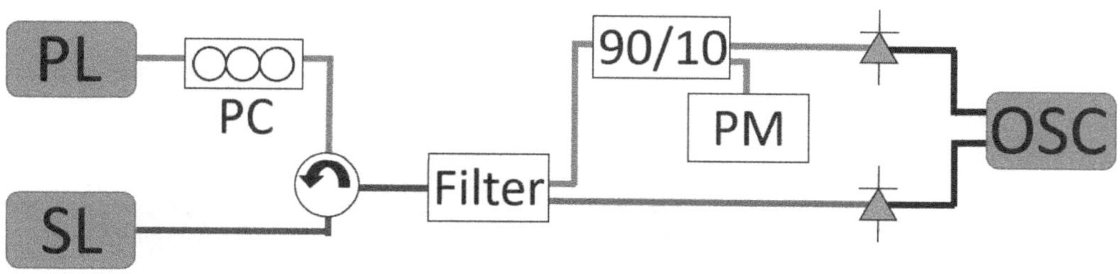

Figure 1. Schematic experimental setup. SL is the secondary (QD) laser. PL is the tunable primary laser. Light from the PL is is sent to a circulator and is then injected into the SL. A polarisation controller (PC) is used to maximise coupling. Light from the SL is sent to the circulator and then to a filter where the ES and GS light are separated. The ES light is sent directly to a 12 GHz detector and the GS light goes to a 90/10 splitter. The 10% is sent to a power meter (PM) for alignment control, and the 90% goes to another 12 GHz detector. Both detectors are connected to a high speed real-time oscilloscope. Red lines are GS light; blue lines are ES light; purple lines contain both GS and ES light; and black lines are high speed electrical cables.

Typically, one can define the experimental injection level as the ratio of the magnitude of the electric field of the injected light to the magnitude of the electric field of the free running laser. However, in this work, when the device is free-running, the GS is never lasing and so this measurement cannot be used. Instead, we define injection strength as the square root of the power of the PL reaching the facet of the SL normalised to the maximum power of the PL reaching the facet of the SL in the experiment, $K = \sqrt{\frac{p^{facet}}{p^{facet}_{max}}}$. Thus, the maximum value of K is 1. Similarly, as the GS mode being injected is always subthreshold, the detuning is also difficult to define. We thus pragmatically define zero detuning to be where the ES is at a minimum power after injection.

The experimental technique to build the stability maps is similar to that of [2], but here we fix the detuning and vary the injection strength for each slice of our map. We record and analyse 20 µs-long time series of both the GS and ES intensities at regular intervals. Initially the frequency of the PL is set and the power of injected light reaching the facet of the secondary laser from the PL is swept from 0.25 mW to 2.47 mW in 120 equal steps of K. Then, a down sweep is performed where the power reaching the facet is swept from to 2.47 mW to 0.25 mW with the same 120 steps. The wavelength is then decreased in steps of 0.2 pm (∼0.0356 GHz) and the power sweeps are repeated after each step.

2.1.1. Stability Maps

The average power from the GS is plotted in Figure 2a,d and that of the ES is plotted in Figure 2b,e. The variance of the ES output is also calculated and plotted in Figure 2c,f. The variance plots are particularly useful for immediate identification of the constant output regions (dark blue) and areas where the intensity is oscillating (green/yellow). When the ES intensity is constant, that of the GS is also constant, and when the ES intensity is oscillating, that of the GS is also oscillating. The upper panels in all stability map figures in this section show the results generated when the injection sweep is upwards (from low strengths to high strengths) and the lower panels show the corresponding maps for the downward sweep (from high strengths to low strengths).

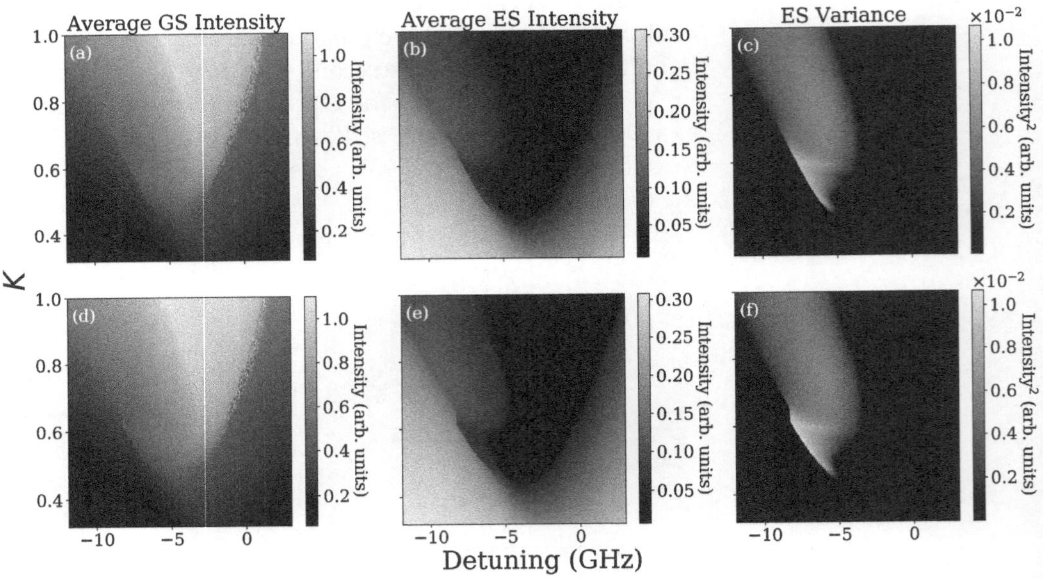

Figure 2. Experimental stability maps. The top panels show results when the injection strength, K, is swept from low to high. The bottom panels show the results when injection strength is swept from high to low. (**a**,**d**) show the average GS intensity. (**b**,**e**) show the average ES intensity. The ES variance is shown in (**c**,**f**). The average ES and GS intensities are calculated for 20 µs long measurements. The phase locked region is identified as the bright yellow/cyan area in (**a**,**d**), where the GS has high output intensity. This corresponds to a quenched ES, shown as dark blue in (**b**,**e**). In (**c**,**f**) green denotes oscillatory behaviour and dark blue corresponds to constant output. The speckled area on the positively detuned boundary of (**a**,**c**) is related to very slow oscillations. A more detailed map of this boundary is shown later.

In Figure 2a, the bright yellow region in the centre is the phase-locked region. Here, only the GS is lasing and the ES is completely quenched as indicated by the corresponding dark blue region in Figure 2b and the zero variance in Figure 2c. To the negative detuning side of the phase locked region there is a broad area of dynamics as indicated by the green and yellow region in the variance in Figure 2c. On the positive boundary of the phase locked region there is a region of very slowly varying GS power, indicated by the non-smooth colouring/speckled pixels in Figure 2a. We defer discussion of this for now and return to it below.

To the positive side of the slowly varying speckled region, the device behaves as a dual state emitter, with both states emitting constant intensities. As the detuning is increased there is a smooth, continuous evolution of both, with the GS power decreasing and the ES power increasing. The same dual state behaviour is mirrored in the negative detuning region to the left of the dynamic oscillatory regime. We interpret this as follows: in the GS only system (or indeed any conventional optical injection system), there are unlocking boundaries outside of which there are only oscillating unlocked solutions. In the limit of large detuning, the oscillations can be physically interpreted as a beating between the injected light and the emission of the secondary laser. However, in our dual state system there is no free-running GS. Thus, this beating cannot arise. The output of our QD SL can then be thought of as a mixture of regenerated injected light [4,25] and ES emission from the secondary laser. We thus distinguish between the phase-locked output (dark blue) in Figure 2b, where the ES is completely off, and the regenerated injected light output of somewhat large detunings where there is dual state emission.

Figure 3 shows just some of traces taken during a single upsweep of the injection strength K for a fixed detuning of -6.9 GHz. The coloured labels correspond to the coloured dots in Figure 4, marking their location on the maps. (The maps in Figure 4 are the same as in Figure 2, with some additional markings to allow for comparison with Figure 3). Figure 3a shows a periodic train of GS dropouts and the corresponding ES pulses as reported in [23,24]. As the injection strength is increased these periodic trains disappear, and in Figure 3b a bursting oscillation is obtained while various mixed mode oscillation (MMO) and bursting MMO regions [26,27] are observed in Figure 3c–e. In Figure 3b there is a switching between a quiescent phase, where the GS is on and the ES is off, and an active phase with oscillations in both states (similar to the dropouts and pulses of Figure 3a). A further increase to $K = 0.61$ leads to a different bursting dynamic as shown in Figure 3c. In this region, the switching is between a quiescent phase with the GS on and the ES off and an evolving bursting phase. The switch to the bursting phase is via an initial period of decreasing amplitude oscillations, followed by a long series of growing oscillations before a switch back to the quiescent phase. Such evolving bursts were previously reported in [15] and shown to arise via an optothermal coupling. The evolution of the bursting state arises via a determinstic thermal sweep of the detuning leading to slow passages through several bifurcations, breaking underlying bistabilities similar to the generation of excitable square waves in the GS only system in [14]. In Figure 3d $K = 0.68$, the trace is qualitatively similar to that in Figure 3c, but in Figure 3d the oscillations following the GS dropout die away quickly and the bursting part of the trace is shorter. Finally, when K is increased to 0.99 as shown in Figure 3e, there are no oscillations following the GS dropout and only a very short region of oscillations preceding the return to the quiescent, GS on phase.

2.1.2. Hysteresis and Bistabilty

In order to identify regions of bistability in the maps, we can compare the upsweep and downsweep figures in Figure 4. The upper panel shows the upsweep with a red line superimposed, marking the boundaries from the downsweep. The lower panel shows the downsweep with a white line superimposed, corresponding to the boundaries of oscillatory behaviour from the upsweep. The boundaries are largely the same, and reveal bistabilities where they differ. The negatively detuned side of the phase-locked region (the red/white line in the centre of each figure) is the same in both cases. At high injection strengths and negative detunings, the two directions match at the boundary between the dynamical region and the regeneration region. However, at moderate injection strengths from approximately $K = 0.5$ to $K = 0.65$, and between detunings of approximately -8.5 GHz and -6 GHz, there is a clear region of hysteresis. This stands out clearly in all of the plots, but most obviously in Figure 4f.

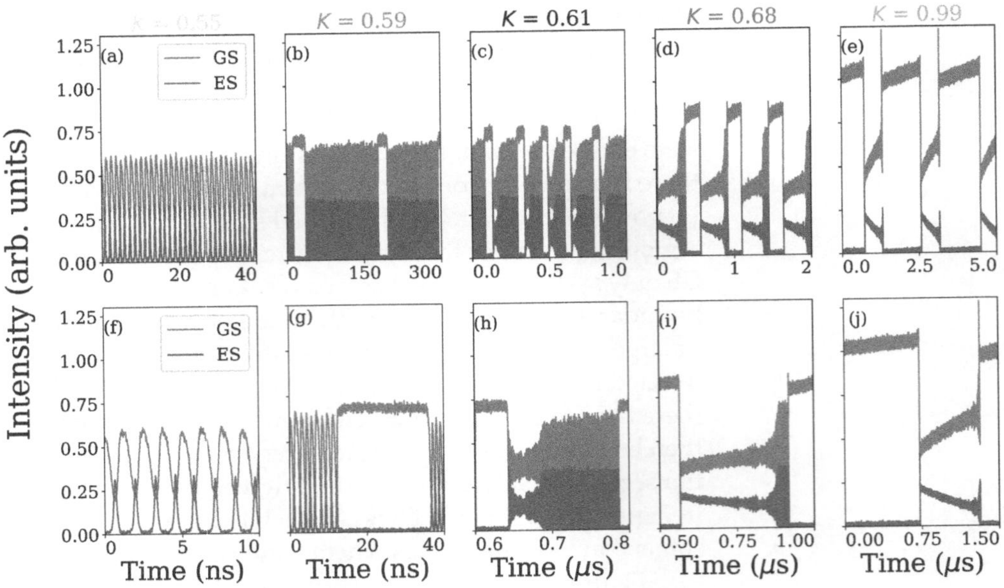

Figure 3. The upper panels, **(a–e)** show representative intensity traces for the different dynamical regimes observed during an upsweep where the detuning remains fixed at −6.9 GHz. The injection strength is given above each subfigure. The lower panels **(f–j)** show zooms of the corresponding upper panels. The colour of the font indicating the injection strength corresponds to the colour of dots in Figure 4 which in turn mark the locations on the map where the intensity traces were recorded.

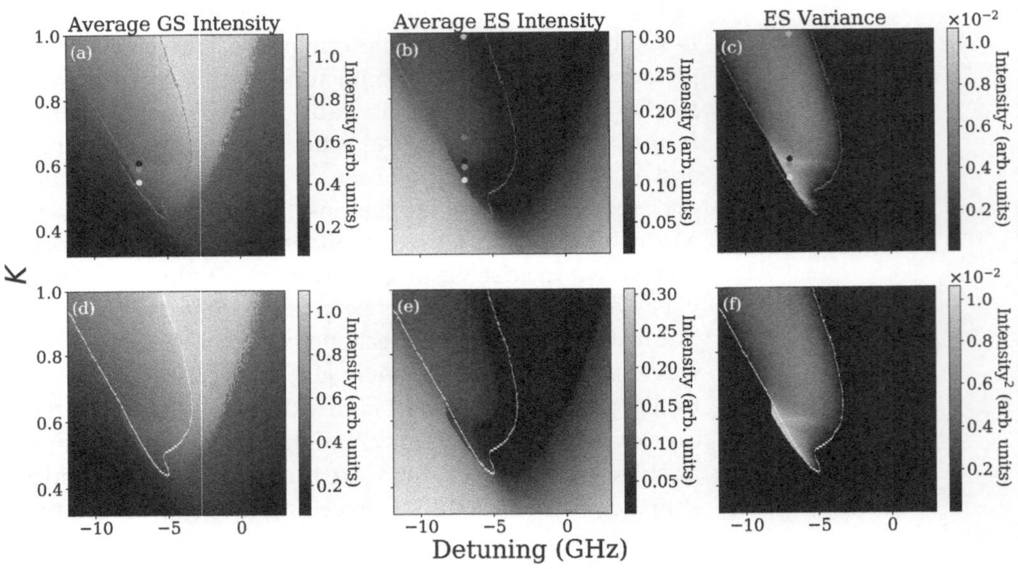

Figure 4. The same experimental stability maps of average GS and ES intensities, and ES variance, previously shown in Figure 2 but shown again here with markings added for direct comparison with the time traces in Figure 3. The injection strength is swept up in **(a–c)** and down in **(d–f)**. The coloured dots correspond to the timetraces in Figure 3. For a more obvious identification of hysteresis the boundary of oscillating dynamics observed in the downsweep is plotted on top of the upsweep data as a red line. Similarly the boundary of oscillating dynamics seen in the upsweep is plotted as a white line on the down sweep map. Hysteresis is particularly evident in **(f)** where the region of oscillating dynamics extends beyond the white line, which marks the boundary observed **(c)**.

Figures 5 and 6 compare the timetraces from an upsweep and a downsweep, respectively, at a detuning of −8.2 GHz. The sweeps move across the widest part of hysteresis region, just to the right of the almost vertical part of the red line shown in the upper panels in Figure 4. It is important to note that the injection strengths are plotted in the same grid position of both figures for easy comparison. Thus, the upsweep shown in Figure 5 was performed from Figure 5a–l, while the timetraces during the downsweep shown in Figure 6 were collected in reverse, from Figure 5l–a.

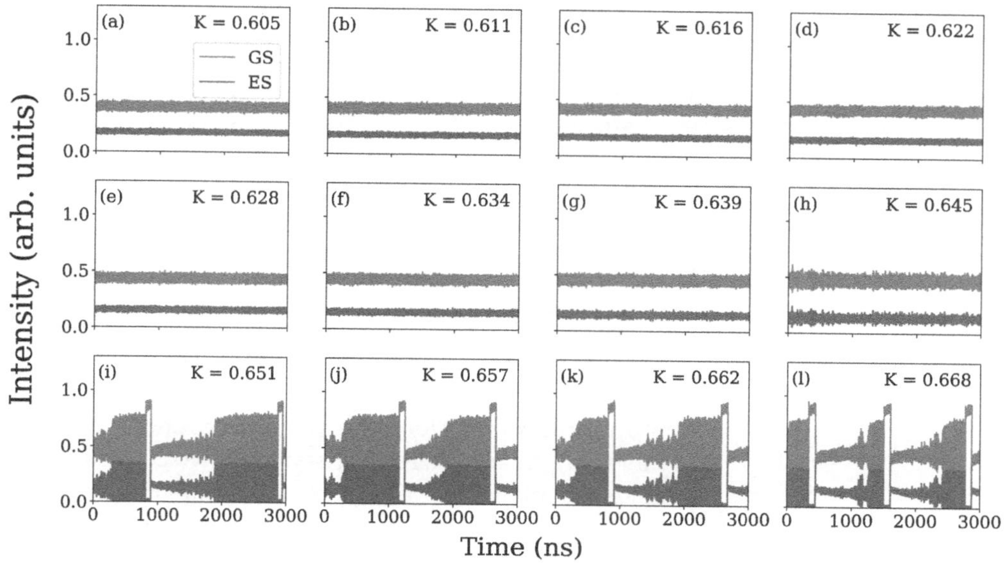

Figure 5. Timetraces from an upsweep in the region of hysteresis at −8.2 GHz detuning. The sweep was performed from panel (**a**) to panel (**l**). The injection strength is indicated in each subfigure. The corresponding downsweep timetraces are shown in Figure 6.

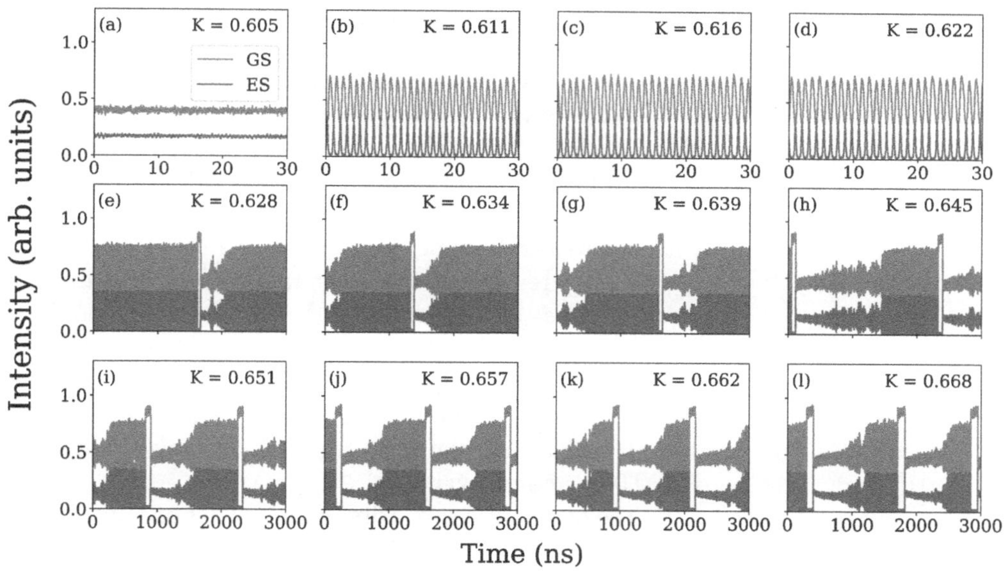

Figure 6. Timetraces from a downsweep in the region of hysteresis at −8.2 GHz detuning. The sweep was performed from panel (l) to panel (**a**). Hysteresis is clear when comparing the subfigures (**b–h**) here with the corresponding subfigures in Figure 5.

In the case of the upsweep in Figure 5a–g, dual state emission is observed (this is in the regenerated region). As K is increased to 0.645, shown in Figure 5h, the GS appears to oscillate with a very small amplitude suggestive of a Hopf bifurcation, before the onset of bursting and MMOs at $K = 0.651$ shown in Figure 5i. The lifetime of the bursting state decreases as the injection strength is further increased.

Similar MMOs are seen for high injection strengths during the downsweep until $K = 0.645$. Then, the presence of hysteresis can be seen by comparing the corresponding subplots of Figures 5 and 6. As K is further decreased below 0.645, we do not find continuous wave dual state emission until $K = 0.605$. Between these two values, the lifetime of the bursting state continues to increase and the switches to the quiescent phase become more rare. Eventually, trains of GS dropouts and ES pulses are observed, as seen in Figure 6b. These trains were observed for the entire 20 μs acquisition time of the oscilloscope. As there are fewer quiescent phases, the overall variance is large and is therefore represented as bright yellow in the variance map (Figure 4f). Clearly, the yellow region extends beyond the white line marking the upsweep dynamical boundary, highlighting the hysteresis cycle. In Figure 6a, where $K = 0.605$, the oscillations disappear, and continuous wave dual state emission is obtained, as is also observed in the upsweep in Figure 5a–g. Thus, we find a bistability between the constant dual state emission (constant ES and regenerated injected light) and dynamic, pulsing behaviour with MMOs.

2.1.3. Slow Oscillations

As mentioned above, on the positive detuning side close to the transition between the phase-locked and regeneration regions, there is an area of dynamics indicated by a speckled, non-uniform colour along the diagonal boundary in Figure 2a,d (and also, of course, in Figure 4a,d). In fact, there are essentially two colours, indicating that there are two distinct average GS intensities. The binomial nature of the colouring suggests a slow dynamic of which we only sample a short, nearly constant part within the 20 μs measurement interval. The sampling window is increased to 1 second and new maps are created and shown in Figure 7. We note that there are also very narrow regions of hysteresis on both the left and right of this region.

Figure 8 shows timetraces from an upsweep at −0.6 GHz marked by white lines in the upper panels of Figure 7. The period of the oscillations is extremely long: approximately 40 ms. The output is strongly dominated by the GS, but very small corresponding oscillations are also observed in the ES as can be seen in Figure 8. The cycle resembles a periodic oscillation between the phase-locked GS output and the dual state regenerated light output.

At the lower injection strength boundary of this region (i.e. at the bottom of the white line), the lower intensity section of the timetrace is longer lived than the upper intensity section, as is clear in Figure 8b. Increasing K, the duty cycle changes, eventually reaching 0.5, as seen in Figure 8h. Continuing to increase the injection strength, the higher GS intensity section becomes longer lived than the lower GS intensity section, as is most clearly seen in Figure 8k. Qualitatively, this behaviour is extremely similar to the optothermal square waves on the negative detuning side in the GS only system [14,16] and to the square wave/bursting phenomena reported here and originally in [15], where optothermal coupling breaks bistabilities. We thus interpret this region as another broken bistability between a low power GS output such as that shown in Figure 8a and a high power GS output such as that shown in Figure 8l. However, and importantly, the oscillations here are ∼100,000 times slower.

The qualitative similarity of the oscillations with the previously investigated optothermal effect is very suggestive. However, even apart from the vast difference in timescales, there is a significant difference that prevents the explanation of the oscillations via the original thermal coupling. Consider the phase locked bistability near the negative detuning boundary of [14]. In that case, the low power solution is on the negatively detuned side of the region and the high power solution on the positive side. The physical source of the coupling in that case is non-radiative carrier recombination. The low-power solution results in high carrier density and an ensuing higher temperature via increased non-radiative recombination. In the high power solution, the carrier density is lower and so the temperature is lower. Thus, higher power leads to lower temperature and vice versa; the deterministic cycle is driven by the low power solution being pushed towards positive detuning and the high power solution pushed towards negative detuning, yielding an anticlockwise phasor cycle as shown in [14,16]. However, in the new region identified in this work, the *high* power solution is the more negatively detuned of the two. Thus, the high power solution would push the system further towards the negative detuning direction and vice versa for the low power solution, and so the cycle would not arise. Thus, heating due to non-radiative recombinations cannot account for the phenomenon, and a coupling with the opposite sign is required. Such a coupling does exist, as can be seen by considering several important ways in which heating can arise in the device. In particular, reabsorption of light in the device can lead to such heating. In [28,29] both recombinative and re-absorptive heating was included. It is also known that self-absorption arises in QD lasers in particular, due to the wide distribution of states and dot sizes [30]. Thus, for re-absorption, a higher optical power leads to an increased temperature, and so the effect results in an effective detuning sweep in the opposite direction to that of the first optothermal effect, as required. Furthermore, the timescale for the second effect should be significantly longer as the time over which re-absorptive heating is distributed through the device can be of the order of tens of milliseconds, or even longer [28,31]. This matches the observed timescale extremely well. We see below that including such an effect in the model allows for excellent agreement with the experiment.

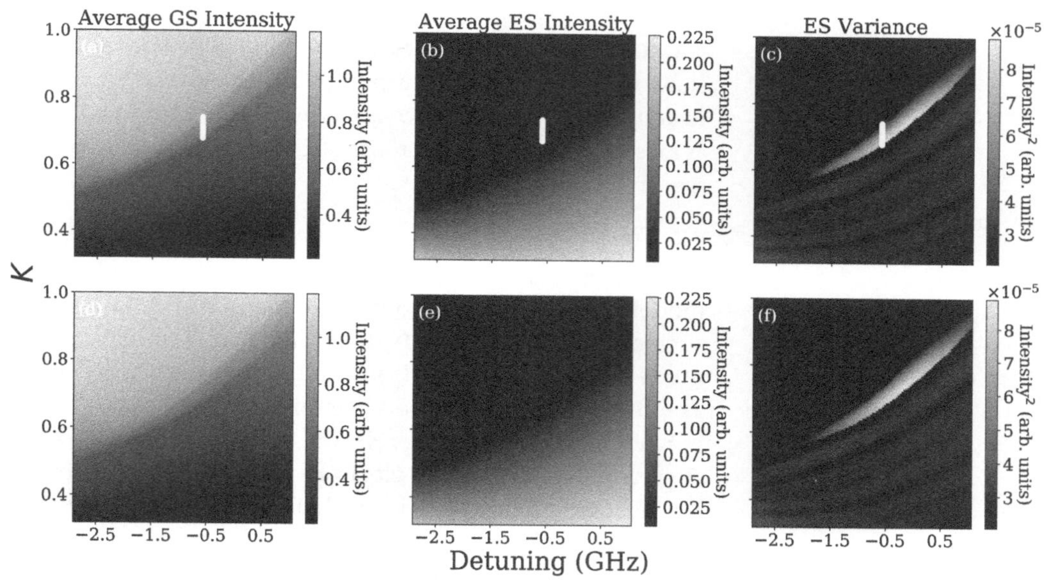

Figure 7. Experimental maps of the positively detuned side of the unlocking boundary. The top row shows an upsweep and the bottom shows a downsweep. The timetraces used to build this map were 1 second long. Three regions are visible in **(a,b)** and in **(d,e)**. In **(a,d)** the large yellow region with high GS intensity is a phase locked region. The blue region is the regenerated light region. The narrow green region in between the yellow and blue regions is where there are slow oscillations between the high and low GS intensity solutions, and thus the average intensity lies somewhere in the middle. **(b,e)** show a completely quenched ES—dark blue—when the laser is phase locked. The region of slow oscillations is represented by the slightly lighter shade of dark blue. For positive detuning values, dual state emission is clear. The variance shows the region where slow oscillations occur more clearly **(c,f)**. There is a small hysteresis cycle of one to two pixels at both sides. The white lines in the region of slow oscillations in **(a–c)** correspond to the timetraces shown in Figure 8.

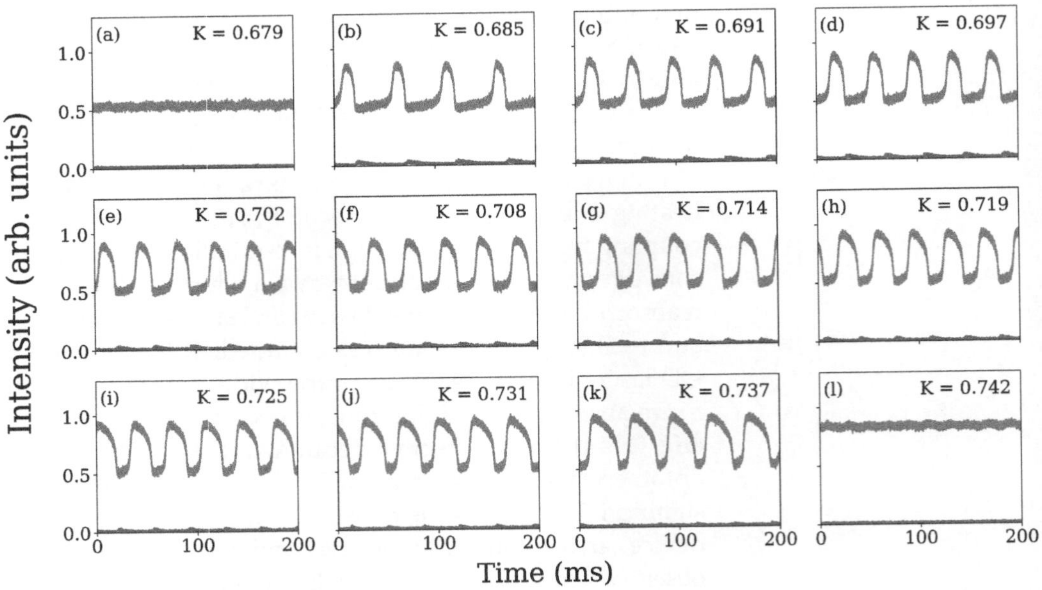

Figure 8. Slow oscillations observed during an upsweep at −0.6 GHz, near the positively detuned side of the unlocking boundary. The sweep is marked by a white line in Figure 7. Comparing **(a)** and **(l)**, there are two very different GS amplitudes at either side of the region of oscillations. The slow oscillations shown in panels **(b–k)** are between these two intensities, suggesting there is an underlying broken bistability. As K is increased the duty cycle moves from longer lived lower sections as most clearly seen in **(b)**, through 50:50 in **(f,g)**, and on to longer lived upper sections as most clearly seen in **(k)**.

2.2. Theory

We use a rate equation model to numerically investigate the system. Rate equation models have been used extensively to great effect in studies of optically injected QD lasers, such as in [3,9,32,33] and further references within. Superficially, there are several different models. Of course, all arise from the same starting point: the Maxwell–Bloch equations. The differences then arise via choices about how to describe carrier capture and escape between the carrier reservoir, the excited state(s), and the ground state of the QDs and even whether to treat the electrons and holes separately or together in an excitonic model. Further, one can also consider whether to include effects such as phase-amplitude coupling and inhomogeneous broadening microscopically or phenomenologically, or indeed at all. We necessarily need to extend any model to include two optothermal effects with extremely slow timescales and so, for purposes of computational efficiency, we use an excitonic model for the carrier dynamics, while for the light dynamics we consider the electric field of the GS and the intensity of the ES. Our model is,

$$\begin{aligned}
\dot{E}_{GS} &= g(1+i\alpha_{GS})[(2\rho_{GS}-1)-\kappa]E_{GS} \\
&\quad + 2ig\alpha_{ES}[(2\rho_{ES}-1)-\kappa]E_{GS} \\
&\quad - 2\pi i(\Delta_0+\Delta_1+\Delta_2)E_{GS} + K, &(1)\\
\dot{I}_{ES} &= [4g(2\rho_{ES}-1)-2\kappa]I_{ES} + \beta\rho_{ES}/\tau, &(2)\\
\dot{\rho}_{GS} &= \mathcal{S}_{rel} - g|E_{GS}|^2(2\rho_{GS}-1) - \rho_{GS}/\tau, &(3)\\
\dot{\rho}_{ES} &= -\tfrac{1}{2}\mathcal{S}_{rel} + \mathcal{S}_{cap} - 2gI_{ES}(2\rho_{ES}-1) - \rho_{ES}/\tau, &(4)\\
\dot{N} &= J - 4\mathcal{S}_{cap} - N/\tau, &(5)
\end{aligned}$$

where E_{GS} is the electric field of the GS, I_{ES} is the power of the ES, ρ_{GS} and ρ_{ES} are the occupation probabilities of the GS and ES, respectively, and N is the normalised charge-carrier number in the quantum well reservoir. The rate equation model in Equations (1)–(5) is thus a two state extension of that used in [3,9] where it was used to model GS only QD lasers under optical injection. The scattering terms in Equations (6)–(8), on the other hand, are of the form used in [32,33] and are given by

$$\mathcal{S}_{\text{rel}} = S_{\text{rel}}\left[(1-\rho_{\text{GS}})\rho_{\text{ES}} - \exp\left(\tfrac{-\varepsilon_{\text{GSES}}}{k_B T}\right)\rho_{\text{GS}}(1-\rho_{\text{ES}})\right], \tag{6}$$

$$\mathcal{S}_{\text{cap}} = S_{\text{cap}}(\rho_{\text{ES}}^{\text{eq}} - \rho_{\text{ES}}), \tag{7}$$

$$\rho_{\text{ES}}^{\text{eq}} = \left(1 + \exp\left(\tfrac{-\varepsilon_{\text{ES}}}{k_B T}\right)\left[\exp\left(\tfrac{N}{D_{\text{2D}}}\right) - 1\right]^{-1}\right)^{-1}. \tag{8}$$

There are three Δ terms controlling the detuning in the system. Δ_0 allows us to choose our reference frame. We define $\Delta_0 \equiv \delta_0 + \delta$ with δ_0 chosen so that the GS frequency of the free-running laser is at zero. Thus, δ_0 compensates the frequency shift due to the phase–amplitude coupling and the optothermal effects in the free-running laser. Then, $\delta = 0$ is at the central GS frequency of the free-running quantum dot laser and it is δ that defines the x-axis in each of our maps below. The other two terms, Δ_1 and Δ_2, are used to implement the aforementioned optothermal coupling mechanisms. Δ_1 accounts for heating of the active region due to recombinative heating, while Δ_2 models overall device heating via re-absorption. We implement these two different effects with individual coupling strengths $c_{1,2}$, and characteristic time scales $\gamma_{1,2}$ in the following two equations,

$$\dot{\Delta}_1 = \gamma_1(c_1[2(\rho_{\text{GS}} - \rho_{\text{GSth}}) + 4(\rho_{\text{ES}} - \rho_{\text{ESth}}) + (N - N_{\text{th}})] - \Delta_1), \tag{9}$$

$$\dot{\Delta}_2 = \gamma_2(c_2[|E_{\text{GS}}|^2 + I_{\text{ES}}] - \Delta_2), \tag{10}$$

where the subscript "th" denotes the value of the corresponding charge-carrier variable at the laser threshold. (We note that in [14,16] the thermal effect was coupled to the power even though the physical explanation is via the carriers. In the Class A case, this is unavoidable, as the carriers are adiabatically eliminated.) The meaning of all the other parameters used in the equations and their values are given in Table 1.

Table 1. Parameter values.

Symbol	Value	Meaning
J	$25\,\text{ns}^{-1}$	Normalised pump current
α_{GS}	2	Phase-amplitude coupling from the GS
α_{ES}	0.5	Phase-amplitude coupling from the ES
g	$80\,\text{ns}^{-1}$	Optical gain coefficient
κ	$76\,\text{ns}^{-1}$	Optical loss coefficient
β	10^{-5}	Spontaneous emission factor
τ	1 ns	Charge-carrier recombination time
S_{cap}	$0.2\,\text{ps}^{-1}$	QD capture rate
S_{rel}	$1\,\text{ps}^{-1}$	QD relaxation rate
ε_{ES}	50 meV	Confinement energy of the ES
$\varepsilon_{\text{GSES}}$	40 meV	Energy separation between GS and ES
D_{2D}	10.9	Normalised 2D density of states
T	300 K	Temperature
γ_1	$40\,\mu\text{s}^{-1}$	Characteristic time scale of non-radiative thermal effects
γ_2	$4\,\mu\text{s}^{-1}$	Characteristic time scale of reabsorption thermal effects

2.2.1. No Optothermal Effects

It turns out to be quite instructive to first consider the system in the absence of any optothermal effects, then with only one effect, and finally with both effects included, thereby allowing for a comprehensive description of how the observed features emerge. There are many features that are common to all three cases. In particular, there is a large, central, phase-locked region; there are large dual state, regenerated, injected light regions for large magnitude detunings; and for low injection strengths at negative detuning, an antiphase dropout and pulsing dynamic is found in each case.

In the absence of any optothermal coupling, there are no square wave phenomena for either negative or positive detuning. Instead, we find multiple bistabilities near the negative detuning boundary. One of these is very clear from Figure 9 in the top left corner of each subplot. Consider the leftmost edge of Figures 9a,d. In the upsweep, the regenerated, two state output persists right up to the top of the figure. Figure 9c shows that the regenerated two state output eventually destabilises via a Hopf bifurcation along the bright line in the top left corner. There is only a very narrow region over which the resulting oscillatory solution exists, after which the system moves to the phase locked output. In the downsweep case, however, the bright yellow phase-locked region extends down to $K{\sim}4.8$ at the extreme left of Figure 9d. Thus, there is a clear bistability between the phase-locked output and the regenerated two state output over a large region and, indeed, another bistability between the phase-locked output and the oscillating two state output over a small region.

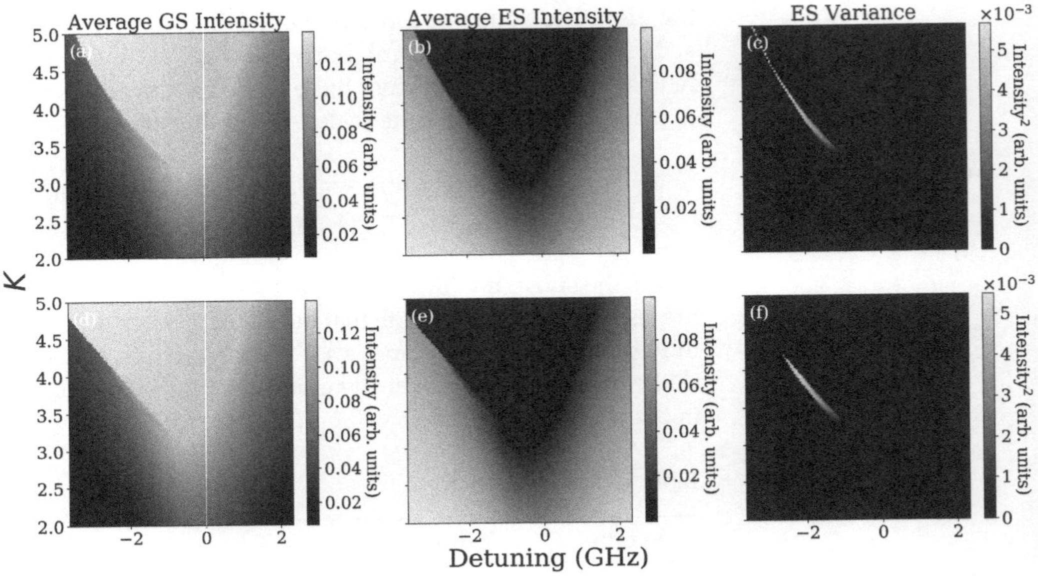

Figure 9. Stability maps in the absence of any optothermal effects. (**a–c**) show the results for the upsweep. (**d–f**) show the corresponding downsweep. Bistabilty 1, the bistability between two phase-locked states, can be seen by comparing the top left corners of (**a**,**d**) at high injection strengths. It can also be seen clearly by comparing the top left corners of (**b**,**e**) at high injection strengths. Bistability 2, the bistability between a phase locked solution and the dual state output at moderate injection strengths, is clear when looking at the top left corners of the variance plots (**c**,**f**). Bistability 3, the low injection strength bistability between a constant (dual state) output and an oscillating (dual state) output, is visible at the bottom of the oscillating regime seen in (**c**,**f**). It is even clearer in the supplemental gif.

There is another region of hysteresis around $K{\sim}4$. This is not as easy to see in the figure as it is over a much smaller area. We include a gif in the supplementary material showing the transition from Figure 9c,f, in which the hysteresis is clear. At this injection strength, the phase-locked solution first destabilises into the oscillating GS dropout/ES pulse regime reported in [24] as the injection strength is decreased. This phase-locked/pulsing regime is bistable with the regenerated two state output and its destabilising Hopf induced cycle discussed above (again, as is clear in the supplemental gif). Notably, though, there is *not* a bistability for positive detuning (where we find the extremely slow dynamic in the experiment).

One might reasonably expect the negative detuning bistabilities from previous GS only analyses [9,17], and some have even already been described in [15,21,32]. In [15], it was shown that the introduction of an optothermal coupling leads to the destruction of high injection strength bistability and that a periodic bursting dynamic is instead obtained. We repeat this analysis here with our model.

2.2.2. First Optothermal Effect

Here, we introduce the first optothermal effect, arising from non-radiative recombinations, and plot the updated maps in Figure 10. As expected, (and again, as was already shown in [15]) this breaks the upper bistability, and yields the periodic bursting square wave regime. The bistability between the phase locked/pulsing output and the dual state/Hopf dynamic is preserved although its precise location and size has changed. However, more intriguing here is the effect that this has on the positive detuning boundary. Now, where before there was no bistability, we find one. It is a bistability between the phase-locked output and the constant dual state output as indicated in Figure 11 near 1 GHz. This arises as the first optothermal effect shears the map so that the two solutions now overlap. It arises in a relatively narrow region precisely where the extremely slow square waves are obtained in the experiment. The appearance of this bistability motivates the introduction of a second optothermal coupling. As discussed in the experimental section above, this coupling must take the opposite sign to the original optothermal effect in order to yield the desired deterministic oscillations.

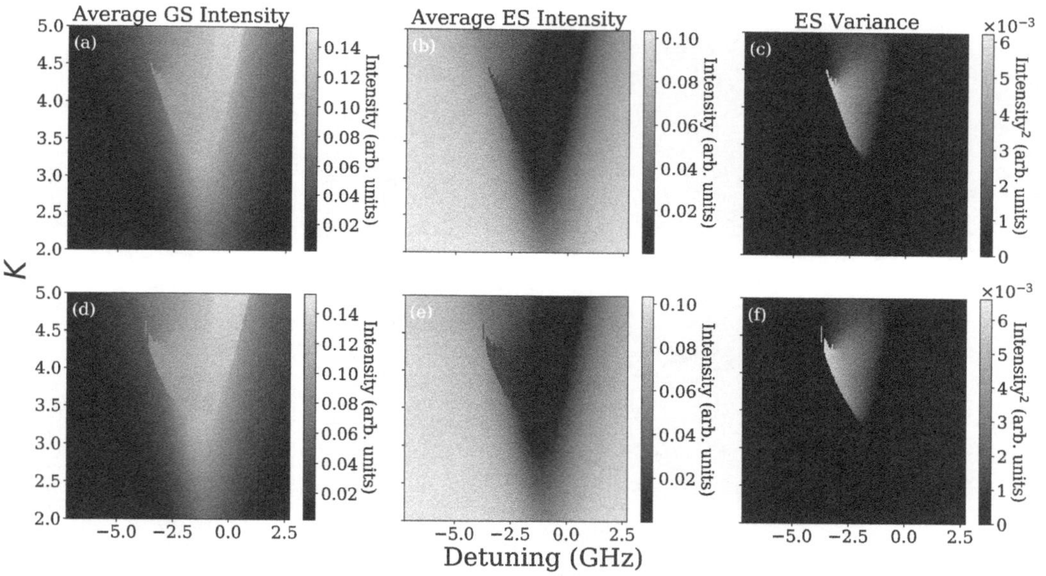

Figure 10. Simulated stability maps with one optothermal effect. The top panels show the results from an upsweep and the lower panels show the corresponding downsweep. (**a**,**d**) show the average GS intensity and (**b**,**e**) show the average ES intensity. (**c**,**f**) show the variance in the ES intensity. The region of high variance seen in (**c**,**f**) is much larger than in Figure 9 as the large phase locked bistability is broken and is replaced with a square wave regime. $\gamma_1 = 40\,\mu s^{-1}$, $c_1 = 3\,ns^{-1}$.

2.2.3. Second Optothermal Effect

We use a characteristic timescale of $\gamma_2 = 4\,\mu s^{-1}$ for the second optothermal effect, which is not quite as slow as the experiment but allows for a reasonable computation time while maintaining a large enough timescale separation with the existing dynamic timescales. We have verified that an even longer timescale for γ_2 reproduces qualitatively identical dynamics. With the addition of the second optothermal effect, the square wave bursting solutions and the experimentally observed bistability are preserved as shown in Figure 12. We note that the value of c_1 when the two effects are included is different to the one we chose for just one optothermal effect. This is because the second optothermal effect counteracts the first one and so, in order to maintain the same overall shape, c_1 must change when the second coupling is introduced. On the positive detuning side, the new thermal effect yields a region of very slow oscillations in the region where there had been a bistability. The region over which these arise is most clearly seen when looking at the GS variance in Figure 12c,f and in the number of maxima as shown in Figure 13c,f in the narrow lines at the positive detuning boundaries. (In fact, there is even still a very small region of hysteresis as can be seen in Figure 11b in agreement with the experiment.)

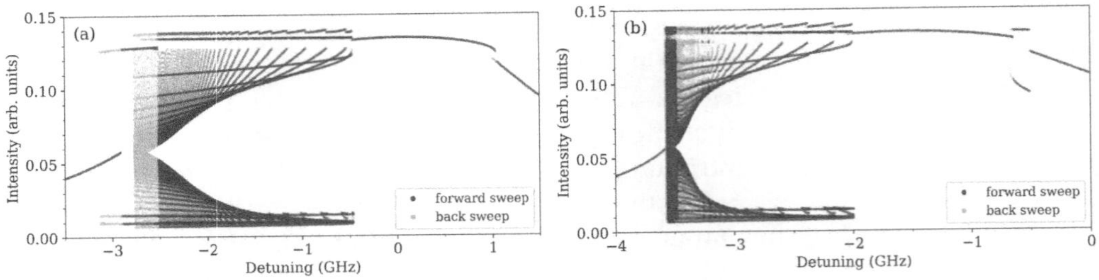

Figure 11. 1D bifurcation diagrams showing the bistabilities. The many diagonal lines are brought about by the bursting phenonema, where the amplitude is growing even though the input parameter for detuning remains fixed. (**a**) only one optothermal effect is included. A bistability is created on the positive unlocking boundary. At the negative unlocking boundary, there is a large cycle between the remnants of a bistability, with some of the bistability remaining intact. (**b**) both optothermal effects are included. The bistability at the positive boundary has been replaced with a limit cycle. In both cases $K = 4.5$.

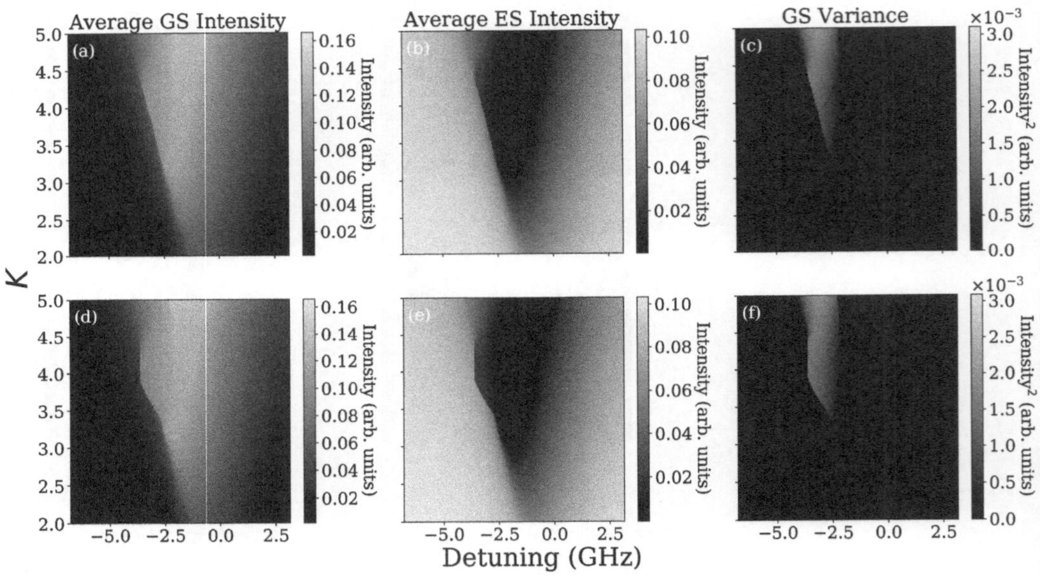

Figure 12. Simulated maps with two optothermal effects. (**a**,**d**) show the average GS intensity and (**b**,**e**) show the average ES intensity. (**c**,**f**) show the variance in the ES intensity. (**a**–**c**) show the results for the upsweep. (**d**–**f**) show the corresponding downsweep. $\gamma_1 = 40\,\mu s^{-1}$, $c_1 = 4.5\,ns^{-1}$, $\gamma_2 = 4\,\mu s^{-1}$, and $c_2 = 0.7\,ns^{-1}$.

Figure 13 is very instructive in highlighting the evolution of dynamical features with the inclusion of optothermal effects. We plot the number of maxima in the time series when there is no optothermal coupling in Figure 13a,d, when there is recombinative optothermal coupling in Figure 13b,e, and when both optothermal couplings are included in Figure 13c,f. Without optothermal coupling, there are very few regions of dynamics and those that exist do so only over small areas. When one optothermal coupling is included as in Figure 13b,e dynamics become much more prevalent in the system and a large region of dynamics appears in the central region of the map. In Figure 13c,f we find that the two competing optothermal effects endow the system with all of the experimentally observed dynamics. Most notably, as already mentioned, the experimentally observed extremely slow oscillations are found near the positive boundary.

Apart from the dynamical features, there are several distinct features in the shape of experimental map. One of these is the near vertical boundary for negative detuning for $K \sim 0.6$ in Figure 1. This feature is recovered with the inclusion of both optothermal effects as seen in the lower panels in Figures 12 and 13c,f for $K \sim 4$. A second feature is the shape of the dynamical region (and consequently the phase locked region). In the experiment, there is a large and very asymmetric region of dynamics for negative detuning and a very narrow region of dynamics along the positive detuning boundary. Figure 13 shows the evolution of the dynamics arising in the model as the optothermal effects are included. There is one very narrow region of dynamics for negative detuning when no optothermal coupling is included. There is a large, somewhat symmetric, central region of dynamics when the recombinative optothermal effect is included. Finally, when both optothermal couplings are included, the map is sheared so that the dynamical region is rendered asymmetric, folding back towards negative detuning as the injection strength is increased, just as in the experiment.

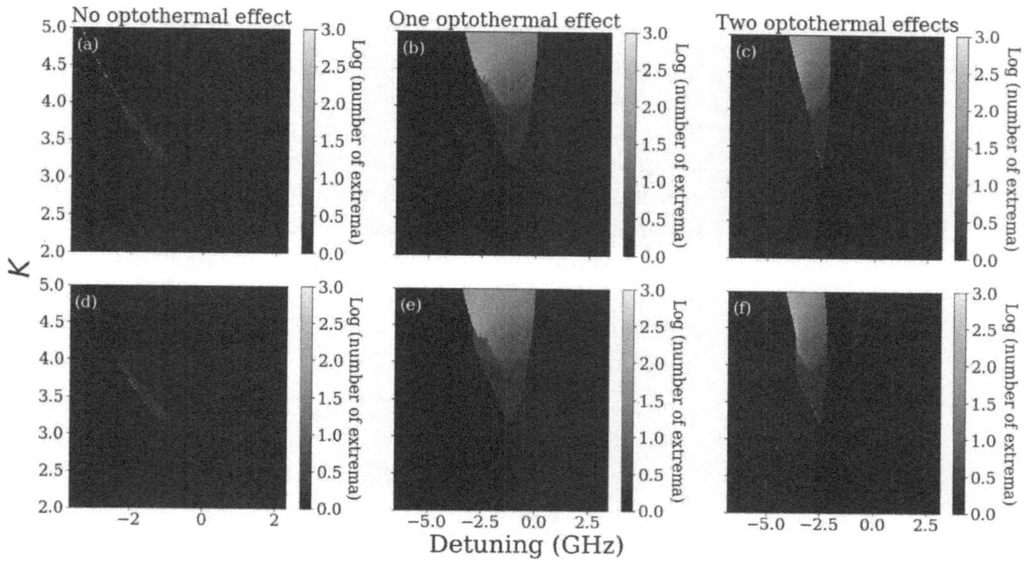

Figure 13. Simulated maps showing the number of maxima in log scale when no optothermal effect is included (**a**,**d**), when the non-radiative optothermal effect is included (**b**,**e**), and when both the non-radiative and light reabsorption optothermal effects are included (**c**,**f**). The top panels show the results from the upsweep and the lower panels show the corresponding downsweep. Chaos is not present in any case.

2.2.4. Absence of Chaos

As mentioned in the introduction, there is a marked absence of chaos when a QD laser emitting from the ES is optically injected near the GS. We have not observed any chaotic regimes in the experiment or in the numerical studies. The presence of chaos in optically injected QD lasers for the GS only case has been reported several times both experimentally and theoretically, such as in [3,32,34] and we have confirmed it arises using our model by adjusting the parameters to allow for GS only lasing. However, it appears to be absent in the dual state system analysed in this work. While there are multiple maxima in large regions of the map as shown in Figure 13c,f, these correspond only to the bursting dynamics, which is complex but not chaotic.

3. Discussion

We have presented a detailed stability map for the optically injected dual state QD system. This is the first comprehensive experimental map in the literature to the best of our knowledge. We also present the first numerical map taking optothermal effects into account. These effects have dramatic effects on the overall system, but allow for excellent agreement between our experimental and numerical results. Unlike with the optically injected conventional semiconductor laser, optothermal effects play a somewhat prominent role in the map, and there is an array of intact and broken bistabilities due to competing optothermal effects. A particularly beautiful route to an extremely slow square wave dynamic near the positive detuning boundary is obtained. As with the implementation of optothermal effects in other systems, a bistability is broken yielding a deterministic optothermal cycle. However, the bistability only exists due to a different, competing, optothermal coupling in the first place! There is no sign or even hint of the dynamic in the bare system.

The need to consider optothermal effects in this system, and indeed in the conventional GS only system, arose entirely due to experimental findings where unexpected square wave and bursting phenomena were observed. However, the inclusion of optothermal effects in the model is shown here to improve the mapping beyond simply recovering the dynamics, with characteristic shapes and features found in the experiment recovered only when the full system is analysed. This suggests that perhaps including such couplings in other injection systems might be worthwhile.

With a view to the future, QD lasers are particularly suited to photonic integrated circuits, given their potential for isolator free operation on chip due to the enhanced stability to optical feedback [35–37]. Novel neuromorphic information processing may be possible using dual state QD lasers, as described recently in [38]. The stability map is an indispensable tool for analysing such applications, revealing the regions and parameter ranges over which dynamics are obtained. In fact, many of the dynamic regions obtained in this system are neuromorphic. The excitable dynamic reported in [24] persists at low injection strength and is of clear relevance while the neuromorphic potential of the optothermal bursting has previously been highlighted in [15].

References

1. Wieczorek, S.; Krauskopf, B.; Simpson, T.; Lenstra, D. The dynamical complexity of optically injected semiconductor lasers. *Phys. Rep.* **2005**, *416*, 1. [CrossRef]
2. O'Shea, D.; Osborne, S.; Blackbeard, N.; Goulding, D.; Kelleher, B.; Amann, A. Experimental classification of dynamical regimes in optically injected lasers. *Opt. Express* **2014**, *22*, 21701. [CrossRef] [PubMed]
3. Kelleher, B.; Goulding, D.; Hegarty, S.; Huyet, G.; Viktorov, E.; Erneux, T. Optically injected single-mode quantum dot lasers. In *Quantum Dot Devices*; Springer: New York, NY, USA, 2012; pp. 1–22.
4. Chan, S.-C. Analysis of an Optically Injected Semiconductor Laser for Microwave Generation. *IEEE J. Quantum Electron.* **2010**, *46*, 421. [CrossRef]
5. Hurtado, A.; Mee, J.; Nami, M.; Henning, I.D.; Adams, M.J.; Lester, L.F. Tunable microwave signal generator with an optically-injected 1310 nm QD-DFB laser. *Opt. Express* **2013**, *21*, 10772. [CrossRef]
6. Li, X.-Z.; Chan, S.-C. Random bit generation using an optically injected semiconductor laser in chaos with oversampling. *Opt. Lett.* **2012**, *37*, 2163. [CrossRef]
7. Prucnal, P.R.; Shastri, B.J.; de Lima, T.F.; Nahmias, M.A.; Tait, A.N. Recent progress in semiconductor excitable lasers for photonic spike processing. *Adv. Opt. Photonics* **2016**, *8*, 228. [CrossRef]

8. Gavrielides, A.; Kovanis, V.; Erneux, T. Analytical stability boundaries for a semiconductor laser subject to optical injection. *Opt. Commun.* **1997**, *136*, 253. [CrossRef]
9. Erneux, T.; Viktorov, E.; Kelleher, B.; Goulding, D.; Hegarty, S.; Huyet, G. Optically injected quantum-dot lasers. *Opt. Lett.* **2010**, *35*, 937. [CrossRef]
10. Mayol, C.; Toral, R.; Mirasso, C.R.; Natiello, M.A. Class-a lasers with injected signal: Bifurcation set and Lyapunov–potential function. *Phys. Rev. A* **2002**, *66*, 013808. [CrossRef]
11. Chlouverakis, K.E.; Adams, M.J. Stability maps of injection-locked laser diodes using the largest Lyapunov exponent. *Opt. Commun.* **2003**, *216*, 405. [CrossRef]
12. Bonatto, C.; Garreau, J.C.; Gallas, J.A.C. Self-similarities in the frequency-amplitude space of a loss-modulated CO_2 laser. *Phys. Rev. Lett.* **2005**, *95*, 143905. [CrossRef] [PubMed]
13. Kelleher, B.; Bonatto, C.; Huyet, G.; Hegarty, S. Excitability in optically injected semiconductor lasers: Contrasting quantum-well- and quantum-dot-based devices. *Phys. Rev. E* **2011**, *83*, 026207. [CrossRef] [PubMed]
14. Dillane, M.; Tykalewicz, B.; Goulding, D.; Garbin, B.; Barland, S.; Kelleher, B. Square wave excitability in quantum dot lasers under optical injection. *Opt. Lett.* **2019**, *44*, 347. [CrossRef]
15. Kelleher, B.; Tykalewicz, B.; Goulding, D.; Fedorov, N.; Dubinkin, I.; Erneux, T. ; Viktorov, E.A. Two-color bursting oscillations. *Sci. Rep.* **2017**, *7*, 8414. [CrossRef]
16. Dillane, M.; Robertson, J.; Peters, M.; Hurtado, A.; Kelleher, B. Neuromorphic dynamics with optically injected quantum dot lasers. *Eur. Phys. J. B* **2019**, *92*, 1. [CrossRef]
17. Lingnau, B.; Lüdge, K.; Chow, W.W.; Schöll, E. Failure of the α factor in describing dynamical instabilities and chaos in quantum-dot lasers. *Phys. Rev. E* **2012**, *86*, 065201. [CrossRef] [PubMed]
18. Lingnau, B.; Dillane, M.; O'Callaghan, J.; Corbett, B.; Kelleher, B. Multimode dynamics and modeling of free-running and optically injected fabry-pérot quantum-dot lasers. *Phys. Rev. A* **2019**, *100*, 063837. [CrossRef]
19. Markus, A.; Chen, J.X.; Paranthoën, C.; Fiore, A.; Platz, C.; Gauthier-Lafaye, O. Simultaneous two-state lasing in quantum-dot lasers. *Appl. Phys. Lett.* **2003**, *82*, 1818–1820. [CrossRef]
20. Bimberg, D.; Kirstaedter, N.; Ledentsov, N.; Alferov, Z.I.; Kop'Ev, P.; Ustinov, V. Ingaas-gaas quantum-dot lasers. *IEEE J. Sel. Top. Quantum Electron.* **1997**, *3*, 196. [CrossRef]
21. Tykalewicz, B.; Goulding, D.; Hegarty, S.P.; Huyet, G.; Dubinkin, I.; Fedorov, N.; Erneux, T.; Viktorov, E.A.; Kelleher, B. Optically induced hysteresis in a two-state quantum dot laser. *Opt. Lett.* **2016**, *41*, 1034. [CrossRef]
22. Tykalewicz, B.; Goulding, D.; Hegarty, S.P.; Huyet, G.; Byrne, D.; Phelan, R.; Kelleher, B. All-optical switching with a dual-state, single-section quantum dot laser via optical injection. *Opt. Lett.* **2014**, *39*, 4607. [CrossRef]
23. Viktorov, E.A.; Dubinkin, I.; Fedorov, N.; Erneux, T.; Tykalewicz, B.; Hegarty, S.P.; Huyet, G.; Goulding, D.; Kelleher, B. Injection-induced, tunable all-optical gating in a two-state quantum dot laser. *Opt. Lett.* **2016**, *41*, 3555. [CrossRef]
24. Dillane, M.; Dubinkin, I.; Fedorov, N.; Erneux, T.; Goulding, D.; Kelleher, B.; Viktorov, E.A. Excitable interplay between lasing quantum dot states. *Phys. Rev. E* **2019**, *100*, 012202. [CrossRef] [PubMed]
25. Simpson, T.B.; Liu, J.M.; Huang, K.F.; Tai, K. Nonlinear dynamics induced by external optical injection in semiconductor lasers. *Quantum Semiclass. Opt.* **1997**, *9*, 765. [CrossRef]
26. Desroches, M.; Guckenheimer, J.; Krauskopf, B.; Kuehn, C.; Osinga, H.M.; Wechselberger, M. Mixed-mode oscillations with multiple time scales. *SIAM Rev.* **2012**, *54*, 211. [CrossRef]
27. Desroches, M.; Kaper, T.J.; Krupa, M. Mixed-mode bursting oscillations: Dynamics created by a slow passage through spike-adding canard explosion in a square-wave burster. *Chaos* **2013**, *23*, 046106. [CrossRef] [PubMed]
28. Rzhanov, Y.A.; Richardson, H.; Hagberg, A.A.; Moloney, J.V. Spatiotemporal oscillations in a semiconductor étalon. *Phys. Rev. A* **1993**, *47*, 1480. [CrossRef]
29. Abraham, E. Modelling of regenerative pulsations in an InSb etalon. *Opt. Commun.* **1987**, *61*, 282. [CrossRef]
30. Bera, D.; Qian, L.; Tseng, T.-K.; Holloway, P.H. Quantum dots and their multimodal applications: A review. *Materials* **2010**, *3*, 2260. [CrossRef]
31. Ziegler, M.; Tomm, J.W.; Elsaesser, T.; Erbert, G.; Bugge, F.; Nakwaski, W.; Sarzała, R.P. Visualization of heat flows in high-power diode lasers by lock-in thermography. *Appl. Phys. Lett.* **2008**, *92*, 103513. [CrossRef]
32. Meinecke, S.; Lingnau, B.; Röhm, A.; Lüdge, K. Stability of Optically Injected Two-State Quantum-Dot Lasers. *Ann. Phys.* **2017**, *529*, 1600279. [CrossRef]
33. Meinecke, S.; Lingnau, B.; Lüdge, K. Increasing stability by two-state lasing in quantum-dot lasers with optical injection. In *Physics and Simulation of Optoelectronic Devices XXV. San Francisco, California, United States*; Witzigmann, B., Osiński, M., Arakawa, Y., Eds.; International Society for Optics and Photonics (SPIE): Bellingham, WA, USA, 2017; Volume 10098, pp. 67–77.
34. Pausch, J.; Otto, C.; Tylaite, E.; Majer, N.; Schöll, E.; Lüdge, K. Optically injected quantum dot lasers: impact of nonlinear carrier lifetimes on frequency-locking dynamics. *New J. Phys.* **2012**, *14*, 053018. [CrossRef]
35. O'Brien, D.; Hegarty, S.; Huyet, G.; McInerney, J.; Kettler, T.; Laemmlin, M.; Bimberg, D.; Ustinov, V.; Zhukov, A.; Mikhrin, S.; et al. Feedback sensitivity of 1.3/spl mu/m inas/gaas quantum dot lasers. *Electron. Lett.* **2003**, *39*, 1819. [CrossRef]
36. O'Brien, D.; Hegarty, S.; Huyet, G.; Uskov, A. Sensitivity of quantum-dot semiconductor lasers to optical feedback. *Opt. Lett.* **2004**, *29*, 1072. [CrossRef] [PubMed]

37. Duan, J.; Huang, H.; Dong, B.; Norman, J.C.; Zhang, Z.; Bowers, J.E.; Grillot, F. Dynamic and nonlinear properties of epitaxial quantum dot lasers on silicon for isolator-free integration. *Photonics Res.* **2019**, *7*, 1222. [CrossRef]
38. Kelleher, B.; Dillane, M.; Viktorov, E.A. Optical information processing using dual state quantum dot lasers: Complexity through simplicity. *Lightw. Sci. Appl.* **2021**, *10*, 238. [CrossRef] [PubMed]

Two Polarization Comb Dynamicsin VCSELs Subject to Optical Injection

Abstract: Optical frequency comb technologies have received intense attention due to their numerous promising applications ranging from optical communications to optical comb spectroscopy. In this study, we experimentally demonstrate a new approach of broadband comb generation based on the polarization mode competition in single-mode VCSELs. More specifically, we analyze nonlinear dynamics and polarization properties in VCSELs when subject of optical injection from a frequency comb. When varying injection parameters (injection strength and detuning frequency) and comb properties (comb spacing), we unveil several bifurcation sequences enabling the excitation of free-running depressed polarization mode. Interestingly, for some injection parameters, the polarization mode competition induces a single or a two polarization comb with controllable properties (repetition rate and power per line). We also show that the performance of the two polarization combs depends crucially on the injection current and on the injected comb spacing. We explain our experimental findings by utilizing the spin-flip VCSEL model (SFM) supplemented with terms for parallel optical injection of frequency comb. We provide a comparison between parallel and orthogonal optical injection in the VCSEL when varying injection parameters and SFM parameters. We show that orthogonal comb dynamics can be observed in a wide range of parameters, as for example dichroism linear dichroism ($\gamma_a = -0.1$ ns^{-1} to $\gamma_a = -0.8$ ns^{-1}), injection current ($\mu = 2.29$ to $\mu = 5.29$) and spin-flip relaxation rate ($\gamma_s = 50$ ns^{-1} to $\gamma_s = 2300$ ns^{-1}).

Keywords: optical frequency comb; polarization switching; optical injection; nonlinear dynamics; VCSEL

---

1. Introduction

The nonlinear dynamics of externally driven semiconductor laser have been extensively studied last decades [1–3]. The polarization dynamics in Vertical-Cavity Surface-Emitting Lasers (VCSELs) have particularly attracted much attention due to their compactness, low cost, low energy consumption and possibility of mass production. These properties have enabled several applications such as optical communications, sensing and computing [4]. The polarization properties of the VCSEL are directly linked to their cavity and active region geometrical properties. The cylindrical geometry of the VCSEL cavity combined with the symmetry of the gain in the plane of quantum wells yields a weak polarization anisotropy giving rise to light emission of two linear orthogonally polarized fundamental modes. In addition, polarization mode instability including polarization switching (PS), polarization bistability and polarization mode hopping can be observed in the VCSELs between the two polarization modes [5–7]. Polarization switching has been first observed in the free-running operation when varying the injection current [8]. More recently, optical injections have been demonstrated as a powerful tool to induce PS in VCSELs [6,9,10]. Depending on the injected power and detuning frequency, the PS has been found to bifurcate to more complex dynamics including periodic dynamics, complex dynamics and chaos [6]. Interestingly, a very recent study has shown that VCSEL output can exhibit complex chaotic dynamics even in free-running [11]. These rich nonlinear polarization dynamics in VCSELs have found several applications such as optical frequency comb generation [12].

Optical frequency combs have been in recent years the focus of intense scientific research from the fundamental viewpoint, as well as from the technological perspectives [13–15]. These works have unveiled several physical systems generating optical frequency combs such as mode-locked lasers [14,16], electro-optic modulator and optically pumped nonlinear microresonators [17]. The development of semiconductor laser based frequency combs has been widely motivated by their numerous properties such as low energy consumption, low size and possibility of mass production and on-chip integration. The semiconductor lasers employed for optical comb generation by mode locking include multisection lasers, Vertical Extended-Cavity Surface-Emitting Lasers (VECSELs) [16,18], quantum cascade lasers [19,20] and quantum dot lasers [21,22]. Unfortunately, the repetition rate of the combs based on these systems depend intrinsically of their cavity properties including cavity length and temperature inside the cavity. Frequency combs with large comb spacing have been proposed using the above systems but achieving a low and stable comb spacing remains a challenge because of the change in the cavity due to temperature variations. For example, 6-millimeter-long devices correspond to a comb spacing of $\approx$7.5 GHz [23] and require an active stabilization by acting on the cavity temperature for practical implementations such as dual comb spectroscopy. The comb generation system based on the electro-optic modulators has been adopted to overcome the limitations of the mode-locked lasers-based frequency combs. Optical comb generation using electro-optic modulators is relatively simple because the comb is obtained by directly modulating the output of a single-frequency CW laser. The physical phenomena underlying these comb generation systems are linked to the electro-optic effect. Electro-optic modulator based-optical combs have shown several limitations including fluctuation in the amplitude of the comb lines and low bandwidth. Insertions of the electro-optic modulator inside an optical resonator and cascade electro-optic modulator have been adopted as alternatives to improve comb bandwidth and flatness. Recent works have controlled the comb properties by injecting a narrow comb into semiconductor lasers [24–27]. Optical injection of comb is used: first as a method to select and amplify a desired line in an optical frequency comb, but more recently as a technique to induce rich nonlinear laser dynamics, such as relaxation oscillation frequency locking [24], harmonics frequency locking [24] and Devil's staircase resonance [25]. However, to date, the full potential of this technique for broadband optical comb generation has only recently been addressed, with an in-depth analysis and mapping of nonlinear comb dynamics [26,27]. In addition, in a very recent work, we have experimentally taken advantage of polarization dynamics in VCSEL to extend the comb's bandwidth and improve the power per comb lines referred to as carrier to noise ratio (CNR) [28]. It is shown that polarization switching induced by orthogonal optical injection plays a key role in the control of the comb properties (bandwidth and CNR) through the generation of two combs with orthogonal polarization. This work was restricted to a single-mode VCSEL with birefringence of $\sim$18 GHz and does not consider the full potential of the VCSEL parameters. The literature provides a lot of research on the optical injection dynamics in VCSEL with orthogonal polarized light [7,10,29–35]. The parallel (orthogonal) injection is the injection of light with linear polarisation parallel (orthogonal) to the free-running VCSEL linear polarisation direction. Although parallel optical injection in VCSELs shows similarities to optical injection in the conventional single-polarization edge-emitting laser, very interesting additional polarization dynamics have been reported recently [36–40].

The excitation of the polarization modes has been found to bifurcate to periodic, complex or chaotic dynamics with orthogonal polarization. The question of how parallel optical frequency comb injection in VCSELs induces additional nonlinear polarization dynamics is of great interest and has not been addressed so far.

In this paper, we analyze, in detail, nonlinear dynamics and polarization properties in VCSELs subject to optical frequency comb injection. More specifically, we provide an in-depth experimental and theoretical description of the bifurcation scenarios when the polarization of injection comb is parallel to that of the VCSELs. We show that a variation of the injection parameters together with the bias current allows enabling the excitation of the normally depressed polarization mode accompanied by the generation of two combs with orthogonal polarization directions. The appearance of the two polarization combs is directly linked to the evolution of the injection current. Indeed, for a fixed injection current, comb generation performance is limited by the increase in injected comb spacing, but the increase in bias current allows overcoming this limitation. Experimental and numerical simulations highlight that the injection current is an addition parameter to control comb properties. Beyond this, the injection parameters are used to tailor the comb properties in a VCSEL in our previous work [28]; here, we take advantage of linear dichroism and birefringence to improve comb properties in the VCSEL. We also contrast experimentally and theoretically the comb dynamics in the cases of parallel and orthogonal optical injection when varying injection parameters, i.e., the comb spacing and injection strength and most importantly the injection current in the VCSEL. More specifically, we provide an in-depth bifurcation analysis of polarization comb dynamics when varying VCSEL parameters. The bifurcation diagrams show that VCSEL parameters can be used to suppress the comb in the normally depressed polarization mode or to tailor power in each polarization comb.

2. Experimental Polarization Dynamics

2.1. Experimental Setup

Figure 1a shows the experimental setup for parallel optical injection in a single-mode VCSEL. The tunable laser (TL) output is first amplified by an Erbium-Doped Fiber Amplifier (EDFA). The output of EDFA is then sent to the RF port of the Mach–Zehnder Modulator (MZM), which is driven with an electrical signal modulation generated by an Arbitrary Waveform Generator (AWG) (Tektronix AWG 700002A). Three optical frequency lines are created at the output of the MZM. The polarization controller at the input of the MZM is used to align its polarization with the tunable laser. A fiber circulator is arranged to provide isolation for the comb injection in the VCSEL. The total power of the injected comb is controlled using a Variable Optical Attenuator (VOA). The VCSEL output is amplified before being sent to a 50/50 coupler to analyze the optical spectra and the corresponding time series separately. Optical spectra are analyzed with a high-resolution optical spectrum analyzer, BOSA 400, which allows monitoring optical spectra with a resolution of about a minimum of 0.1 pm (12 MHz) at the operating wavelength of 1550 nm. Figure 1b–d present the optical spectra of the injected comb for a fixed comb spacing of $\Omega = 1$ GHz, $\Omega = 2$ GHz and $\Omega = 4$ GHz, respectively. The ratio between the power of the central comb line and the side comb lines is around 12 dB. The VCSEL (Raycan) used is a single-mode device with a threshold current equal to I = 3 mA. In free-running conditions, the dominant polarization mode (normally depressed polarization mode) emits along the X (Y) axis, as shown in Figure 1e. The ratio between the power of the linear polarization modes (X-PM and Y-PM) is around 43 dB. Figure 1e is obtained for a bias current at I = 6 mA, which corresponds to 2-times the threshold current. At that current, the total output power of the VCSEL has been measured to be $P_{inj} = 330$ µW. The dominant polarization mode (X-PM) emits at 1553.8 nm at I = 6 mA. The difference in frequency between linear polarization modes (VCSEL birefringence) is around 17.71 GHz at 23 °C. The x-axes of the optical spectra will show the relative frequency with respect to the X-Polarization mode (X-PM), i.e., the zero value will correspond to the frequency position of the dominant X-Polarization mode (X-PM). In the following, the frequency detuning $\Delta\nu$ will be defined from the frequency of the central injected comb line to the frequency of dominant polarization mode of the VCSEL (X-PM), i.e., $\Delta\nu = \nu_0 - \nu_x$, where ν_0 and ν_x are the frequencies of the central injected comb line and the X-polarization mode of VCSEL (X-PM), respectively.

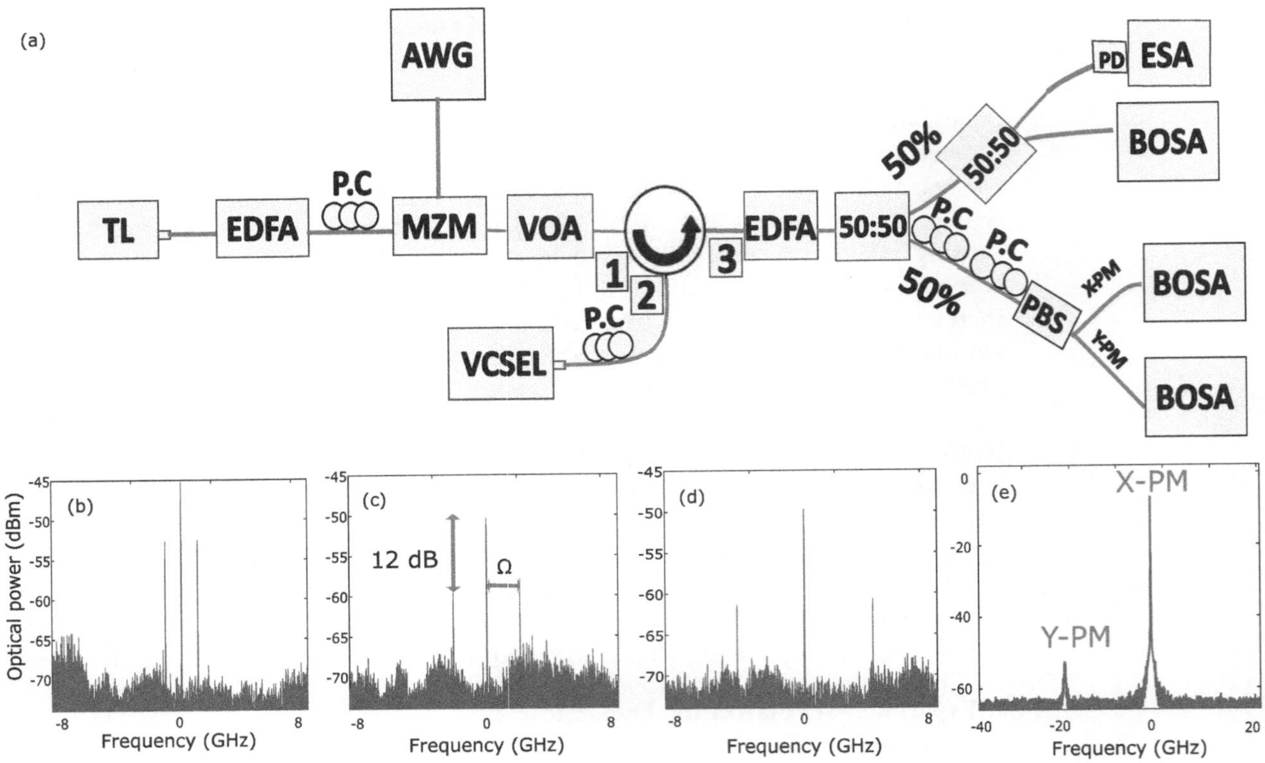

Figure 1. (**a**) Setup for frequency comb injection into a single-mode VCSEL. TL: Tunable Laser, EDFA: amplifier, P.C: Polarization Controller, AWG: Arbitrary Waveform Generator, MZM: Mach–Zehnder Modulator, VOA: Variable Optical Attenuator, OSA: Optical Spectrum Analyser, PD: photodiode, ESA: Electrical spectrum analyzer. (**b–d**) correspond to the optical spectra of the injected comb for comb spacing of $\Omega = 1$ GHz, $\Omega = 2$ GHz and $\Omega = 4$ GHz, respectively. (**e**) shows the optical spectrum of the VCSEL in free running.

2.2. *Impact of the Injected Comb Spacing*

We first analyze the nonlinear polarization dynamics of a VCSEL under a narrow optical comb injection. To this end, we fix the bias current at I = 6 mA and then scan the plane of injection parameters, namely the detuning frequency and injected power that leads to various polarization mode competition. That bias current corresponds to a relaxation oscillation frequency of $\omega_{RO} = 4.2$ GHz. Figure 2 describes the sequence of bifurcations leading to the excitation of the normally depressed polarization mode (Y-PM) accompanied by unlocked time-periodic dynamics, which extends the injected comb to a much broader optical spectrum. Figure 2 is obtained for a small injected comb spacing, $\Omega = 1$ GHz and fixed detuning frequency $\Delta\nu = 1.6$ GHz when varying the injected power, P_{inj}. The VCSEL output shows stable free-running operation at a very low injected power (see Figure 2a). A small increase in the injected power destabilizes the VCSEL output to modulated signal as a result of nonlinear wave-mixing in the dominant polarization mode (X-PM) (Figure 2b). In agreement with our previous publications [26,27], this nonlinear wave mixing takes place at the detuning frequency and a new frequency that depends on the injected comb spacing. The number of frequency lines involved in the nonlinear wave-mixing decreases with the injected power, as shown in Figure 2c. The VCSEL output is then an unstable comb with a high noise pedestal in X-PM at the injected comb repetition rate. By fine-tuning the injected power, instead of stabilizing the comb in a single polarization mode, a new bifurcation scenario results in the excitation of the free-running normally depressed polarization mode of the VCSEL. Interestingly, this polarization mode competition induces in the VCSEL two optical frequency combs with orthogonal polarization directions, as seen in Figure 2d. The two polarization combs have comparable power and the same repetition rate as the injected one with nine frequency lines in the X-PM comb and eight in the Y-PM comb at −30 dB from the maximum. The strongest comb line in the optical spectrum of Y-PM appears at the frequency position of the free-running Y-PM, i.e., its frequency position is ruled by the VCSEL linear birefringence. When we keep increasing the injected power, harmonic frequency comb dynamics takes place simultaneously in the two polarization combs, as shown in Figure 2e,f. The harmonic comb dynamics result in a reduction in comb repetition rate. Subharmonic bifurcations are well studied and understood phenomena that occur naturally in periodically forced oscillators [41]. Subharmonic dynamics have been expected to be a form of frequency locking different from the Adler type in diode laser optically injected with optical comb [24,25]. They appear when the detuning frequency is close to the rational fraction of the injected comb spacing. When we increase the injected power, the harmonic polarization comb destabilizes to complex polarization dynamics in X-PM and Y-PM (Figure 2g). It is worth observing that the X-PM still shows some remarkable comb lines at the repetition rate of the injected comb. A further increase in injected power results in a progressive suppression of all the power in the depressed polarization mode Y-PM accompanied by the stabilization of the comb dynamics in the dominant polarization mode X-PM, as seen in Figure 2h,i. Such single-polarization comb has the same repetition rate as the injected one.

Figure 3 shows the polarization resolved optical spectra for comb dynamics in Figure 2d–f. Figure 3 confirms that the total VCSEL output combs shown in Figure 2d–f are formed by two combs with orthogonal polarization. In the optical spectrum of X-PM (Y-PM), we observed some comb lines at the frequency position of Y-PM (X-PM). This is due to the extension ratio of the polarization controllers.

We have shown in Figure 2 the possibility of two orthogonal combs generation from polarization mode competition in a single-mode VCSEL. As discussed earlier in the introduction, the tunability of the optical frequency comb is a key property for many applications. In order to demonstrate tuning capability, we show in Figure 4 bifurcation scenarios inducing two polarization comb dynamics for the same detuning $\Delta\nu = 1.6$ GHz when increasing the injected comb spacing to $\Omega = 2$ GHz. Similarly for $\Omega = 1$ GHz comb injection, the VCSEL output at low injected power shows the free-running operation, which bifurcates to nonlinear wave mixing due to a modulation involving detuning and injected comb spacing. In agreement again with Figure 2, wave-mixing results in a two polarization combs generation. Most importantly, despite the comparable power of the two polarization combs, the numbers of lines in X-PM and Y-PM have significantly decreased compared to $\Omega = 1$ GHz comb injection case. Indeed, the number of frequency lines in each polarization mode has decreased to achieve 3 at −30 dB from the maximum. It is important to notice that the two polarization comb dynamics are observed in a very small area of injection parameters for $\Omega = 2$ GHz comb injection. We have checked by using polarization resolved optical spectra that these total combs are formed by two micro-comb with orthogonal polarization in single-mode VCSEL, as shown in Figure 5. A small increase in the injected power destabilizes VCSEL to more complex dynamics in each polarization mode, as shown in Figure 4d. We have checked that the size of the two polarization comb region shrinks when increasing the injected comb spacing until it disappears at $\Omega = 4$ GHz. Figure 6 shows how to overcome this limitation. Similarly to $\Omega = 1$ GHz comb injection, complex polarization dynamics bifurcates to the single-polarization comb in X-PM. This polarized comb has an important noise pedestal and, therefore, deteriorating the comb performances. Interestingly, we have also checked that the two polarization comb dynamics in the single-mode VCSEL under parallel optical injection are not observed when the bias current is below $I = 6$ mA. In contrast, when the polarization of the injected comb is orthogonal to that of VCSEL, the two polarization comb dynamics is always observed, whatever the bias current is.

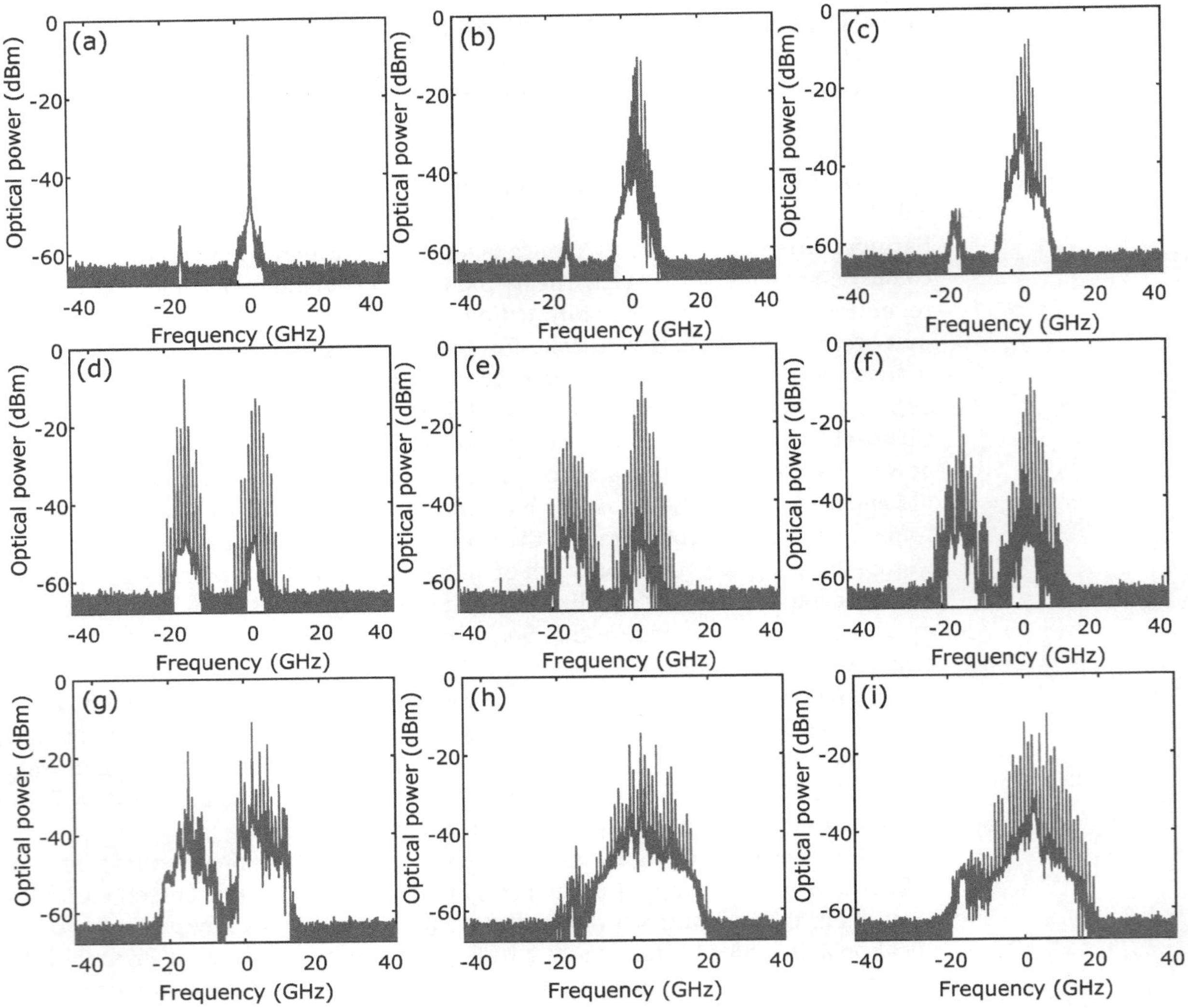

Figure 2. Bifurcation scenarios resulting in the excitation of the depressed polarization mode. These optical spectra are obtained for detuning $\Delta\nu = 1.6$ GHz and injected comb spacing $\Omega = 1$ GHz when increasing the injected power, P_{inj}. (**a**) Stable output at $P_{inj} = 3$ µW, (**b**,**c**) wave mixing at $P_{inj} = 16$ µW and $P_{inj} = 32$ µW, respectively, (**d**) two polarization comb at $P_{inj} = 48$ µW, (**e**,**f**) two polarization harmonics comb at $P_{inj} = 80$ µW and $P_{inj} = 96$ µW, respectively, (**g**) two polarization complex dynamics at $P_{inj} = 128$ µW and (**h**,**i**) X-polarization comb at $P_{inj} = 144$ µW and $P_{inj} = 240$ µW, respectively.

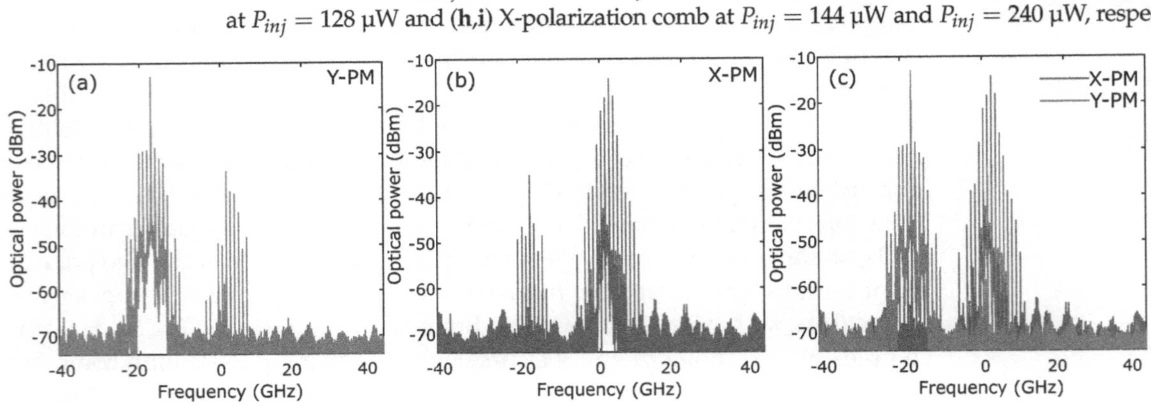

Figure 3. Polarization resolved-optical spectra corresponding to an example of two polarization comb dynamics similar to Figure 2d–f for detuning $\Delta\nu = 1.6$ GHz and injected power $P_{inj} = 20$ µW. (**a**–**c**) correspond to the optical spectra of Y-PM, X-PM and the superposition of Y-PM and X-PM, respectively.

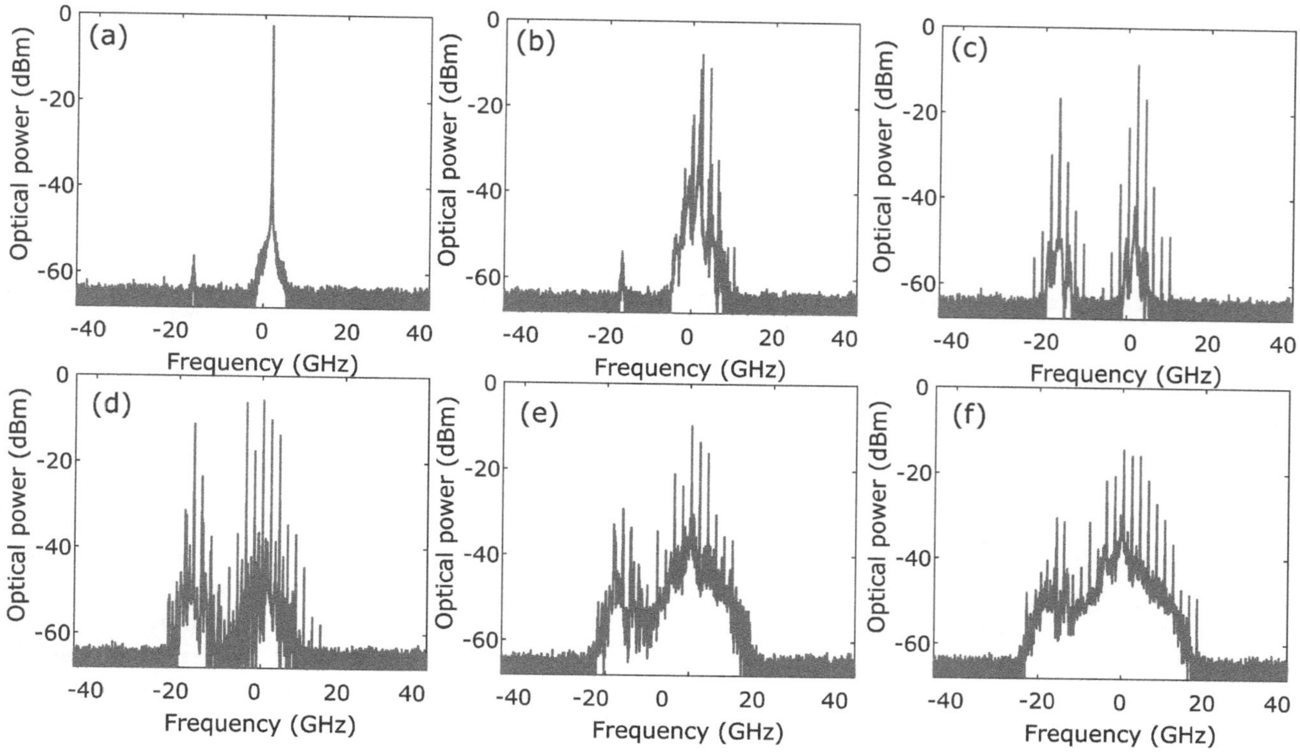

Figure 4. Bifurcation scenarios resulting in the excitation of the depressed polarization mode. These optical spectra are obtained for detuning $\Delta\nu = 1.6$ GHz and injected comb spacing $\Omega = 2$ GHz. (**a**) Stable output at $P_{inj} = 3$ µW, (**b**) wave mixing at $P_{inj} = 16$ µW, (**c**) two polarizations comb at $P_{inj} = 48$ µW, (**d**) two polarization harmonics comb at $P_{inj} = 112$ µW, (**e**) two polarization complex dynamics at $P_{inj} = 208$ µW and (**f**) X-polarization comb at $P_{inj} = 240$ µW.

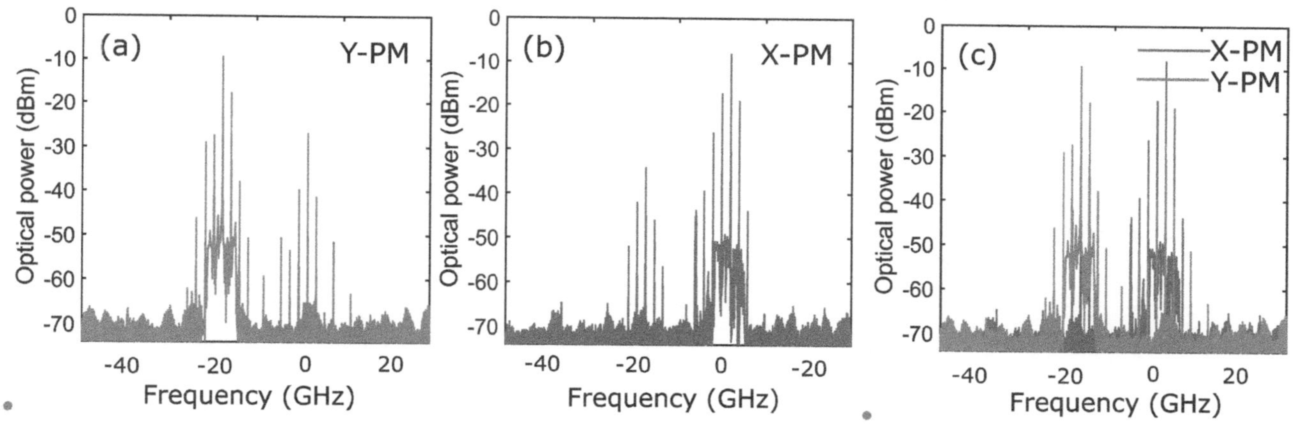

Figure 5. Polarization resolved-optical spectra corresponding to an example of two polarization comb dynamics in Figure 4c. (**a–c**) are obtained for detuning $\Delta\nu = 1.6$ GHz and injected power $P_{inj} = 20$ µW. (**a–c**) correspond to the optical spectra of Y-PM, X-PM and the superposition of Y-PM and X-PM, respectively.

2.3. *Influence of the Injection Current*

We next increase the injected comb spacing to $\Omega = 4$ GHz and the bias current to $I = 8$ mA, which is 2.67 times the threshold current. Such bias currents have been used in VCSELs to observe various nonlinear dynamics in the presence of single-mode injection with parallel polarization [37,39]. Compared to the $I = 6$ mA bias current injection where nonlinear wave-mixing is accompanied by two polarization comb as a result of excitation of the normally depressed polarization mode, the VCSEL output bifurcates to selective amplification of the central injected lines in X-PM, as shown in Figure 6a–c. As observed in

Figure 6c,d, selective amplification smoothly induces a broad optical comb characterized by the appearance of new frequency lines. When we increase the injected power, polarization mode competition results in the extension of the polarized comb to a much broader optical spectrum, as shown in Figure 6e. We checked through polarization resolved-optical spectra in Figure 7 that this overall comb corresponds to two combs based on the linear orthogonal polarization of the VCSEL. It is also possible to control the power in each polarization mode by fine-tuning the detuning. The two polarization combs and the polarized comb have the same repetition rate as the injected comb. By fine-tuning the injected power, the noise pedestal smoothly increases accompanied by harmonics comb line generation simultaneously in the two polarization combs as shown in Figure 6e. The harmonic comb significantly increases the number of resulting output lines and gives rise to a new comb with a low repetition rate ($\Omega = 2$ GHz in Figure 6e). Most importantly, Figure 6e highlight a new possibility of broadband $\Omega = 2$ GHz comb generation using the same injection technique as in Figure 4c,d. When we keep increasing the injected power, the harmonics comb destabilizes to complex dynamics with remarkable frequency comb lines mainly around the dominant polarization mode (X-PM), as seen in Figure 6g. Interestingly, when further increasing injected power, VCSEL bifurcates again to the new broadband comb encompassing the two polarization modes with a significant increase in noise pedestal, as shown in Figure 6h,i. Similarly to Figure 6e, this overall comb is formed by two combs with the orthogonal polarization direction. The noise pedestal increases again with the injected power to deteriorate comb performance, such as CNR and comb bandwidths. We have shown that the properties (bandwidth and CNR) of the two polarization combs induced by parallel optical injection are limited by increasing the injected comb spacing for fixed bias current. However, the increase in bias current is, therefore, an alternative to improve polarization combs properties.

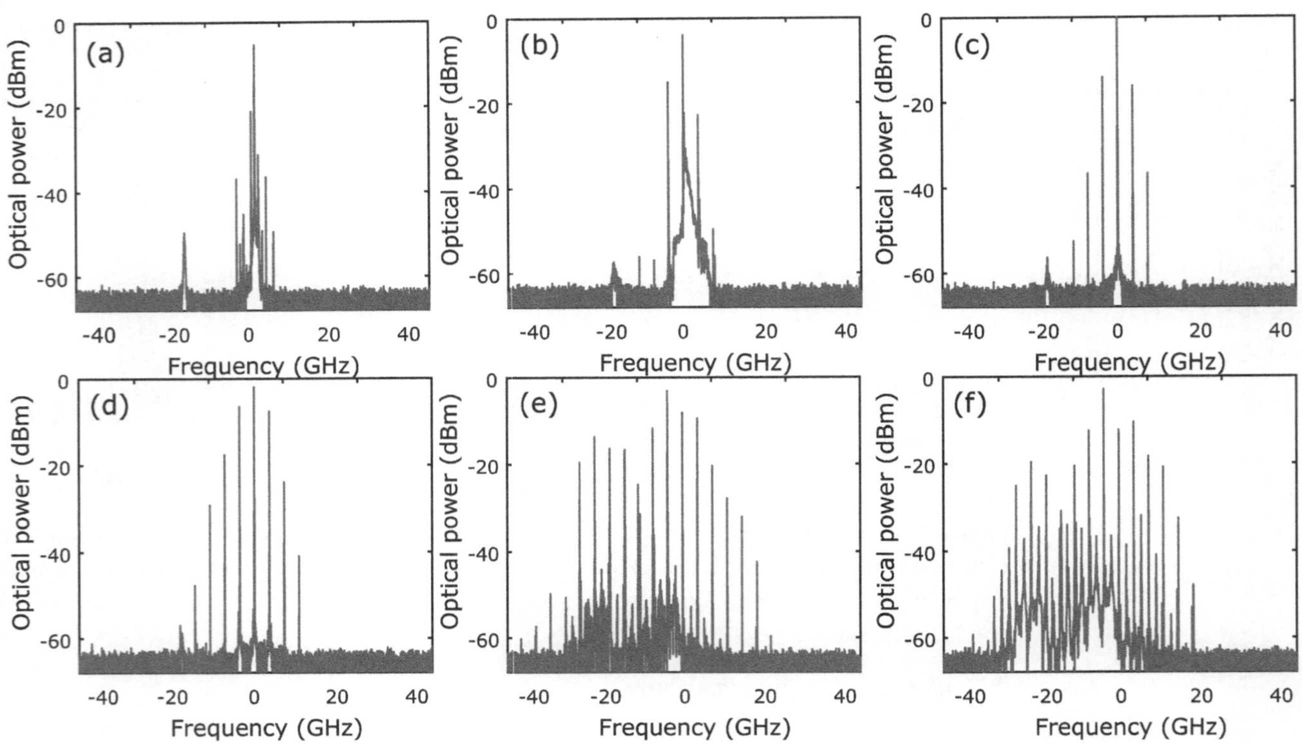

Figure 6. *Cont.*

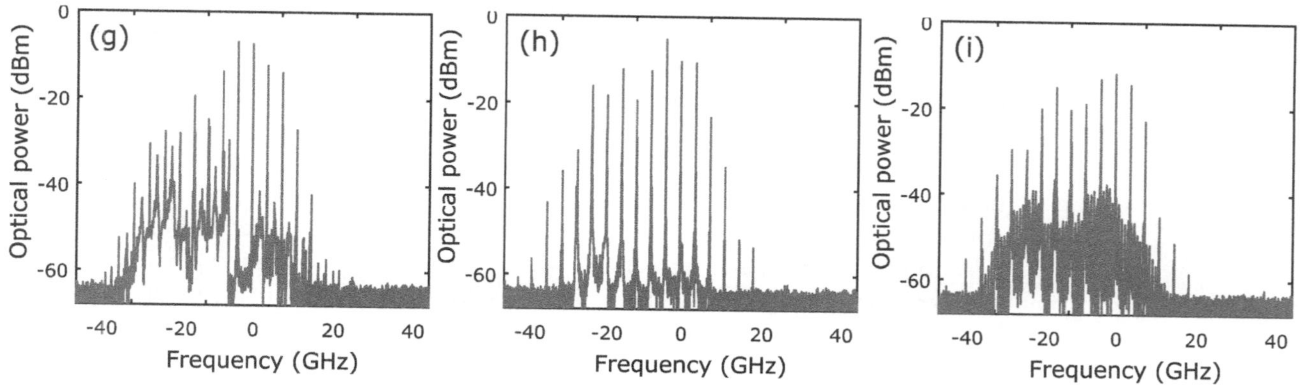

Figure 6. Bifurcation scenarios resulting in the excitation of the depressed polarization mode. These optical spectra are obtained for detuning $\Delta\nu = -0.9$ GHz and injected comb spacing $\Omega = 4$ GHz. (**a**,**b**) Wave mixing at $P_{inj} = 3$ µW and $P_{inj} = 32$ µW, respectively, (**c**,**d**) single polarization comb at $P_{inj} = 48$ µW and $P_{inj} = 228$ µW, respectively, (**e**,**h**,**i**) two polarizations comb at $P_{inj} = 304$ µW, $P_{inj} = 560$ µW and $P_{inj} = 704$ µW, respectively, (**f**) two polarization harmonics comb at $P_{inj} = 376$ µW and (**g**) two polarization complex dynamics at $P_{inj} = 448$ µW.

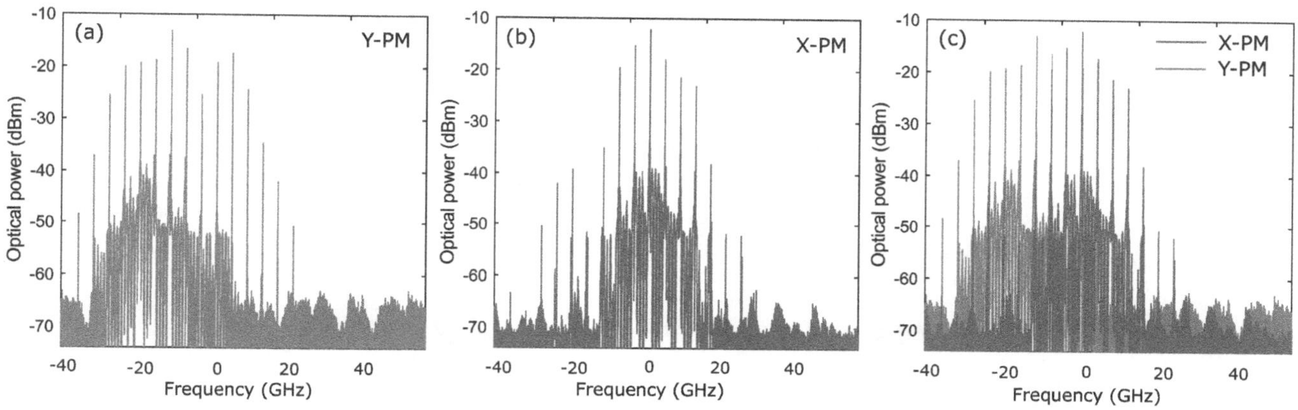

Figure 7. Polarization resolved-optical spectra corresponding to an example of two polarization comb dynamics in Figure 6e–i. (**a**–**c**) are obtained for detuning $\Delta\nu = -0.9$ GHz and injected power $P_{inj} = 432.8$ µW. (**a**–**c**) correspond to the optical spectra of Y-PM, X-PM and the superposition of Y-PM and X-PM, respectively.

2.4. Tailoring Comb Properties

Figure 8 provides further insight in comb dynamics when varying injection parameters, the injection current in the VCSEL and the polarization of injected light. Figure 8a,b analyze the comb properties for an injected comb spacing of $\Omega = 2$ GHz and fixed detuning, $\Delta\nu = 1.6$ GHz. For the measurement of comb bandwidth, we consider an output comb line when its amplitude lies above -30 dB from the maximum amplitude in the optical spectrum. When the spectrum total optical comb shows a big dip, the total comb bandwidth is estimated from the separate bandwidth of X-PM and Y-PM spectra. The polarization of the injected comb is parallel to that of the free-running VCSEL. When increasing injected power, the comb bandwidth for $I = 6$ mA and $I = 8$ mA current reaches a maximum of around $P_{inj} = 100$ µW and $P_{inj} = 200$ µW, respectively, as shown Figure 8a. The decrease in comb bandwidth is due to the bifurcation of the two polarization comb to a single polarization comb generation in X-PM, as shown in Figures 2i and 4f. Figure 8b shows that the best CNR is found for low injected power P_{inj}. CNR decreases with injected power due to the increase in the noise pedestal, which destabilizes comb properties. Figure 8a,b also show that the comb bandwidth and CNR increase with the injection current for the same injected comb properties. We next highlight the impact of polarization of the injected comb lines on VCSEL output comb dynamics in Figure 8c,d. In Figure 8c,d, blue and red correspond to parallel and orthogonal optical injection cases, respectively. As discussed in [28], the two polarization comb dynamics are mainly observed when VCSEL is injected between the two linear polarizations. However, for parallel optical injection, the two polarizations combs are observed only when the central injected comb lines are close to the dominant polarization mode of the VCSEL (detuning close to zero) at low injection currents, as shown in Figure 8a,b. When the injection current increases, the two comb dynamics are observed in a large portion of the plane of the injection parameters. Figure 8c shows that comb bandwidth increases with the injected power in the parallel and orthogonal optical injection case, which is in agreement with our previous publication [26–28]. Interestingly, the comb bandwidth for parallel optical injection is better than orthogonal optical injection at high injection currents, as shown in Figure 8c. When comparing Figure 8a to [28], we can observe that the comb bandwidth at low injection current is better for orthogonal than parallel optical injection. We also checked that optical comb performances do not change with the increase in injection currents. In the case of parallel optical injection, the improvement in the comb performance with the increase in the injection current is due to polarization mode instability, i.e., the excitation of the depressed polarization mode in the VCSEL at a large injection current. The excitation of the free-running depressed polarization mode in the VCSEL has been recently analyzed in several studies [37,39,40]. In these works, VCSEL was injected with a single frequency line master laser, and the injection current was higher than two-times the threshold current of VCSEL. The influence of the polarization of the injected comb is more impactful when we consider CNR. Figure 8d shows the CNR of comb dynamics induced by parallel (blue) and orthogonal (red) optical injections. We observe that, for parallel optical injection, whatever is the injected comb spacing and the injection current, the best CNR is found at low injected power and then decreases due to the increase in noise pedestal, as shown in Figure 6d–f. As shown in Figure 8b,d, the best values of CNR are found at low injections for orthogonal and parallel optical injections, which is in agreement with [28].

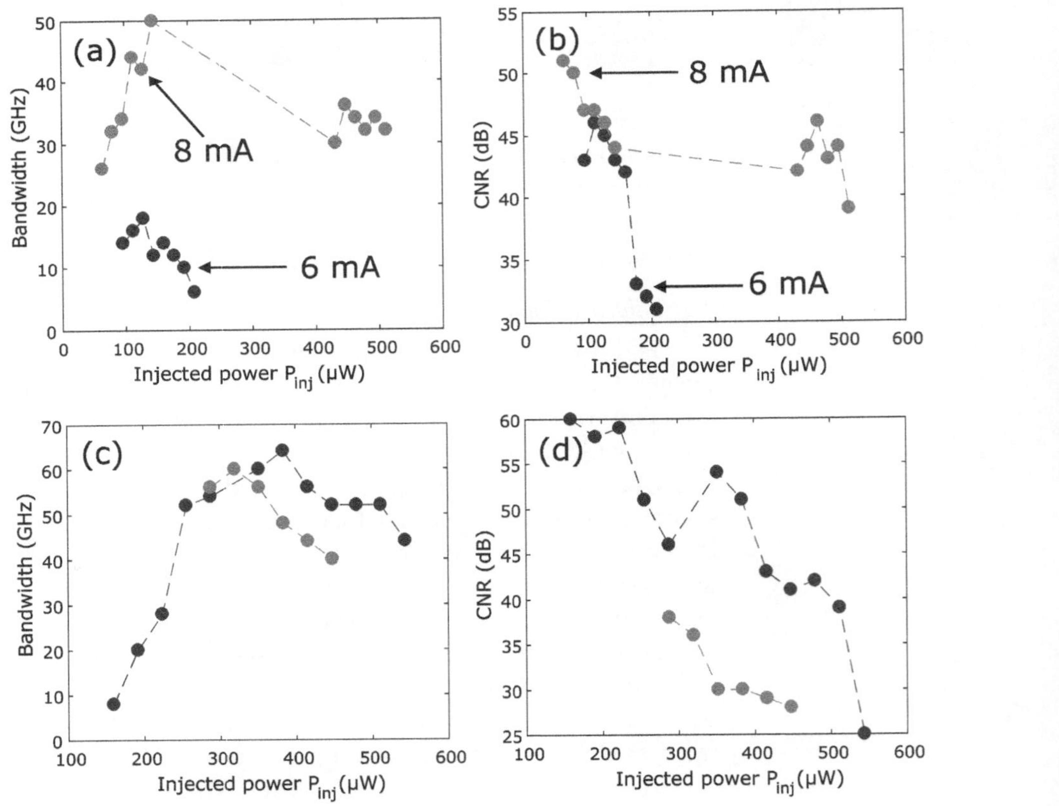

Figure 8. Control of comb properties using the injection parameters and polarization of the injected light. (**a**,**b**) are obtained for fixed detuning and comb spacing $\Delta\nu = 1.6$ GHz and $\Omega = 2$ GHz, respectively. (**c**,**d**) are obtained for fixed comb spacing ($\Omega = 4$ GHz) and injection current $I = 8$ mA.

The blue and red curves correspond to the parallel and orthogonal optical injection, respectively. The comb dynamics in parallel and orthogonal optical injection are obtained for fixed detuning $\Delta\nu = -0.9$ GHz and $\Delta\nu = -11.6$ GHz, respectively.

3. Theoretical Bifurcation Analysis

In addition to experimental results, theoretical studies have been performed to provide further insight into the polarization dynamics of comb injected VCSELs. We have used the Spin-Flip Model (SFM) [42], which is widely used to describe polarization dynamics in free-running VCSELs and supplemented it with additional term describing optical comb injection. SFM parameters are chosen such that the dominant polarization mode (normally depressed polarization mode) of the VCSEL in free-running emits along the X-axis ((Y-axis). Model equations are given in Equations (1)–(4). In these equations, E_x and E_y are the two linearly polarized slowly varying components of the complex fields in the X and Y directions, respectively. D and n are the sum and the difference between the population inversions for spin-up and spin-down radiation channels. μ is the normalized current injection.

$$\frac{dE_x}{dt} = -(\kappa + \gamma_a)E_x - i(\kappa\alpha + \gamma_p)E_x + \kappa(1 + i\alpha)(DE_x + inE_y) + E_M, \tag{1}$$

$$\frac{dE_y}{dt} = -(\kappa - \gamma_a)E_y - i(\kappa\alpha - \gamma_p)E_y + \kappa(1 + i\alpha)(DE_y - inE_x), \tag{2}$$

$$\frac{dD}{dt} = -\gamma[D(1 + |E_x|^2 + |E_y|^2) - \mu + in(E_yE_x^* - E_xE_y^*)], \tag{3}$$

$$\frac{dn}{dt} = -\gamma_s n - \gamma[n(|E_x|^2 + |E_y|^2) + iD(E_yE_x^* - E_xE_y^*)]. \tag{4}$$

The meaning of the remaining SFM parameters is the following: γ_p and γ_a correspond to linear birefringence and linear dichroism, respectively; γ_s is the spin-flip relaxation rate, γ is the decay rate of D, κ is the field decay rate and α is the linewidth enhancement factor.

The complex electrical field of the injected comb is written as follows:

$$E_M = \kappa_{inj} \sum_j E_j(t) e^{i(2\pi\nu_j t + \varphi_j(t))} \tag{5}$$

with angular frequency ω_j and amplitude E_j corresponding to j_{th} comb lines and coupling coefficient κ_{inj}. We consider an optimal coupling between the VCSEL output and the injected comb, which correspond to $\kappa_{inj} = \kappa$. We consider a comb with three frequency lines and we simplify the calculations by supposing $\varphi_j = 0$, i.e., the phase of the individual injected comb lines is zero. The detuning frequency $\Delta\nu_j$ is the difference between ν_j and the intermediate frequency between those of X and Y polarizations, $\frac{2\pi\nu_x + 2\pi\nu_y}{2}$, with $2\pi\nu_x = \alpha\gamma_a - \gamma_p$ and $2\pi\nu_y = \gamma_p - \alpha\gamma_a$, the frequency corresponding to the linear polarization mode X and Y, respectively.

SFM parameters are taken as $\kappa = 33$ ns^{-1}, $\gamma_a = -0.1$ ns^{-1}, $\alpha = 2.8$, $\mu = 2.29$, $\gamma_p = 9$ GHz, $\gamma = 2.08$ ns^{-1} and $\gamma_s = 2100$ ns^{-1}. These numerical values are taken from [38]. In the following, we shall use κ_{inj} for injection strength, with $\kappa_{inj} = \frac{E_{inj}}{E_0}$, where E_{inj} and E_0 are the total amplitude of the injected field and the total amplitude of the VCSEL in free-running mode, respectively. Rate Equations (1)–(4) are numerically integrated using a fourth-order Runge–Kutta method. Numerical simulations are typically performed with a times series of 200 ns and a time step equal to 1.2 ps. The detuning frequency value is defined from the central injected comb line and is referred to as $\Delta\nu$. In this section, we model the experiment by considering that the difference between the amplitude of the central injected comb line and the side comb lines is 12 dB.

3.1. Y-PM Comb Dynamics

Experimental results have demonstrated the possibility of generating two combs with orthogonal polarization in a VCSEL subject to parallel optical injection. We have also shown that this polarization comb performance is limited by the increase in injected comb spacing. In Figure 9, we analyze the two polarization comb numerically by using bifurcation diagrams. Figure 9 is obtained for fixed injected comb spacing $\Omega = 2$ GHz and detuning $\Delta\nu_x = -9$ GHz. These bifurcation diagrams are plotted by selecting the minima and the maxima of polarization-resolved intensities, $I_{x,y} = |E_{x,y}|^2$, for each injection strength. In Figure 9, the left and right panels correspond to the bifurcation diagrams of X-PM and Y-PM, respectively. Figure 9(a_1,b_1) show the bifurcation diagrams for fixed linear dichroism $\gamma_a = -0.1$ ns^{-1}. When increasing the injection strength, the VCSEL output first shows a nonlinear wave-mixing involving the detuning and the injected comb spacing in X-PM. When we kept increasing the injection strength, wave mixing bifurcates to periodic dynamics corresponding to the optical frequency comb in the two linear orthogonal polarization modes. The two polarization comb region is referred to as "comb" in Figure 9. The X-PM shows a larger number of lines than the Y-PM, but the number of lines in Y-PM can be controlled with injection parameters. The two comb dynamics remain stable with a simultaneous increase in the number of lines over a large range of injection strengths. Interestingly, by finetuning the injection strength, the VCSEL output shows harmonic comb dynamics simultaneously in the two orthogonal combs at the boundary of the comb regions. A further increase in injection strength results in complex dynamics in X-PM and Y-PM accompanied by the abrupt suppression of all power in Y-PM. Cascade comb dynamics resulting in a very low comb repetition rate are observed in between complex dynamics depending on injection parameters. The complex polarization dynamics bifurcate to a single polarization comb in the dominant polarization mode (X-PM). We next vary linear dichroism to highlight its impact on the two polarization comb dynamics. Interestingly, when decreasing linear dichroism, the size of the two polarization comb region decreases, as observed in Figure 9(a_2,a_3,b_2,b_3). The double horizontal arrows indicate how much the size of the combs' regions decreases with the decrease in linear dichroism. It is worth observing that all bifurcation sequences are similar to the case of $\gamma_a = -0.1$ ns^{-1}. When linear dichroism reaches $\gamma_a = -0.8$ ns^{-1}, the two polarization comb disappear and Y-PM is suppressed, as shown in Figure 9(a_4,b_4). Therefore, with the exception of the disappearance of comb lines in Y-PM, the bifurcation scenarios remain similar for X-PM. The bifurcation scenarios described in Figure 9 are in a very good agreement with the experimental ones in Figures 2 and 4.

Figure 10 shows an example comb and complex dynamics in the two polarization modes. Figure 10(a_1,b_1) show the optical spectra of comb dynamics in X-PM and Y-PM, respectively. We observe clearly in these spectra that the bandwidth of the frequency comb in the injected polarization mode (X-PM) is larger than the comb in the depressed polarization mode, Y-PM. As discussed early in the experimental section, these X-PM and Y-PM combs can combine to form a broad comb in the total output power of the VCSEL. Figure 10(a_2,b_2) show an example of complex polarization dynamics. Depending on injection parameters and injected comb properties, these complex polarization dynamics can simultaneously bifurcate to a two polarization harmonic comb or single-polarization comb.

Figure 9 has shown that the decrease in linear dichroism results in the complete suppression of the comb lines in the normally depressed polarization mode (Y-PM). In Figure 11, we provide a further insight on comb generation, whatever the linear dichroism is. To this end, we keep the parameters used to obtain Figure 9(a_4,b_4), and then we vary the bias current. When the normalized bias current reaches 2.9 times the threshold, i.e., $\mu = 4.2$, comb dynamics start to take place again in the two polarization modes, as observed in Figure 11a,b. Most importantly, the bifurcation scenarios giving rise to the two polarization comb dynamics remain similar to the case of $\mu = 2.29$. When we increase the bias current to $\mu = 5.29$ in Figure 11c,d, the bifurcation to the two polarization comb dynamics remains similar to the case of $\mu = 4.2$. Interestingly, unlike the case of $\mu = 4.2$, where the two polarization becomes abruptly chaotic, the VCSEL output bifurcates to cascade harmonic comb dynamics in the two polarization modes. The two polarization harmonics comb results in the suppression of all the power in Y-PM and bifurcates and, therefore, to polarized comb generation in the X-PM. Interestingly, we checked that when the polarization of the injected comb is orthogonal to that of the VCSEL, the size of the two polarization combs area increases slightly when linear dichroism, γ_a, is decreased.

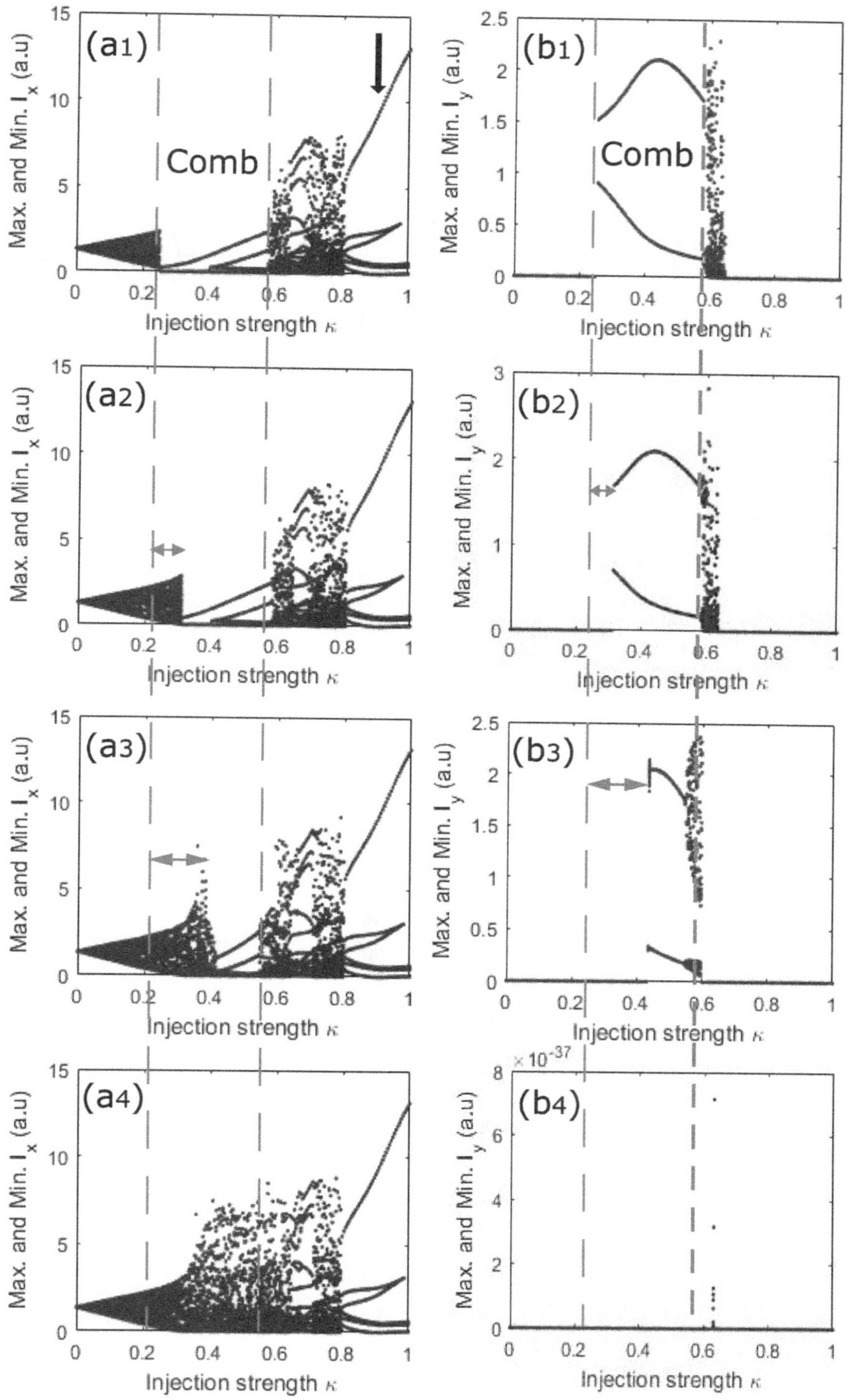

Figure 9. Bifurcation diagrams for fixed injected comb spacing $\Omega = 2$ GHz and detuning $\Delta\nu_x = -9$ GHz. The left and right panels correspond to X-polarization mode (X-PM) and Y-polarization mode (Y-PM). ($\mathbf{a_1}$,$\mathbf{b_1}$), ($\mathbf{a_2}$,$\mathbf{b_2}$), ($\mathbf{a_3}$,$\mathbf{b_3}$) and ($\mathbf{a_4}$,$\mathbf{b_4}$) are obtained for $\gamma_a = -0.1$ ns^{-1}, $\gamma_a = -0.2$ ns^{-1}, $\gamma_a = -0.6$ ns^{-1} and $\gamma_a = -0.8$ ns^{-1}.

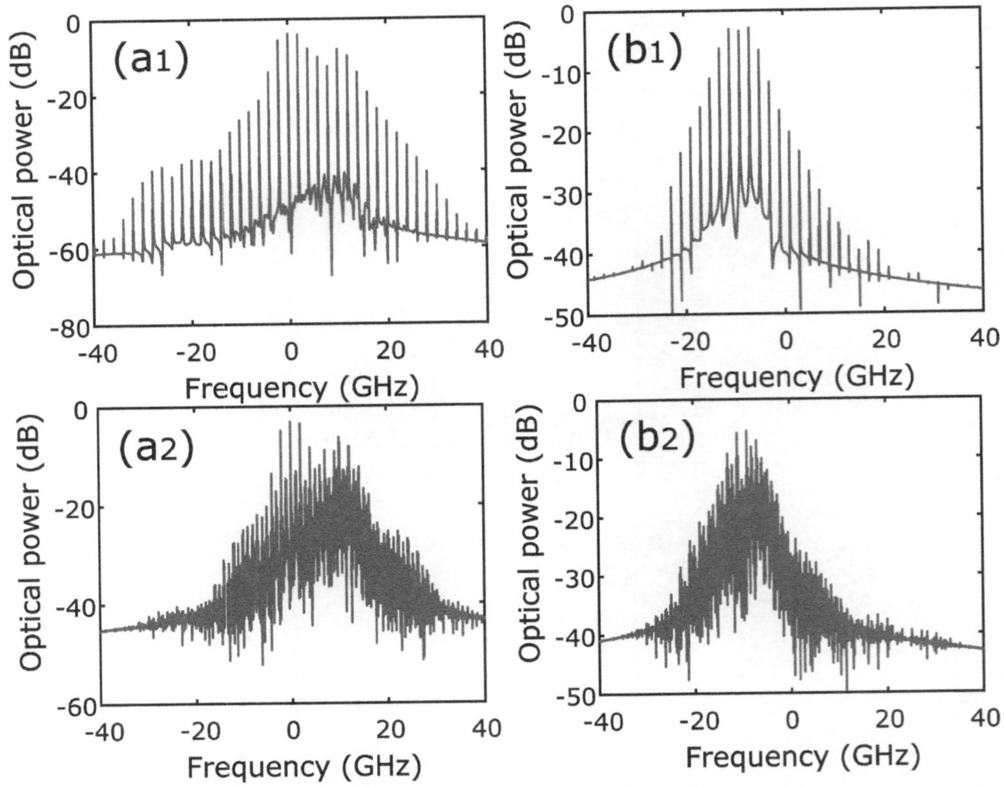

Figure 10. Optical spectra for fixed $\Omega = 2$ GHz, $\Delta\nu_x = -9$ GHz and $\gamma_a = -0.6$ ns^{-1}. The left (**a**) and right (**b**) panels correspond to X-PM and Y-PM, respectively. The top figures (**a$_1$**,**b$_1$**) are obtained for $\boldsymbol{\kappa = 0.525}$ and the bottom figures (**a$_2$**,**b$_2$**) are obtained for $\boldsymbol{\kappa = 0.6}$.

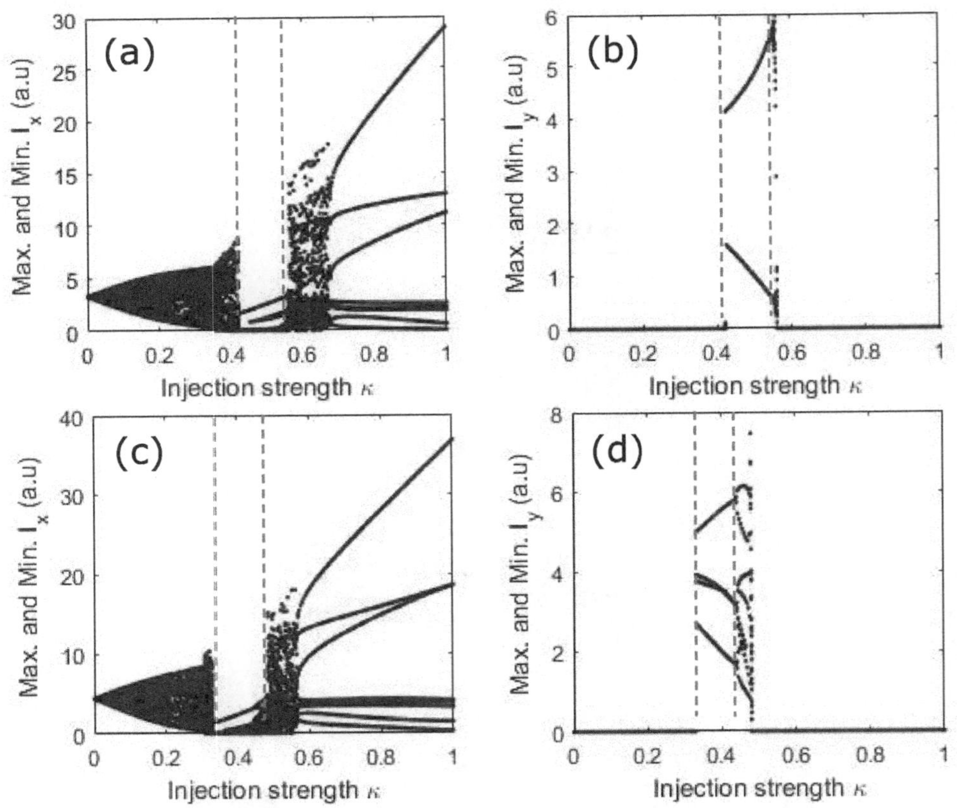

Figure 11. Bifurcation diagrams for fixed injected comb spacing $\Omega = 2$ GHz and $\gamma_a = -0.8$ ns^{-1}. The left and right panels correspond to X-PM and Y-PM, respectively. (**a**,**b**) are obtained for $\mu = 4.2$ and (**c**,**d**) for $\mu = 5.29$.

3.2. Influence of Spin-Flip Relaxation Rate

Further insight into the two polarization comb dynamics is provided in Figure 12, where bifurcation diagrams are shown for fixed $\Delta\nu_x = -9$ GHz when varying the injection strength together with the spin-flip relaxation rate γ_s. The blue and red colors in Figure 12 correspond to the bifurcation diagrams of X-PM and Y-PM, respectively. When γ_s is small (Figure 12a), the two polarization modes are excited at very low injection strengths. Most importantly, the comb lines start to progressively appear in each polarization mode as injection strength increases. Interestingly, when we keep increasing the spin-flip relaxation rate γ_s, the VCSEL output first shows the complex dynamics in the dominant polarization mode, which bifurcates to the excitation of the depressed polarization mode accompanied by comb generation, as shown in Figure 12b. The bifurcation sequence remains quite similar when further increasing the spin-flip relaxation rate. Still, the size of the complex dynamics region in X-PM increases, therefore resulting in a decrease in the size of the two combs region, as seen in Figure 12b–d. The size of the comb regions remains fixed at a high value of γ_s, as seen in Figure 12c,d. Unlike the case of Figure 12a,b, where the two polarization combs take place smoothly, the two comb appearance is abrupt when the spin-flip relaxation rate is large in Figure 12c,d. The two polarization comb dynamics bifurcates to a single polarization comb at large injection strength for each bifurcation diagram. It is worth observing that cascade harmonics characterizes bifurcation to single polarization comb dynamics at low γ_s. In contrast, the two complex polarization dynamics result in the single polarization comb at significant γ_s. Once the single polarization mode operation is achieved in VCSEL, bifurcation scenarios remain similar whatever the value the spin-flip relaxation rate is. In addition to the spin-flip relaxation rate, we have checked that the two polarization comb dynamics are observed in a wide range of SFM parameters including linear dichroism ($\gamma_a = -0.1$ ns^{-1} to $\gamma_a = -0.8$ ns^{-1}), injection current ($\mu = 2.29$ to $\mu = 5.29$), linewidth enhancement factor ($\alpha = 1$ to $\alpha = 5$) and linear birefringence ($\gamma_p = 1$ GHz to several hundred of GHz).

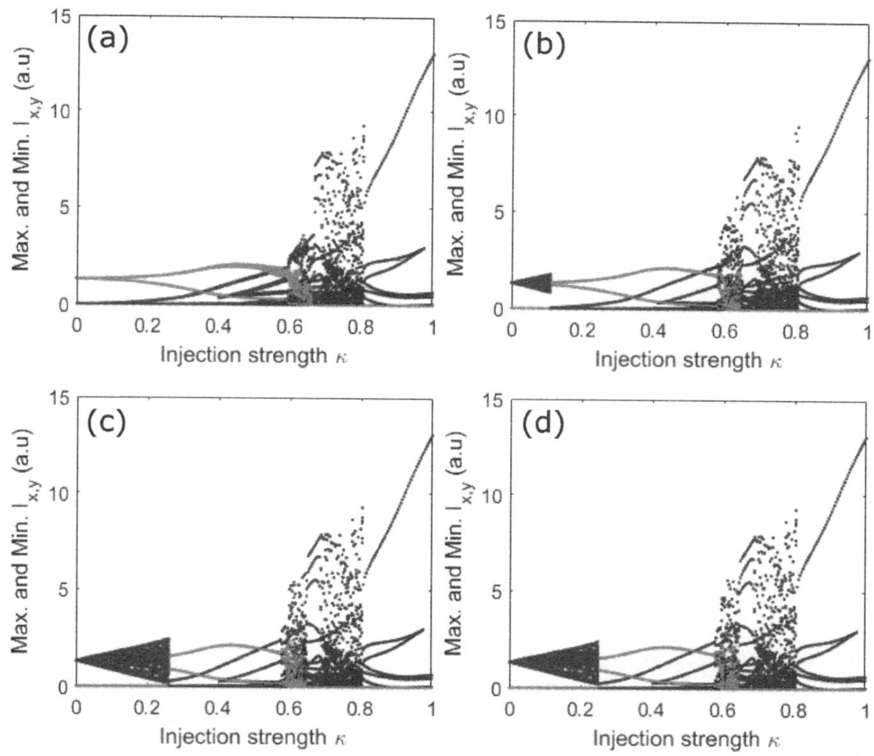

Figure 12. Bifurcation diagrams for fixed injected comb spacing $\Omega = 2$ GHz and detuning $\Delta\nu_x = -9$ GHz when varying γ_s. (**a**–**d**) are obtained for $\gamma_s = 50$ ns^{-1}, $\gamma_s = 200$ ns^{-1}, $\gamma_s = 1000$ ns^{-1} and $\gamma_s = 2300$ ns^{-1}, respectively.

4. Conclusions

We have shown experimentally and theoretically that VCSEL can bifurcate to interesting polarization dynamics under optical frequency comb injection. Interestingly, we have demonstrated that when the polarization of the injected comb is parallel to that of the VCSEL, polarization mode competition can induce two combs in a single-mode VCSEL. These combs are based on the two linear orthogonal polarizations in the VCSEL. We have also shown that these polarization combs are limited by an increase in the injected comb spacing, especially in the normally depressed polarization mode, Y-PM. Our results have shown that comb performance can be regained when increasing the bias current in VCSEL. We have performed numerical simulations to highlight a strong agreement with experimental results by using bifurcation diagrams. We have also numerically analyzed the influence of linear dichroism and the spin-flip relaxation rate on the two polarization comb dynamics. Interestingly, the two polarization comb area becomes narrower when linear dichroism decreases until it disappears at $\gamma_a = -0.8\ \text{ns}^{-1}$. By contrast, the size of the region of two polarizations comb dynamics increases with the decrease in linear dichroism when VCSEL is under an orthogonal optical injection. The single or two polarization combs induced by parallel optical injection are not as efficient in terms of bandwidth and CNR as the case of orthogonal comb injection investigated in [28] in terms of the number of lines at low bias current. Furthermore, two polarization combs are observed only if the bias current is significantly large, at least twice the threshold current for the VCSEL considered here. In the case of orthogonal optical injection, single and two polarization combs dynamics show the same performance (bandwidth and CNR), whatever the injection current is.

References

1. Sciamanna, M.; Shore, K.A. Physics and applications of laser diode chaos. *Nat. Photonics* **2015**, *9*, 151–162. [CrossRef]
2. Wieczorek, S.; Krauskopf, B.; Simpson, T.B.; Lenstra, D. The dynamical complexity of optically injected semiconductor lasers. *Phys. Rep.* **2005**, *416*, 1–128. [CrossRef]
3. Mogensen, F.; Olesen, H.; Jacobsen, G. Locking conditions and stability properties for a semiconductor laser with external light injection. *IEEE J. Quantum Electron.* **1985**, *21*, 784–793. [CrossRef]
4. Larsson, A. Advances in VCSELs for Communication and Sensing. *IEEE J. Sel. Top. Quantum Electron.* **2011**, *17*, 1552–1567. [CrossRef]
5. Sciamanna, M.; Panajotov, K.; Thienpont, H.; Veretennicoff, I.; Mégret, P.; Blondel, M. Optical feedback induces polarization mode hopping in vertical-cavity surface-emitting lasers. *Opt. Lett.* **2003**, *28*, 1543–1545. [CrossRef] [PubMed]
6. Sciamanna, M.; Panajotov, K. Route to polarization switching induced by optical injection in vertical-cavity surface-emitting lasers. *Phys. Rev. A* **2006**, *73*, 023811. [CrossRef]
7. Altés, J.B.; Gatare, I.; Panajotov, K.; Thienpont, H.; Sciamanna, M. Mapping of the dynamics induced by orthogonal optical injection in vertical-cavity surface-emitting lasers. *IEEE J. Quantum Electron.* **2006**, *42*, 198–207.
8. Choquette, K.D.; Schneider, R.P.; Lear, K.L.; Leibenguth, R.E. Gain-dependent polarization properties of vertical-cavity lasers. *IEEE J. Sel. Top. Quantum Electron.* **1995**, *1*, 661–666. [CrossRef]

9. Hurtado, A.; Quirce, A.; Valle, A.; Pesquera, L.; Adams, M.J. Nonlinear dynamics induced by parallel and orthogonal optical injection in 1550 nm vertical-cavity surface-emitting lasers (VCSELs). *Opt. Express* **2010**, *18*, 9423–9428. [CrossRef] [PubMed]
10. Gatare, I.; Buesa, J.; Thienpont, H.; Panajotov, K.; Sciamanna, M. Polarization switching bistability and dynamics in vertical-cavity surface-emitting laser under orthogonal optical injection. *Opt. Quantum Electron.* **2006**, *38*, 429–443. [CrossRef]
11. Virte, M.; Panajotov, K.; Thienpont, H.; Sciamanna, M. Deterministic polarization chaos from a laser diode. *Nat. Photonics* **2013**, *7*, 60–65. [CrossRef]
12. Quirce, A.; de Dios, C.; Valle, A.; Acedo, P. VCSEL-based optical frequency combs expansion induced by polarized optical injection. *IEEE J. Sel. Top. Quantum Electron.* **2018**, *25*, 1–9. [CrossRef]
13. Minoshima, K.; Matsumoto, H. High-accuracy measurement of 240-m distance in an optical tunnel by use of a compact femtosecond laser. *Appl. Opt.* **2000**, *39*, 5512–5517. [CrossRef] [PubMed]
14. Link, S.M.; Maas, D.; Waldburger, D.; Keller, U. Dual-comb spectroscopy of water vapor with a free-running semiconductor disk laser. *Science* **2017**, *356*, 1164–1168. [CrossRef] [PubMed]
15. Marin-Palomo, P.; Kemal, J.N.; Karpov, M.; Kordts, A.; Pfeifle, J.; Pfeiffer, M.H.; Trocha, P.; Wolf, S.; Brasch, V.; Anderson, M.H.; et al. Microresonator-based solitons for massively parallel coherent optical communications. *Nature* **2017**, *546*, 274–279. [CrossRef] [PubMed]
16. Hargrove, L.; Fork, R.L.; Pollack, M. Locking of He–Ne laser modes induced by synchronous intracavity modulation. *Appl. Phys. Lett.* **1964**, *5*, 4–5. [CrossRef]
17. Del'Haye, P.; Schliesser, A.; Arcizet, O.; Wilken, T.; Holzwarth, R.; Kippenberg, T.J. Optical frequency comb generation from a monolithic microresonator. *Nature* **2007**, *450*, 1214–1217. [CrossRef]
18. Tilma, B.W.; Mangold, M.; Zaugg, C.A.; Link, S.M.; Waldburger, D.; Klenner, A.; Mayer, A.S.; Gini, E.; Golling, M.; Keller, U. Recent advances in ultrafast semiconductor disk lasers. *Light. Sci. Appl.* **2015**, *4*, e310. [CrossRef]
19. Hugi, A.; Villares, G.; Blaser, S.; Liu, H.; Faist, J. Mid-infrared frequency comb based on a quantum cascade laser. *Nature* **2012**, *492*, 229–233. [CrossRef]
20. Silvestri, C.; Columbo, L.L.; Brambilla, M.; Gioannini, M. Coherent multi-mode dynamics in a quantum cascade laser: Amplitude- and frequency-modulated optical frequency combs. *Opt. Express* **2020**, *28*, 23846–23861. [CrossRef] [PubMed]
21. Weber, C.; Columbo, L.L.; Gioannini, M.; Breuer, S.; Bardella, P. Threshold behavior of optical frequency comb self-generation in an InAs/InGaAs quantum dot laser. *Opt. Lett.* **2019**, *44*, 3478–3481. [CrossRef] [PubMed]
22. Grillot, F.; Duan, J.; Dong, B.; Huang, H. Uncovering recent progress in nanostructured light-emitters for information and communication technologies. *Light. Sci. Appl.* **2021**, *10*, 156. [CrossRef] [PubMed]
23. Villares, G.; Hugi, A.; Blaser, S.; Faist, J. Dual-comb spectroscopy based on quantum-cascade-laser frequency combs. *Nat. Commun.* **2014**, *5*, 5192. [CrossRef] [PubMed]
24. Shortiss, K.; Lingnau, B.; Dubois, F.; Kelleher, B.; Peters, F.H. Harmonic frequency locking and tuning of comb frequency spacing through optical injection. *Opt. Express* **2019**, *27*, 36976–36989. [CrossRef]
25. Lingnau, B.; Shortiss, K.; Dubois, F.; Peters, F.H.; Kelleher, B. Universal generation of devil's staircases near Hopf bifurcations via modulated forcing of nonlinear systems. *Phys. Rev. E* **2020**, *102*, 030201. [CrossRef] [PubMed]
26. Doumbia, Y.; Malica, T.; Wolfersberger, D.; Panajotov, K.; Sciamanna, M. Nonlinear dynamics of a laser diode with an injection of an optical frequency comb. *Opt. Express* **2020**, *28*, 30379–30390. [CrossRef]
27. Doumbia, Y.; Malica, T.; Wolfersberger, D.; Panajotov, K.; Sciamanna, M. Optical injection dynamics of frequency combs. *Opt. Lett.* **2020**, *45*, 435–438. [CrossRef]
28. Doumbia, Y.; Wolfersberger, D.; Panajotov, K.; Sciamanna, M. Tailoring frequency combs through VCSEL polarization dynamics. *Opt. Express* **2021**, *29*, 33976–33991. [CrossRef]
29. Gatare, I.; Sciamanna, M.; Nizette, M.; Panajotov, K. Bifurcation to polarization switching and locking in vertical-cavity surface-emitting lasers with optical injection. *Phys. Rev. A* **2007**, *76*, 031803. [CrossRef]
30. Gatare, I.; Sciamanna, M.; Buesa, J.; Thienpont, H.; Panajotov, K. Nonlinear dynamics accompanying polarization switching in vertical-cavity surface-emitting lasers with orthogonal optical injection. *Appl. Phys. Lett.* **2006**, *88*, 101106. [CrossRef]
31. Valle, A.; Gatare, I.; Panajotov, K.; Sciamanna, M. Transverse mode switching and locking in vertical-cavity surface-emitting lasers subject to orthogonal optical injection. *IEEE J. Quantum Electron.* **2007**, *43*, 322–333. [CrossRef]
32. Panajotov, K.; Gatare, I.; Valle, A.; Thienpont, H.; Sciamanna, M. Polarization-and transverse-mode dynamics in optically injected and gain-switched vertical-cavity surface-emitting lasers. *IEEE J. Quantum Electron.* **2009**, *45*, 1473–1481. [CrossRef]
33. Gatare, I.; Panajotov, K.; Sciamanna, M. Frequency-induced polarization bistability in vertical-cavity surface-emitting lasers with orthogonal optical injection. *Phys. Rev. A* **2007**, *75*, 023804. [CrossRef]
34. Nizette, M.; Sciamanna, M.; Gatare, I.; Thienpont, H.; Panajotov, K. Dynamics of vertical-cavity surface-emitting lasers with optical injection: A two-mode model approach. *JOSA B* **2009**, *26*, 1603–1613. [CrossRef]
35. Sciamanna, M.; Panajotov, K. Two-mode injection locking in vertical-cavity surface-emitting lasers. *Opt. Lett.* **2005**, *30*, 2903–2905. [CrossRef] [PubMed]
36. Hong, Y.; Spencer, P.S.; Rees, P.; Shore, K.A. Optical injection dynamics of two-mode vertical cavity surface-emitting semiconductor lasers. *IEEE J. Quantum Electron.* **2002**, *38*, 274–278. [CrossRef]
37. Denis-le Coarer, F.; Quirce, A.; Valle, Á.; Pesquera, L.; Sciamanna, M.; Thienpont, H.; Panajotov, K. Polarization dynamics induced by parallel optical injection in a single-mode VCSEL. *Opt. Lett.* **2017**, *42*, 2130–2133. [CrossRef] [PubMed]

38. Quirce, A.; Pérez, P.; Popp, A.; Valle, Á.; Pesquera, L.; Hong, Y.; Thienpont, H.; Panajotov, K. Polarization switching and injection locking in vertical-cavity surface-emitting lasers subject to parallel optical injection. *Opt. Lett.* **2016**, *41*, 2664–2667. [CrossRef]
39. Quirce, A.; Popp, A.; Denis-le Coarer, F.; Pérez, P.; Valle, Á.; Pesquera, L.; Hong, Y.; Thienpont, H.; Panajotov, K.; Sciamanna, M. Analysis of the polarization of single-mode vertical-cavity surface-emitting lasers subject to parallel optical injection. *JOSA B* **2017**, *34*, 447–455. [CrossRef]
40. Denis-le Coarer, F.; Quirce, A.; Pérez, P.; Valle, A.; Pesquera, L.; Sciamanna, M.; Thienpont, H.; Panajotov, K. Injection locking and polarization switching bistability in a 1550 nm VCSEL subject to parallel optical injection. *IEEE J. Sel. Top. Quantum Electron.* **2017**, *23*, 1–10. [CrossRef]
41. Guckenheimer, J.; Holmes, P. *Nonlinear Oscillations, Dynamical Systems, and Bifurcations of Vector Fields*; Springer Science & Business Media: Berlin/Heidelberg, Germany, 2013; Volume 42.
42. San Miguel, M.; Feng, Q.; Moloney, J.V. Light-polarization dynamics in surface-emitting semiconductor lasers. *Phys. Rev. A* **1995**, *52*, 1728. [CrossRef]

High-Resolution Simulation of Externally Injected Lasers Revealinga Large Regime of Noise-Induced Chaos

Abstract: We present comprehensive numerically simulated scans of the spectral evolution of the output from a single-mode semiconductor laser diode undergoing external light injection. The spectral scans are helpful to understand the different regimes of operation as well as the system evolution between each state: i.e., locked state, four-wave mixing, pulsations, chaos. We find that, when under strong injection, when the injected power equals about half of the laser power, two distinct regions of chaotic behaviour are observed. One of the chaotic regions arises due to the usual period-doubling route to chaos; the other chaotic region is a blurring of what would be higher-order period pulsations whose periodicity is broken by spontaneous emission and the laser spectrum is chaotic. Eliminating spontaneous emission in our simulations confirms the latter chaotic region becomes a region with higher-order pulsations.

Keywords: semiconductor lasers; injection-locking; noise; simulation; pulsation; chaos

---

1. Introduction

External laser injection-locking is a curious field of study due to the range of dynamical states that are produced: laser synchronisation; chaos; four-wave mixing (FWM); self-pulsation and associated period-doubling routes to chaos [1–10]. Recently, it was discovered that lasers acting under injection-locking exhibit a strong increase in the potential direct modulation bandwidth, with subsequent studies to unlock the potential of using such strong modulation capability [9,10], with algorithms developed to achieve the optimal injection-locking point [11]. A recent detailed review, in article [4], details optical injection applications including laser synchronisation for phase-sensitive applications and frequency distribution. Injection-locking is also important for stabilizing gain-switched optical pulse sources and for phase stabilization of optical frequency combs [12,13]. External injection of a semiconductor is highlighted in Figure 1, where two lasers are involved and light from the master laser is injected into the slave laser. The optical isolator (Iso.) ensures that no light from the slave laser is injected back into the master laser. In order to observe the phenomena associated with external injection, the central lasing frequency (or wavelength) of both lasers should be similar, within ± 25 GHz of each other.

Laser rate equations have been shown to capture the various phenomena of externally injected lasers [1–3,6]. Despite their simplicity, the equations reveal all of the relevant dynamical phenomena of externally injected lasers. An extended treatise in [6], analysing the dynamical regimes of externally injected lasers, is given in this paper, identifying chaotic, pulsation states including Hopf-bifurcations of the pulsation into general period-N (P-N) pulsation states [6]. One aspect that we notice when solving laser rate equations when spontaneous emission (SE) is included [3] is the absence of higher pulsation states beyond the P-2 pulsation states; this would be predicted when SE is omitted in the analysis. Previous studies attributed SE to be the cause of an observed absence of a period-doubling route to chaos [14]; a more formal analysis [15] examined the behaviour of noise-induced chaos when the external-injected laser system was in a delicate high order pulsation state and that study found that SE could induce chaotic behaviour.

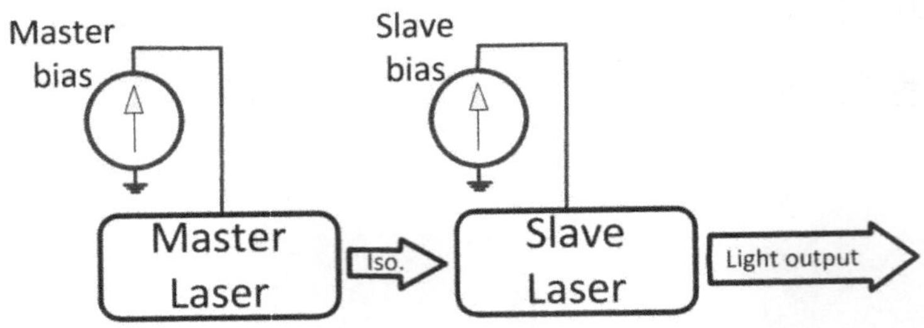

Figure 1. Schematic of external injection laser system. Light from the master laser is coupled to the slave laser, the optical isolator (Iso.) ensures that light from the slave laser does not couple to the master. Depending on the master–slave detuning, master power lever, a whole range of dynamical effects are observed, from injection-locking, four-wave mixing, pulsation and chaos.

In this paper, we present a detailed study of the evolution of the optical spectral output from the injected laser system. We concentrate on varying the detuning frequency between the master and slave lasers and building up the output spectra. The plots are characterised by constant master power, and show FWM, injection-locking, self-pulsations, period-doubling and chaos. We find that only period-2 (P-2) pulsations (i.e., pulsations with double the period of the self-pulsation) are present when we solve the complete rate equations with stochastic SE included. For the first time, to the best of our knowledge, we find a chaotic regime solely created from noise-induced chaos, when SE is removed from the simulation that the region is entirely comprised of high-order pulsations of periodicity P-2 and greater. The conditions to observe this noise-induced chaos regime is when the injection level is quite high, close to 50% of the slave laser power. The noise-induced chaos regime is entirely separate from the chaotic regime caused by the usual period-doubling route to chaos that remains even when the SE is set to zero. In order to quantify how much SE is needed for noise-induced chaos to exist, we use a value for the SE coupling coefficient into the lasing mode to be 1×10^{-4}; this value would be at the lower end of the observed values for this parameter [16]. We run our simulations with hypothetical lower values of SE coupling coefficient and that for values $< 1 \times 10^{-7}$ that noise-induced chaos still exists.

2. Materials and Methods

To conduct this study, we employ the field version of the rate equations for semiconductor lasers. The fundamental derivation of the model is given in [3], though we keep with the complex-valued envelope of the optical field to avoid numerical instability issues when the slave power tends briefly to zero, as can happen with external injection. We implemented the field version of the rate equations before in relation to laser gain-switching [17]; here, we just concentrate on DC biasing of the lasers.

$$\frac{dN}{dt} = \frac{I_{bias}}{eV} - R(N) - \frac{a(N-N_0)}{1+\varepsilon_{NL}|E|^2}|E|^2 + F_N \tag{1}$$

$$\frac{dE}{dt} = \frac{(1-j\alpha_H)}{2}\left[\frac{a(N-N_0)}{1+\varepsilon_{NL}|E|^2} - \frac{1}{\tau_p}\right]E + k_c E_{inj}\exp(j2\pi f_D t) + F_E \tag{2}$$

where all of the symbols have their usual meaning and are defined in Table 1. The carrier density is given by N and E is the envelope of the optical field and is related to the photon density in the laser. The first term on the right-hand side of (1) is given by the electric bias current flowing into the laser, where I_{bias} = 60 mA throughout for this study. Carrier recombination is given by $R(N) = AN + BN^2 + CN^3$ for nonradiative, bimolecular and Auger recombination, respectively; the third term represents stimulated emission. E is a complex-valued quantity describing the envelope of the optical field and encompasses all amplitude and phase modulation effects imposed by the injected-laser system; E is normalised such that $|E|^2$ represents the photon density of the laser field. F_N denotes stochastic carrier recombination (which will be defined later). The first term on the right-hand side of (2) describes the complex gain of the laser field. The gain coefficient is given by $a(N - N_0)$, where a is the differential gain and N_0 is the carrier density at transparency. The gain coefficient is modified by the nonlinear gain ε_{NL}. τ_P is the cavity lifetime. Note that lumping the photon lifetime with the gain-phase coupling allows for the slave laser to be centred at the zero frequency within the simulation; this is beneficial so that master–slave detuning can be easily controlled via the $\exp(j2\pi f_D t)$ frequency-translation operation term for the external injection. E_{inj} is the envelope of the optical field of the master laser; in this paper, we take a constant amplitude (noiseless) for each simulation run. The final term, F_E, is the random addition of SE due to bimolecular recombination into the lasing field, denoting spontaneous emission into the lasing field.

Table 1. Parameters and values used in the simulation.

Symbol	Definition	Value and/or Unit
N	Carrier density	m^{-3}
E	Optical field	$m^{\frac{3}{2}}$
I	Laser bias current	60 mA
e	Quantum of electronic charge	1.6×10^{-19} C
V	Volume of active region	6×10^{-17} m^3
A	Non-radiative carrier recombination rate coefficient	1×10^{9} s^{-1}
B	Bimolecular recombination rate coefficient	1×10^{-16} m^3 s^{-1}
C	Auger recombination rate coefficient	1×10^{-41} m^6 s^{-1}
a	Differential gain	9×10^{-13} s^{-1} m^3
Γ	Confinement factor	0.3
α_H	Linewidth enhancement factor	4
N_0	Carrier density at transparency	1×10^{24} m^{-3}
τ_P	Photon lifetime	3 ps
k_c	Coupling of external injection into the slave laser	2×10^{12} s^{-1}
β	Fraction of spontaneous emission into the lasing mode	1×10^{-4}
ε_{NL}	Nonlinear gain coefficient	1×10^{-23} m^3
Δf_D	Master –slave detuning	Hz
Δt	Simulation timestep	1 ps
B_{sim}	Simulation bandwidth [1]	1 THz
ν_0	Lasing frequency	193 THz
A	Area of lasing mode	1×10^{-13} m^2
n_g	Group index	3.5

[1] Inverse of the simulation timestep.

The stochastic terms are appropriately scaled for numerical computations (1) and (2) with $B_{sim} = t_s^{-1}$, where t_s is the step time. We solve the system of equations using Huen's predictor–corrector method.

$$F_N = \sqrt{2R(N)B_{sim}}e_N(t) \tag{3}$$

$$F_E = \sqrt{\beta B N^2 B_{sim}}\left(e_{EI}(t) + je_{EQ}(t)\right) \tag{4}$$

Each e term is an independent identically distributed random sample taken from a Gaussian random number generator with unity variance. The current is held constant; the only sweeping parameters that we consider is the power of the master laser and the master–slave detuning. We define the detuning $f_D = \nu_{master} - \nu_{slave}$. When solving the equations, we are adjusting f_D directly in (2) and the spectrum is always centred for the slave; however, centring the spectrum for the master laser makes it easier to unequivocally show the injection-locking. To centre the spectrum for the master, one needs to take the complex-conjugate of the $\exp(2\pi f_D t)$ array (calculated when constructing the injection term in (4)) and multiply by the output from solving the differential equations $E(t)$ to frequency translate $E(t)$ to be centred at the master laser.

Converting from photon density to optical power P_W is given by

$$\frac{P_W}{|E|^2} = \frac{c}{n_g} h\upsilon_0 A \tag{5}$$

where c is the speed of light in a vacuum, n_g is the group index, h is Planck's constant, and A is the area of the laser mode. One can interpret the spectral scan as essentially tuning the slave laser into the master laser and noting the spectrum. Practically, one would tune the slave laser using temperature control, taking advantage of the available ~0.1 nm/K (12 GHz/K at 1550 nm) thermal tuning of semiconductor lasers. In all of our simulations, we simulate at a 1 ps timestep, using an initial condition of $N_{IC} = 1 \times 10^{24}$ and $E_{IC} = 1 \times 10^{12}$. In order to lessen the strength and duration of the transient, we are simulating E by taking 110,000 sample points (unless otherwise stated), and the first 10,000 samples are discarded to remove any transient. The remaining 100,000 data points are used to calculate the spectrum using fast Fourier transforms (FFT). The number of data points and the sampling time allows us a resolution bandwidth of 10 MHz, and this is within the range of high-resolution optical spectral analysers. The spectrum we are calculating is the squared magnitude of the (complex valued) FFT array, and here we are circularly shifting by half of the number of samples in the FFT array such that the spectrum is centred at the zero frequency. In order to minimise randomness within the spectra, we perform the following: for each set of parameters the simulation is run twenty times and the spectrum is averaged over those twenty runs. This averages the noise to the average spectral power within each frequency bin in the FFT array.

Preliminary simulations showing the standard laser power of the slave, without external injection, versus bias current are shown in Figure 2a; the threshold current is about 15 mA. The optical spectrum of the slave is shown in Figure 2b with a linewidth of 2 MHz. The relative intensity noise is show in Figure 2c; the noise at low frequencies is -165 dB/Hz and the relaxation oscillation frequency (ROF) occurs at 9.4 GHz.

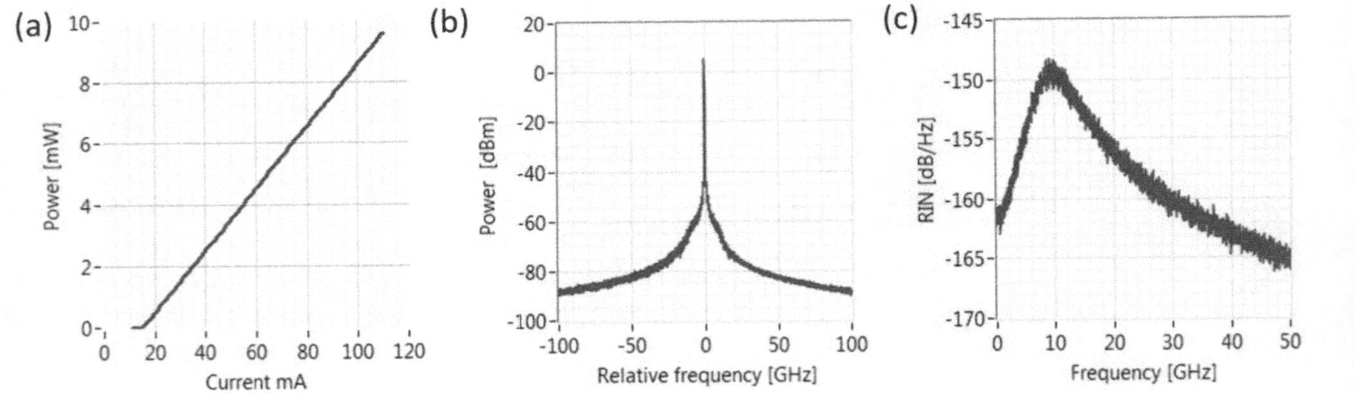

Figure 2. Plot of (**a**) laser L-I curve of the slave without external injection. (**b**) The optical spectrum of the free-running slave laser showing Lorentzian broadening. (**c**) RIN spectrum of the slave laser without external injection showing a relaxation oscillation frequency of 9.4 GHz.

3. High-Resolution Spectral Scans Due to External-Injection

We show the detailed scans of the output spectra as the slave laser is tuned across the master laser, as was performed experimentally in [18] (Figure 4). To build up the scans, we simulate the averaged spectrum for a given set of laser parameters; then, we adjust one of the parameters (here, we adjust the detuning) and then stack all the calculated spectra and display the spectra as a 3D colour map. Even though our calculated spectra run from -500 GHz to 500 GHz, we only show the spectra running from -100 GHz to 100 GHz for clarity because the most interesting features of the spectra are located within ± 100 GHz of the master laser. We plot the spectral scan with the injected power of the master taken to be constant within each scan. The spectral maps are shown in Figures 3 and 4. Different regimes are identified in Figures 3 and 4 by letters A–H; individual spectra corresponding to the of the different regimes are shown in Figure 5. Trajectories of the photon density and normalised carrier density (N–P trajectories) for the same points A–H are shown in Figure 6. Figure 7 presents the calculated RF spectrum after photodetection of the laser output for the different regimes A–H. We will explain the dynamics in more detail later; first, we qualitatively describe the regions labelled A–H in Figures 3 and 4. In Figure 3a, the injection power is 10 μW, and we can clearly see that there is a short locking range from about −3 GHz to 3 GHz, when the output of the slave laser is locked to the master laser; when locking has not been achieved, obvious four wave mixing (FWM) products appear in the spectrum. Note that all of the spectra shown in this section are qualitatively similar to the experimentally taken spectra of external injection dynamics in [11].

Case A: FWM. Region A corresponds to FWM between the master and slave, and the master and slave are clearly not locked; the beating between both lasers modulates the slave laser to create additional frequency products in the spectrum, each spaced by f_d about the slave laser. This is clearly shown in the optical spectrum in Figure 3a. The N–P trajectory in Figure 6A for FWM shows a slight modulation of the carrier density (as expected), and the RF spectrum in Figure 7A shows a few peaks each spaced by f_d.

Case B: Injection-locking. When the detuning between the two lasers is reduced, the magnitude of the varying carrier density increases and becomes sufficient to synchronise the slave laser to the master laser. The lasing frequency of the slave also reduces because the carrier density for threshold of the injected laser system is smaller than that for the solitary free-running laser system. Injection-locking has been extensively studied previously [1–13] and we shall only briefly describe the results. The N–P trajectories (Figure 6B) indicate a single point for the injection-locked case and negligible modulation of the photon and carrier densities. We can deduce the carrier density at threshold of N_{th} 1.42×10^{24} m^{-3} (injection power of 10 μW), which is lower than the carrier density at threshold for the solitary slave laser of 1.422×10^{24} m^{-3}. When N exceeds (goes below) the threshold, the slave becomes amplifying (attenuating). The optical spectrum shows a single lasing mode, and the absence of strong mixing tones in the RF spectrum (Figure 7) indicates that the system is locked.

Increasing the injection power exacerbates the modulation of the photon and carrier densities; therefore, we expect more complex and interesting behaviour. As the injection power is increased, the locking range is increased, as is the case in Figure 3b. We notice, here, that for detuning values close to the locking range, the slave is pulled more strongly towards the master. The next dynamical feature that appears is FWM-induced period-doubling.

Case C: FWM-induced period-doubling. This feature appears when the FWM goes into a period-doubling type oscillation for detuning close to the locking range before injection-locking occurs. In this example, there is detuning of about −18 GHz (Figure 4a and it is also observed in [18] (Figure 4). The cause of this behaviour is that the RO frequency of the slave is midway between the master and slave detuning; hence, the pulsation arising from the FWM exacerbates period-doubling oscillation. As the detuning is increased, the slave laser emerges into a FWM regime without the laser going into self-pulsation.

Case D: Self-pulsation. As the master power is increased further, regions of self-pulsation appear, as shown at point D in Figure 4a. This happens because the system becomes unstable as the damping can no longer suppress random fluctuations in the photon density; hence, optical pulsations grow. One can clearly see that the spectral lines of the self-pulsation coincide with the ROF. The laser system cannot support continuously growing oscillation and the oscillating pulses deplete the carrier density and the pulses thus decay, thereby keeping the self-pulsation stable. The process keeps repeating because the carrier density builds up in the absence of photons, and once the pulse builds up, this depletes the carrier density. In Figure 6D, by looking at the N–P trajectory, one can see a large swing in the carrier density in addition to a large swing in the photon density.

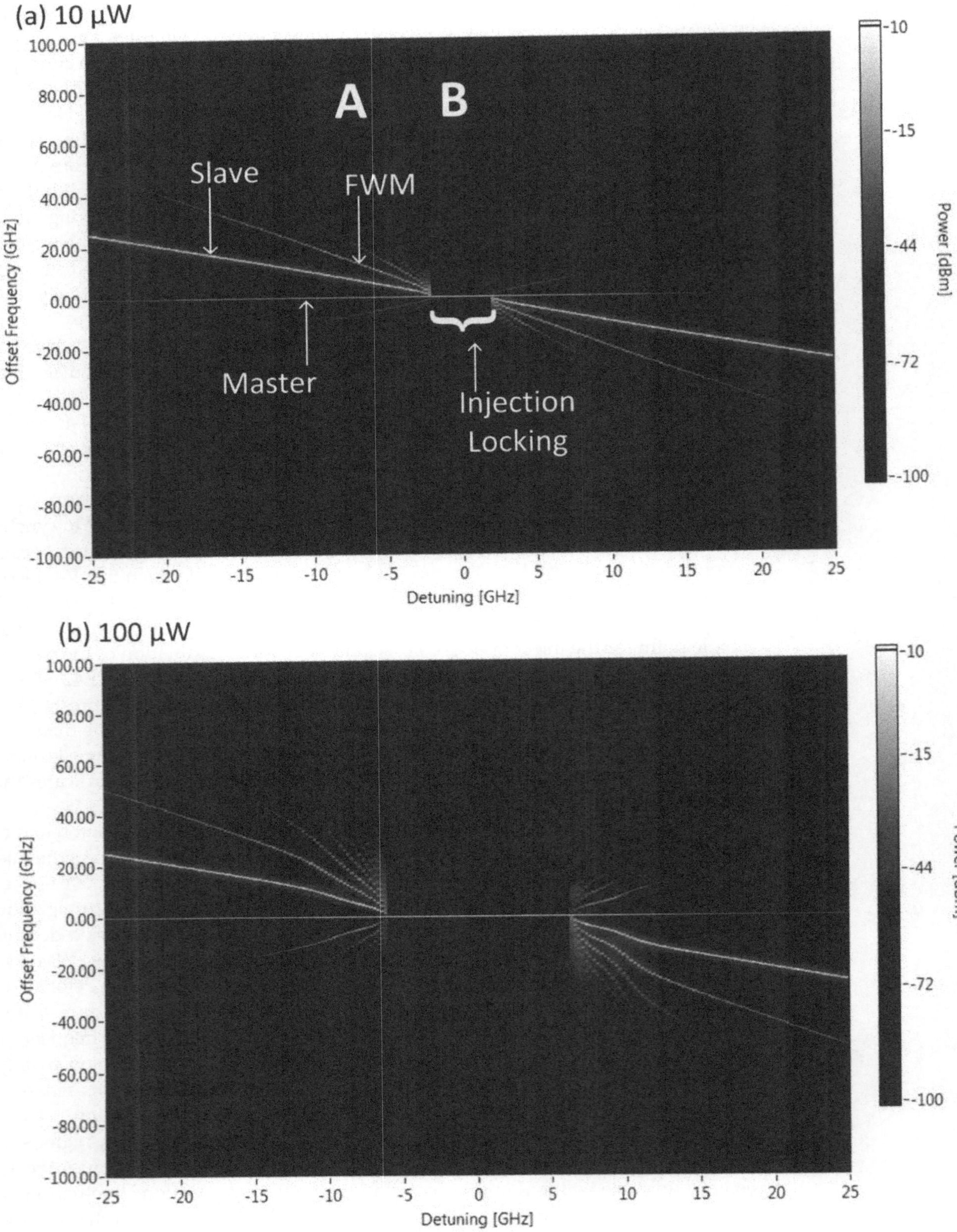

Figure 3. Simulation of the spectra as the master slave detuning is increased from −25 GHz to 25 GHz. Each subplot shows the scan for a different master power. The master power is set to (**a**) 10 μW, (**b**) 100 μW. The different regions are identified in yellow capitals: A is FWM; B is injection-locking.

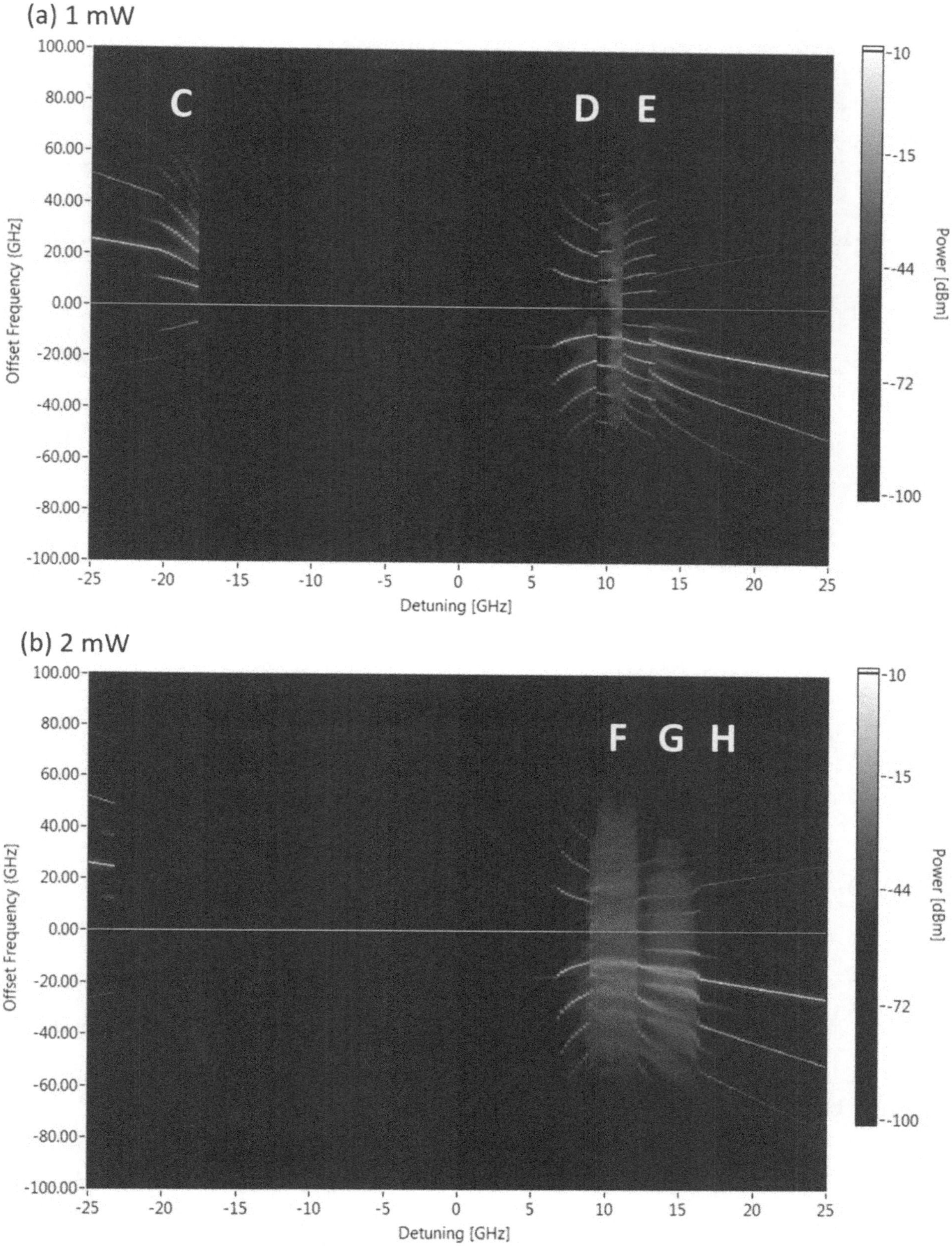

Figure 4. Simulated detuning scan of the spectral evolution of the externally injected laser as the injection power is increased to (**a**) 1 mW and (**b**) 2 mW. In these plots, the labelled regions are: (C) is a frequency doubled pulsation due to FWM as the laser nears injection-locking regime, self-pulsation (D), period-doubling (E), chaos (F), (G) is chaos with period-doubled oscillations and subject to a more detailed study in Section 4; (H) is another FWM regime.

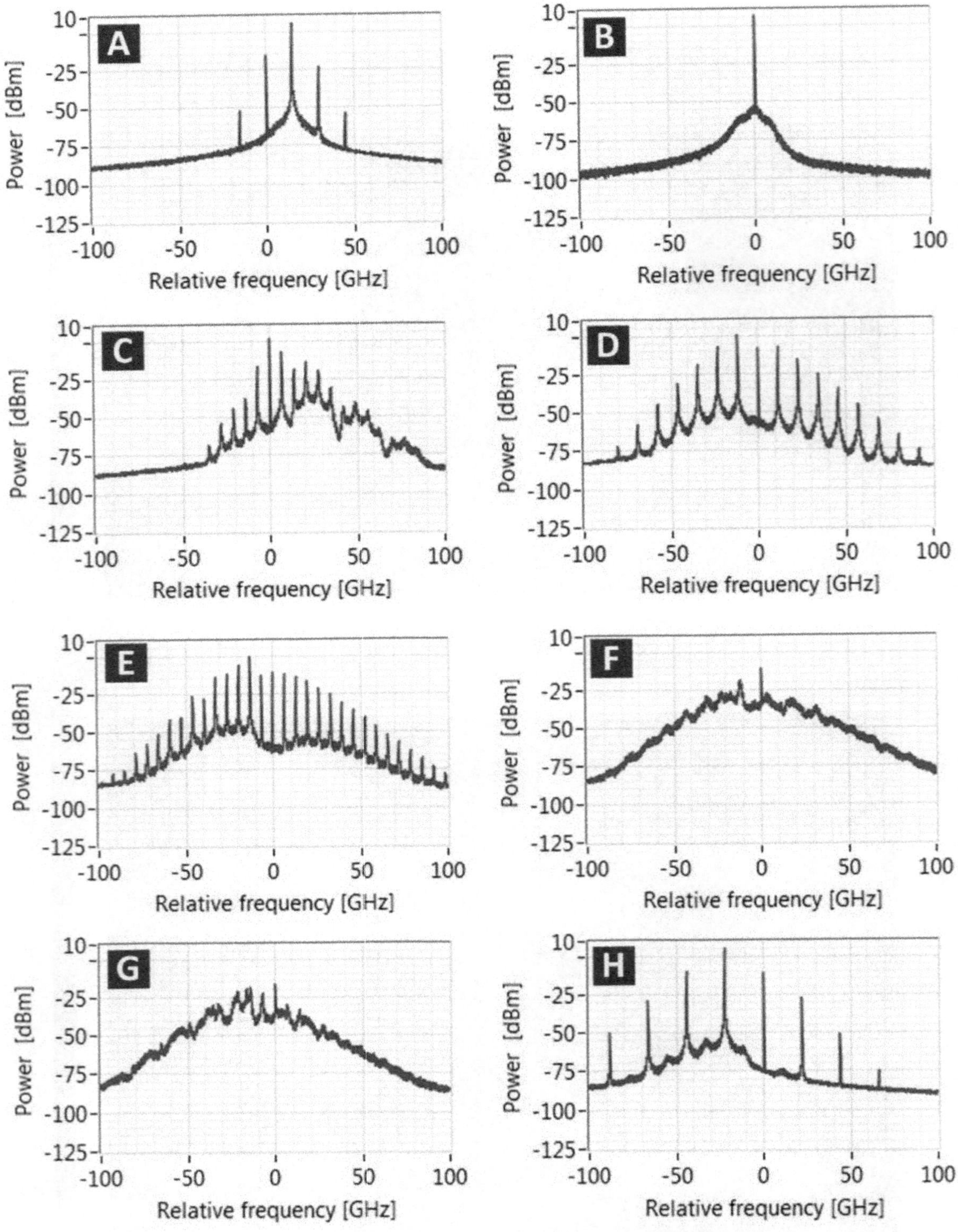

Figure 5. Averaged optical spectra corresponding to the regimes (**A–H**) in Figures 3 and 4. One could consider Figures 3 and 4 to be a stacked collection of these spectra. The pulsation regimes (**D**) are the P-1 oscillation regimes, whereas (**E**) is the P-2 oscillation regime. The chaotic regimes in (**F,G**) show wide spectrum though a clear absence of any spectral line structure that would indicate regular pulsation.

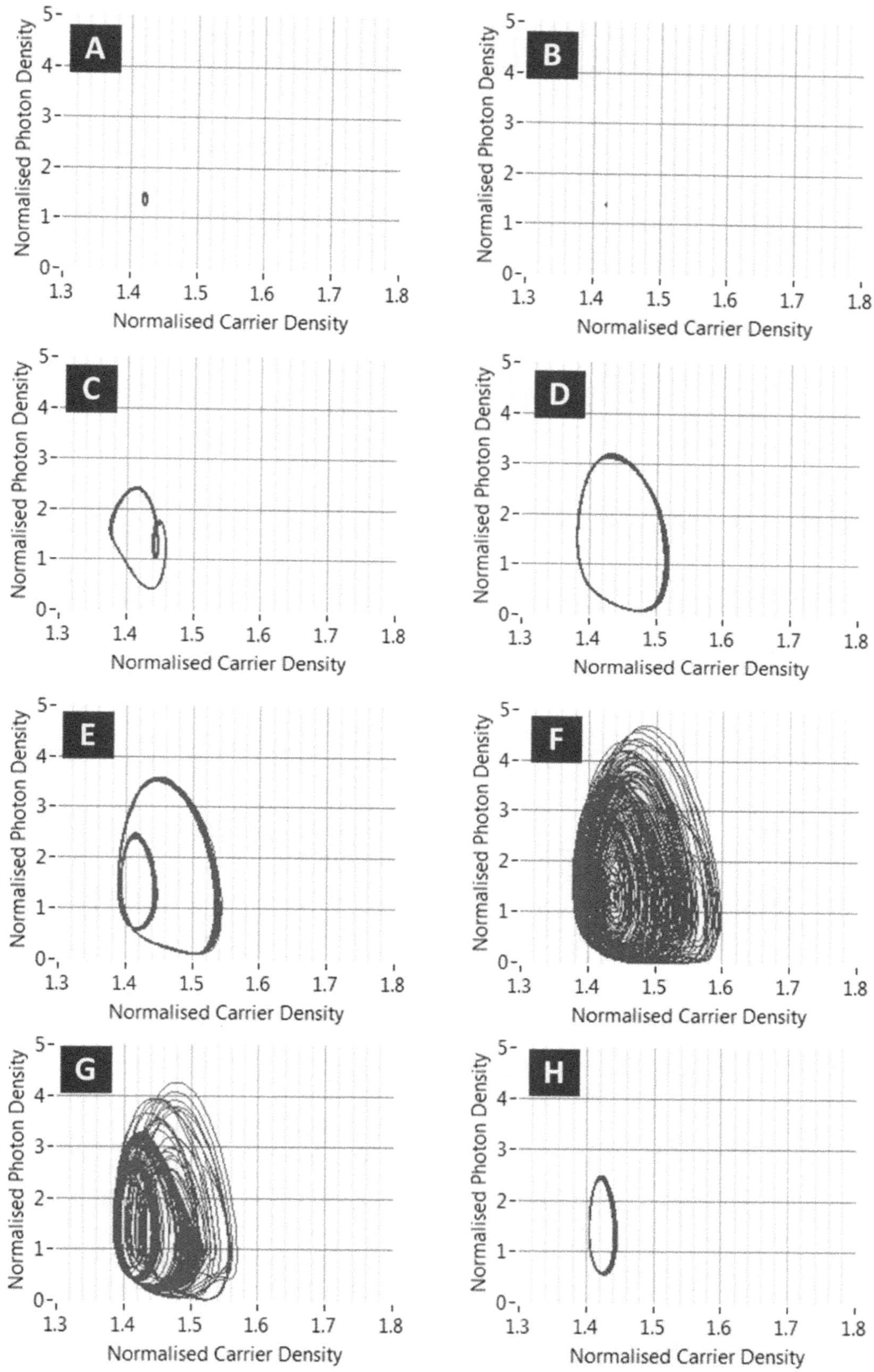

Figure 6. Phase–space portraits of the trajectory of the carrier and photon densities for the regimes outlined in Figure 1. Slight modulation for the FWM states in (**A**,**H**), a static point when injection-locked in (**B**). Period-doubling in (**C**,**E**); note the position of the extra loop in both cases for P-1 pulsations (**D**) and chaos in (**F**,**G**).

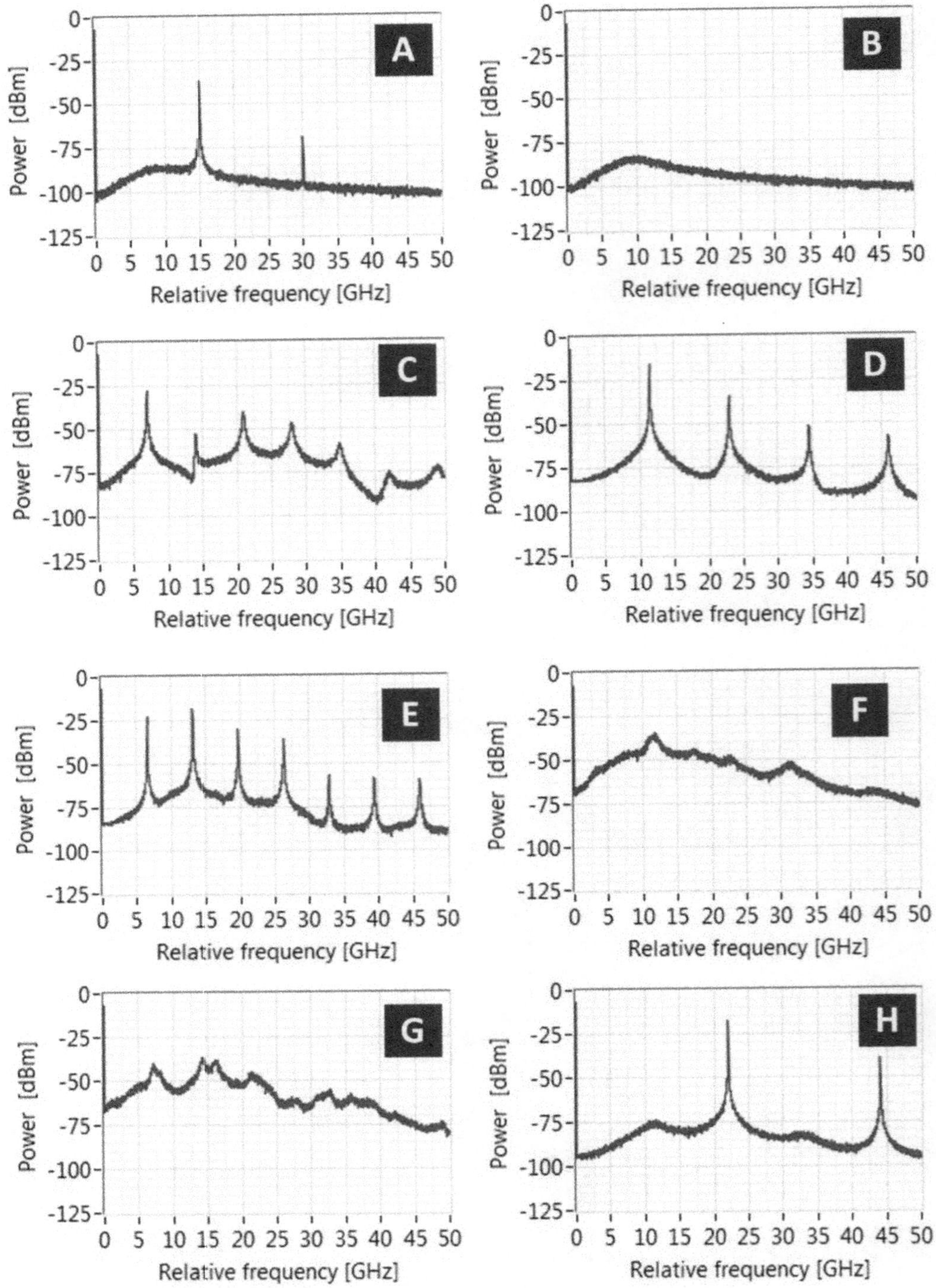

Figure 7. Calculated RF spectra of the regimes (**A–H**) identified in Figures 3 and 4. Injection-locking is achieved when the RF spectrum is minimised. The RF spectrum can yield information about the pulsation regimes, allowing one to distinguish between pulsation regimes (**C–E**) showing clear spectral lines and chaotic pulsation regimes (**F**,**G**) showing a broad continuum. The RF spectra for the FWM cases (**A**,**H**) show a few tones in the spectrum with the tone spacing equalling the master-slave detuning frequency.

Case E: Period-doubling. When the photon and carrier densities require two oscillation periods to revert to the same position, extra spectral lines, which are spaced at half the distance for the case of self-pulsation spectral lines, also appear in the spectrum in region E in Figure 4a and within the optical and RF spectra in Figures 5E and 7E, respectively. In the trajectory in Figure 6E, a small loop appears, indicating a pattern of consecutive larger and smaller pulses in the pulse train. One interesting aspect to note that is indicated by the simulations is that there is an obvious narrowing of the spectral lines for period-doubling (Figure 5E compared to the spectral lines self-pulsation case (Figure 5D). This can be explained by looking at their N–P traces in Figure 6E,D, respectively. One can clearly see that the photon density tends towards zero in both scenarios; large phase changes occur when the value of the laser field is small compared to the SE [19]. For the self-pulsation case, the photon density goes to zero once per pulsation cycle; however, for the period-doubling case, the photon density goes to zero once at half the rate for self-pulsation. This explains why the spectral line broadening is smaller for period-doubling compared to self-pulsation

Case F: Chaos. Chaos has been studied in detail before [6,7], so we will just briefly describe the effect here. The main characteristic of chaos is the broad continuous spectrum; this is clearly seen in Figures 4b and 5F. As the injection power is increased to 2 mW, then it is possible for the pulsations to become completely chaotic, as there is no way to regularise the photon density in the cavity at the instant the carrier density reaches above the threshold. Looking at the detuning range at which chaotic behaviour is observed in Figure 4b (~10 GHz detuning), and noting the corresponding region in Figure 4a when less injection power is used: in that detuning region (~10 GHz) in Figure 4a, notice there is an abrupt change in the spectral line spacing when increasing the detuning in region D in Figure 4a; it is within this region of abrupt pulsation frequency change that chaos occurs when the injection power is increased. It is interesting to note from the N–P trajectory of chaotic regime in Figure 6F that there is no regular pulsation, as is the case for the trajectory for self-pulsation; however, in the coherence case, photon density can go towards zero, and the time taken to build up a lasing field again depends on the value of the photon density in the slave laser as the laser goes above the threshold. One curious aspect of chaos is that the optical spectrum is the same irrespective of whether or not SE is included in the simulation. We have described in previous work how regular pulse trains can have continuous optical spectra without any 'comb'-like structure when there is no memory of the optical phase from pulse to pulse [20]. We rule out any such phenomenon being the cause of the continuous spectrum for chaos by looking at the RF spectrum in Figure 7F; there are clearly no distinct lines that would indicate pulsation with jitter [21].

Case G: Noise-Induced Chaos Regime. This is the regime that we are identifying for the first time in this paper. It is clear in Figure 5G that the laser system is going into a different pulsation regime to that of chaos in Case F (Figure 5F). The extra peaks in the spectrum, similar to those of period-doubling, are visible, though the laser is still in a chaotic state. As we show in the following section, without SE, this region is comprised of a higher-order pulsation regime. Clearly, the N–P trajectory is chaotic (Figure 6G); moreover, the chaotic trajectory looks to have the same form for period-doubling in Figure 6E, though there is no regular trajectory path and it is clearly taking a chaotic trajectory.

Case H: Four-wave mixing. As the master–slave detuning is increased further, the lasers no longer pulsate because of the limited carrier density dynamics, and only FWM-products are created, which is similar to case A.

Now that we have described all of the operating regimes of the externally injected laser system, we concentrate on case G, the period-doubling regime, in more detail.

4. Influence of Spontaneous Emission

In this section, we justify the claim made in the previous section that the SE is responsible for keeping the injected laser system, for high injection powers, in a chaotic state. To enact this, we repeat the scans of Figure 4a,b with the value of the spontaneous coefficient β set to zero. The results are plotted in Figure 8 for injection powers of (a) 1 mW and (b) 2 mW. For the case of 1 mW injection without SE, a very thin region of chaos is clearly identifiable by a spectral continuum at a detuning of about 10.5 GHz. Note this was present in Figure 4a, though the detuning range over which chaos occurs here is narrow; thus, we concentrated on chaos at the higher power of 2 mW for clarity. In Figure 8a, clear evidence of P-2, P-3 and P-4 oscillations appears either side of this narrow chaotic state (in the vicinity of detuning of 11 GHz), which were not present when SE was included in Figure 4a; this finding is consistent with [14], where a lack of a period-doubling route to chaos was found. Within this detuning window in Figure 8a, periodic-pulsation states are now observable due to the removal of SE in the model. This noise-induced chaos behaviour was formally studied in [15]. The impact of SE is starker for the second chaotic region in Figure 4b. There are now clearly observable P-2, P-3 and P-4 oscillations within the detuning window from 13 GHz to 17 GHz; these are washed away by the SE in Figure 4a, where a chaotic state with the hallmarks of a P-2 state exists over the detuning range from 13 to 17 GHz (approx.).

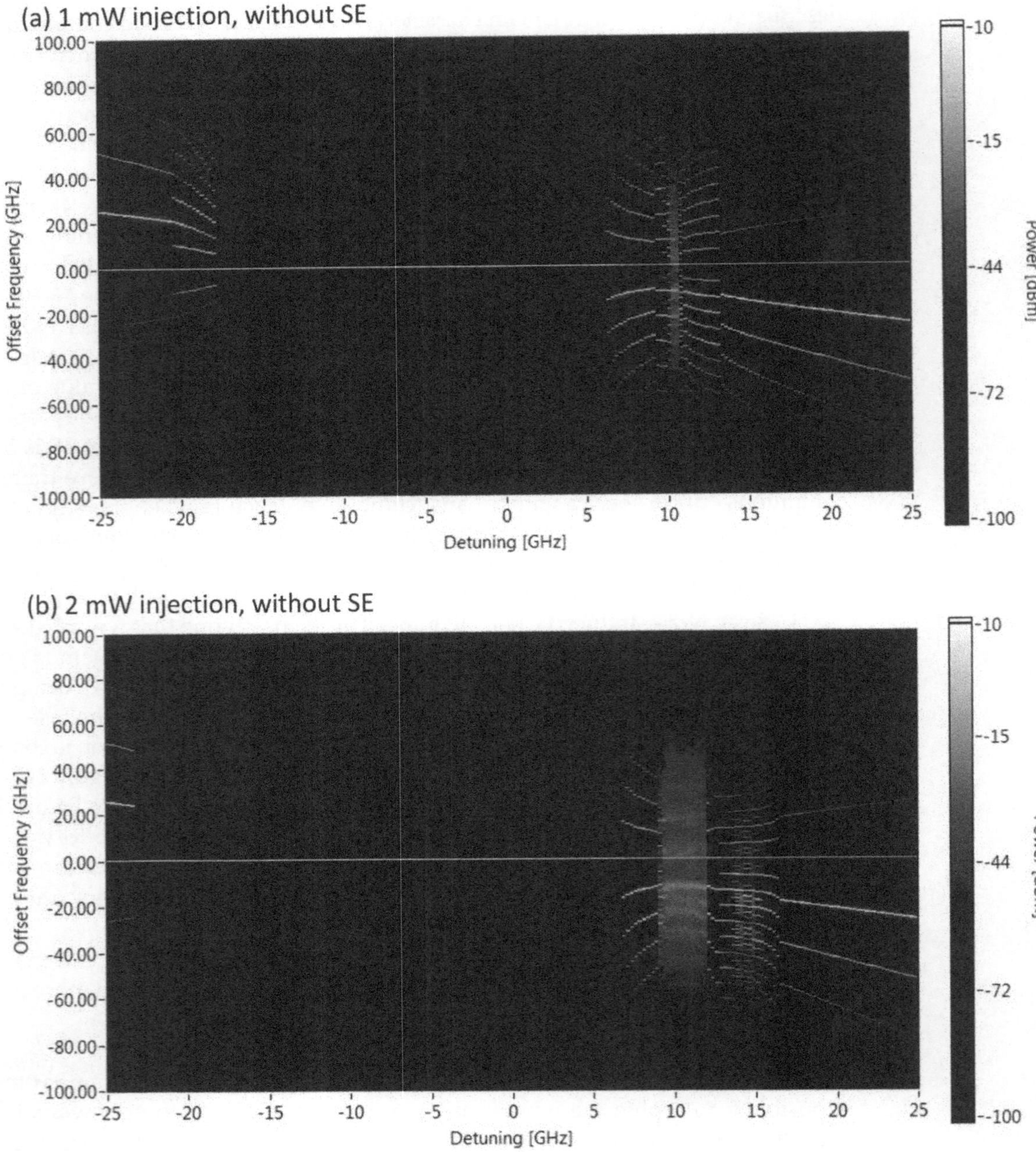

Figure 8. Scan of the injection-locking for (**a**) 100 uW and (**b**) 200 uW. In each case, the value of the spontaneous emission coefficient is set to zero. These plots show more dynamical pulsation states than in Figure 4, especially for the case of 2 mW in the region from 6 to 20 GHz detuning.

The detail in these scans is only as good as the detuning granularity when constructing the spectral scans with detuning. A more in-depth examination at the spectral plots is repeated by exploring the detuning over smaller detuning ranges, as well as zooming into the central portion of the spectrum. Spectral resolution is also enhanced by increasing the number of samples taken when solving the system equations; the number of samples taken was 1,010,000 (the first 10,000 samples are discarded). The spectral resolution is now 1 MHz. These were conducted over the detuning range from 6 to 18 GHz with the injection power equal to 2 mW; the spectral scan results are shown in Figure 9. This figure clarifies the stark differences when SE is included or omitted in the simulations, and higher-order P5 (and above) oscillations are revealed when SE is omitted.

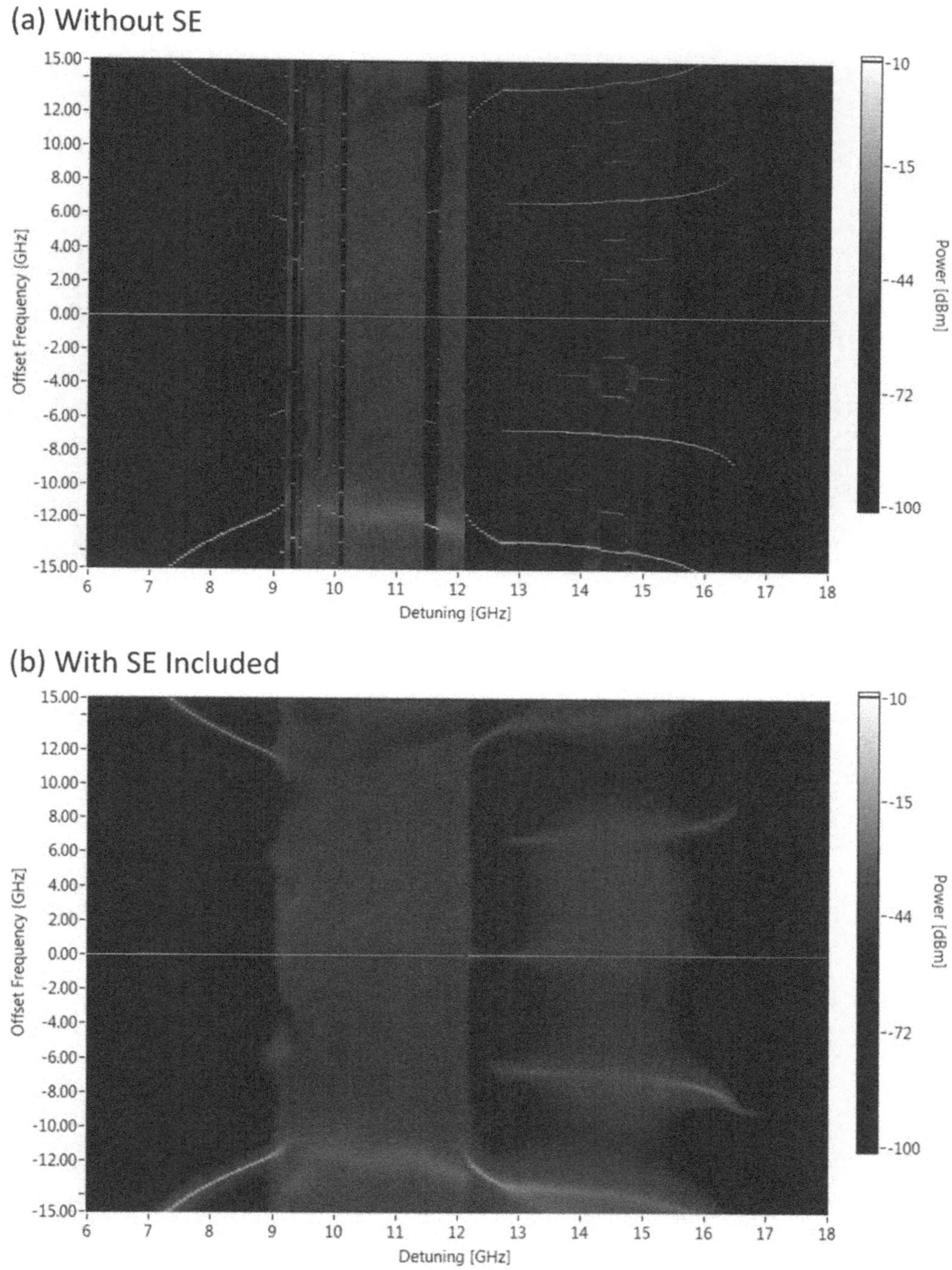

Figure 9. Detailed scan of the region around the chaotic regime showing the changes from regular pulsations to chaotic pulsations for the case (**a**) without SE and (**b**) including SE. The injection power is 2 mW. Note the absence of period-N oscillations when SE is included, leaving just the amalgamated chaotic version of the P-2 state for detunings around 14 GHz to 15 GHz.

For completeness, we investigate the level of SE coupling required to observe oscillations of a higher order than P-2 oscillations. We take the injection power level to be 2 mW and the master–slave detuning to be 14.55 GHz. In Figure 9, we find that there is a P-6 oscillation without SE and a noise-induced chaotic state with SE included. We find that the SE coupling needs to be reduced below the value of 1×10^{-7} to see the P-6 oscillation; the scan of the spectra with the value of SE coupling increased from 0 to 1×10^{-7}, as shown in Figure 10.

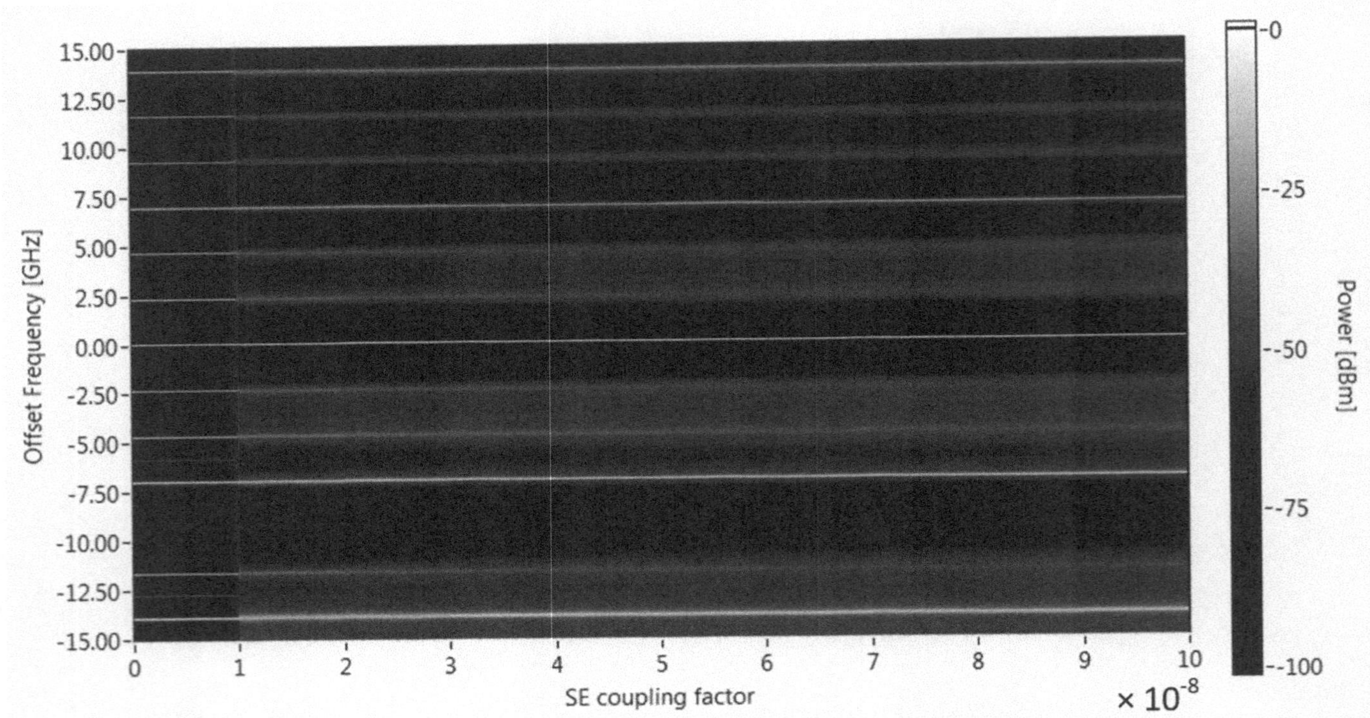

Figure 10. Spectral scan of the influence when very low SE coupling for the external injection laser system when the detuning is 14.55 GHz and the injection power is 2 mW. At the right-hand side of the plot, we note a continuum in the spectrum with two prominent horizontal lines corresponding to the noise-induced chaos within a P2 pulsation. As the SE coupling is reduced towards zero (far left), we see extra horizontal lines emerge, which signify emergence of a higher-order pulsation period. When SE coupling equals zero, there are three extra spectral lines between the P2 pulsation lines; therefore, the system is in a P6 regime. The values of SE coupling are at least 3 orders of magnitude below typical value of SE coupling.

5. Conclusions

We have shown, through numerical solutions, the spectral evolution of an externally injected laser system and the role that SE plays in washing out higher-order pulsation regimes in the dynamics of externally injected lasers. We identify an operating regime comprised entirely of noise-induced chaos destroying higher-order pulsations, especially under strong injection powers. The spectral scans by themselves would be helpful to those operating externally injected lasers to understand the regimes of operation as they are tuning the lasers to achieve injection-locking. The level of employed SE in the slave laser would be considered to be at the lower end, as we chose a value of SE coupling coefficient to be an order of magnitude lower than a typical value for this parameter. We show that a value of SE coupling needs to be as low as 1×10^{-7}, many orders of magnitude smaller than typical values in order to observe higher-order pulsations. There is sufficient experimental evidence to show existence of P3 oscillations in [6,22] and P4 oscillations in [5,14], though much work is needed to understand the laser parameters and operating conditions of injection lasers to achieve higher-order pulsations than the P2 pulsations in the presence of SE.

References

1. Lang, R. Injection locking properties of a semiconductor laser. *IEEE J. Quantum. Electron.* **1983**, *18*, 976–983. [CrossRef]
2. Annovazzi-Lodi, V.; Scire, A.; Sorel, M.; Donati, S. Dynamic behavior and locking of a semiconductor laser subjected to external injection. *IEEE J. Quantum Electron.* **1998**, *32*, 2350–2357. [CrossRef]
3. Schunk, N.; Petermann, K. Noise analysis of injection-locked semiconductor injection lasers. *IEEE J. Quantum Electron.* **1986**, *22*, 642–650. [CrossRef]
4. Liu, Z.; Slavík, R. Optical injection locking: From principle to applications. *IEEE J. Lightw. Technol.* **2019**, *38*, 43–59. [CrossRef]
5. Blin, S.; Guignard, C.; Besnard, P.; Gabet, R.; Stéphan, G.M.; Bondiou, M. Phase and spectral properties of optically injected semiconductor lasers. *C. R. Phys.* **2003**, *4*, 687–699. [CrossRef]
6. Wieczorek, S.; Krauskopf, B.; Simpson, T.B.; Lenstra, D. The dynamical complexity of optically injected semiconductor lasers. *Phys. Rep.* **2005**, *416*, 1–128. [CrossRef]
7. Lenstra, D.; Verbeek, D.H.; den Boef, A.J. Coherence Collapse in Single-Mode Semiconductor Lasers Due to Optical Feedback Coherence collapse in single-mode semiconductor lasers due to optical feedback. *IEEE J. Quantum Electron.* **1985**, *21*, 674–679. [CrossRef]
8. Daly, A.; Roycroft, B.; Corbett, B. Stable locking phase limits of optically injected semiconductor lasers. *OSA Opt. Expr.* **2013**, *21*, 30126–30139. [CrossRef]
9. Simpson, T.B.; Liu, J.M.; Gavrielides, A. Bandwidth enhancement and broadband noise reduction in injection-locked semiconductor lasers. *IEEE Photon. Technol. Lett.* **1995**, *7*, 709–711. [CrossRef]
10. Lau, E.K.; Zhao, X.; Sung, H.K.; Parekh, D.; Hasnain, C.C.; Wu, M.C. Strong optical injection-locked semiconductor lasers demonstrating 100-GHz resonance frequencies and 80-GHz intrinsic bandwidths. *OSA Opt. Exp.* **2008**, *16*, 6609–6618. [CrossRef]
11. Herrera, D.J.; Kovanis, V.; Lester, L.F. Using transitional points in the optical injection locking behavior of a semiconductor laser to extract its dimensionless operating parameters. *IEEE J. Sel. Top. Quantum Electron.* **2022**, *28*, 1800109. [CrossRef]
12. Gunning, P.; Lucek, J.K.; Moodie, D.G.; Smith, K.; Davey, R.P.; Chernikov, S.V.; Guy, M.J.; Taylor, J.R.; Siddiqui, A.S. Gainswitched DFB laser diode pulse source using continuous wave light injection for jitter suppression and an electroabsorption modulator for pedestal suppression. *Electron. Lett.* **1996**, *32*, 1010–1011. [CrossRef]
13. Anandarajah, P.M.; Maher, R.; Xu, Y.Q.; Latkowski, S.; O'Carroll, J.; Murdoch, S.G.; Phelan, R.; O'Gorman, J.; Barry, L.P. Generation of Coherent Multicarrier Signals by Gain Switching of Discrete Mode Lasers. *IEEE Photon. J.* **2011**, *3*, 112–122. [CrossRef]
14. Kovanis, V.; Gavrielides, A.; Simpson, T.B.; Liu, J.M. Instabilities and chaos in optically injected semiconductor lasers. *Appl. Phys. Lett.* **1995**, *67*, 2780–2782. [CrossRef]
15. Hwang, S.K.; Gao, J.B.; Liu, M.; Liu, J. Noise-induced chaos in an optically injected semiconductor laser mode. *Phys. Rev. E* **2000**, *61*, 5162. [CrossRef]
16. Kajiyama, K.; Hata, S.; Sakata, S. Effects on spontaneous emission on dynamic characteristics of semiconductor lasers. *Appl. Phys.* **1977**, *12*, 209–210. [CrossRef]
17. Dúill, S.P.Ó.; Anandarajah, P.M.; Zhou, R.; Barry, L.P. Numerical investigation into the injection-locking phenomena of gain switched lasers for optical frequency comb generation. *Appl. Phys. Lett.* **2015**, *106*, 211105. [CrossRef]
18. Zou, L.-X.; Huang, Y.-Z.; Liu, B.-W.; Lv, X.-M.; Long, H.; Yang, Y.-D.; Xiao, J.-L.; Du, Y. Nonlinear dynamics for semiconductor microdisk laser subject to optical injection. *IEEE J. Sel. Top. Quantum Electron.* **2015**, *21*, 1800408. [CrossRef]
19. Schawlow, A.L.; Townes, C.H. Infrared and optical masers. *Phys. Rev.* **1958**, *112*, 1940. [CrossRef]
20. Dúill, S.P.Ó.; Zhou, R.; Anandarajah, P.M.; Barry, L.P. Analytical approach to assess the impact of pulse-to-pulse phase coherence of optical frequency combs. *IEEE J. Quantum Electron.* **2015**, *51*, 1200208. [CrossRef]

21. Leep, D.A.; Holm, D.A. Spectral measurement of timing jitter in gain-switched semiconductor lasers. *Appl. Phys. Lett.* **1992**, *60*, 2451–2453. [CrossRef]
22. Gavrielides, A.; Kovanis, V.; Nizette, M.; Erneux, T.; Simpson, T.B. Period three limit-cycles in injected semiconductor lasers. *J. Opt. B Quantum Semiclass.* **2002**, *4*, 20. [CrossRef]

Statistics of the Optical Phase of a Gain-Switched Semiconductor Laser for Fast Quantum Randomness Generation

Abstract: The statistics of the optical phase of the light emitted by a semiconductor laser diode when subject to periodic modulation of the applied bias current are theoretically analyzed. Numerical simulations of the stochastic rate equations describing the previous system are performed to describe the temporal dependence of the phase statistics. These simulations are performed by considering two cases corresponding to random and deterministic initial conditions. In contrast to the Gaussian character of the phase that has been assumed in previous works, we show that the phase is not distributed as a Gaussian during the initial stages of evolution. We characterize the time it takes the phase to become Gaussian by calculating the dynamical evolution of the kurtosis coefficient of the phase. We show that, under the typical gain-switching with square-wave modulation used for quantum random number generation, quantity is in the ns time scale; that corresponds to the time it takes the system to lose the memory of the distribution of the initial conditions. We compare the standard deviation of the phase obtained with random and deterministic initial conditions to show that their differences become more important as the modulation speed is increased.

Keywords: semiconductor laser; optical phase; gain-switching; spontaneous emission noise; quantum random number generation

---

1. Introduction

Experimental and theoretical understanding of the fluctuations of laser light began shortly after the invention of the laser [1–5]. Special attention has been devoted to fluctuations of the light emitted by semiconductor lasers [6–10] due to their vast variety of applications. The best available theoretical description of these fluctuations is based on the Fokker–Planck equation, or alternativelly on Langevin's stochastic rate equations [3,6–8,11]. The phase of the laser electrical field is a random quantity, mainly due to the effect of the spontaneous emission noise. The random character of this phase is precisely the basis of some of the available methods for quantum random number generation (QRNG).

Random numbers are a vital resource for numerous applications including cryptography, statistical analysis, stochastic simulations, decission making in engineering processes, quantitative finance, gambling, massive data processing, etc. [12,13]. Random number generators (RNG) use software algorithms (pseudorandom number generators) or hardware physical devices. Typical physical processes used to generate random numbers are radioactive decay, Johnson or Zener's noise, chaos noise [13,14] and quantum phenomena [12,13]. QRNGs are a particular case of physical RNGs that can generate truly random numbers from the fundamentally probabilistic nature of quantum events [13]. The advantage of using QRNGs relies on its unpredictibility, which can be proven to be based on quantum physics laws. Typical QRNGs are based on quantum optics [13]. These generators can be divided in (i) generators that use single-photon sources, and (ii) generators that use multi-photon sources, typically semiconductor lasers or LEDs. QRNGs based on single-photon detection methods include: Branching path generators [15], generators measuring the time of arrival of photons [16], photon counting generators [17], and attenuated pulse generators [18]. These methods have been experimentally compared in [19]. Multi-photon QRNGs include: Generators based on quantum vacuum fluctuations [20], on amplified spontaneous emission (ASE) signals [21,22], and on phase noise in continuous wave [23–25] and in pulsed laser diodes [26–31].

Spontaneous emission is a useful mechanism to generate quantum fluctuations, as it can be ascribed to the vacuum fluctuations of the optical field [26]. Randomness due to spontaneous emission is the basis of QRNGs based on pulsed single-mode laser diodes [26–28,30,31]. These generators have several advantages. They are made of commercially available components: For instance, standard photodetectors can be used due to the high signal levels. They are simple, low-cost, robust, and fast: Generation rates up to 43 Gbps quantum random bit generation have been experimentally demonstrated [27]. In these QRNGs the laser diode is periodically modulated from below to above threshold in such a way that gain-switching operation is obtained, typically at Gbps rates. While the laser is below threshold the optical phase becomes random due to the spontaneous emission noise. Gain-switching operation produces a periodic train of laser pulses with random phases. Phase fluctuations are then converted into amplitude fluctuations by using interferometric setups [27,31]. Detection and filtering of the amplitude fluctuations provides the generation of random values with an almost uniform distribution.

The applications of QRNGs, for instance in cryptography [31,32], require that the physical processes underlying their operation must be properly understood and described. For QRNGs based on pulsed semiconductor lasers, it is essential an accurate description of the phase diffusion process, that is, laser phase fluctuations must be qualitatively and quantitatively characterized. Modelling of these fluctuations has been performed by numerical integration of the laser stochastic rate equations [27,30,31,33–36]. Good quantitative agreement between experiments and theory is achieved when the complete set of parameters of the rate equations is known for the specific laser diode. Good agreement between experimental and theoretical phase fluctuations has been recently reported for a discrete mode laser (DML) [36] for which a complete extraction of the intrinsic parameters was performed [35,37]. This permits a quantitative description of the dependence of phase diffusion on the laser and modulation parameters. On the qualitative side, statistics of optical phase has been described as Gaussian in numerical simulations [27,33,34] since spontaneous emission noise has also Gaussian distribution. However, in these generators the bias current is periodically modulated in such a way that the evolution is mainly in a transient regime, specially when operating at fast bit rates. It is then expected that the choice of initial conditions in the simulations must have impact on the statistics of the optical phase and on the time it takes the phase to be distributed as a Gaussian. This is in fact the main objective of this work: The investigation of the conditions for which the phase becomes distributed as a Gaussian.

In this paper we report a theoretical study of the phase diffusion in gain-switched semiconductor lasers. This is done by performing numerical simulations of the stochastic rate equations for the complex electrical field and carrier number. In our modelling we use the set of parameters recently extracted for a DML device. With these parameters we first analyze the impact of the carrier noise on the phase statistics. In the rest of the paper we focus on the calculation of the temporal dependence of the statistical moments and distribution of the phase. We first consider random initial conditions that contrast to previous analysis in which deterministic fixed initial conditions were chosen [34]. We compare the phase statistics obtained for both types of initial conditions. For both cases we show that the phase is not distributed as a Gaussian because of the non-Gaussianity of the initial conditions. This contrasts to the Gaussian character of the phase that has been assumed in previous works [27,33,34]. We characterize the time it takes the phase becomes approximately Gaussian by calculating the temporal evolution of the kurtosis coefficient of the phase. Our calculations indicate that under the typical gain-switching with square-wave modulation used for QRNG, the time it takes to the phase to become Gaussian is in the ns scale. These are the typical times for which the memory of the distribution of the initial conditions is lost. The comparison between the variance of the phase obtained with random and fixed initial conditions show that their differences become more important as the modulation speed is increased.

Our paper is organized as follows. In Section 2, we present our theoretical model. Section 3 is devoted to analyze the dynamical evolution of the relevant variables. In Section 4, we study the temporal evolution of the phase statistics. Finally, in Section 5 we discuss and summarize our results.

2. Theoretical Model

Gain-switched single-mode semiconductor laser dynamics can be modelled by using a set of stochastic rate-equations that read (in Ito's sense) [6,37,38]

$$\frac{dP}{dt} = \left[\frac{G_N(N-N_t)}{1+\epsilon P} - \frac{1}{\tau_p}\right]P + R_{sp}(N) + \sqrt{2R_{sp}(N)P}F_p(t) \quad (1)$$

$$\frac{d\phi}{dt} = \frac{\alpha}{2}\left[G_N(N-N_t) - \frac{1}{\tau_p}\right] + \sqrt{\frac{R_{sp}(N)}{2P}}F_\phi(t) \quad (2)$$

$$\frac{dN}{dt} = \frac{I(t)}{e} - (AN + BN^2 + CN^3) - \frac{G_N(N-N_t)P}{1+\epsilon P} \quad (3)$$

where $P(t)$ is the number of photons inside the laser, $\phi(t)$ is the optical phase, and $N(t)$ is the number of carriers in the active region. The parameters appearing in these equations are the following: G_N is the differential gain, N_t is the number of carriers at transparency, ϵ is the non-linear gain coefficient, τ_p is the photon lifetime, α is the linewidth enhancement factor, e is the electron charge, and A, B and C are the non-radiative, spontaneous, and Auger recombination coefficients, respectively. In these equations we consider a temporal dependence of the injected current, $I(t)$, and a rate of the spontaneous emission given by $R_{sp}(N) = \beta BN^2$ where β is the fraction of spontaneous emission coupled into the lasing mode. The Langevin terms $F_P(t)$ and $F_\phi(t)$ in Equations (1) and (2), represent fluctuations due to spontaneous emission, with the following correlation properties, $< F_i(t)F_j(t') >= \delta_{ij}\delta(t-t')$, where $\delta(t)$ is the Dirac delta function and δ_{ij} the Kronecker delta function with the subindexes i and j referring to the variables P and ϕ.

QRNG systems based on gain-switching of single-mode laser diodes are such that a large signal modulation of the bias current is applied to the device. We consider an injected current following a square-wave modulation of period T with $I(t) = I_{on}$ during $T/2$, and $I(t) = I_{off}$ during the rest of the period. This modulation is such that $I_{off} < I_{th}$, where I_{th} is the threshold current of the laser, for obtaining a random evolution of the optical phase induced by the spontaneous emission noise. Numerical integration of the previous stochastic rate equations by usual Euler–Maruyama [3,39] or Heun's predictor-corrector algorithms [37] present instabilities when the photon number is very small, a situation always present in this type of QRNGs: some spontaneous noise events cause negative values of P that lead to numerical instabilities due to the square root factor multipliying the noise term in Equations (1) and (2). Very recently a set of rate equations for the complex electrical field, $E(t)$, instead of equations for P and ϕ has been proposed to solve this problem [35]. These equations are the following:

$$\frac{dE}{dt} = \left[\left(\frac{1}{1+\epsilon \mid E\mid^2} + i\alpha\right)G_N(N-N_t) - \frac{1+i\alpha}{\tau_p}\right]\frac{E}{2} + \sqrt{\beta B}N\xi(t) \quad (4)$$

$$\frac{dN}{dt} = \frac{I(t)}{e} - (AN + BN^2 + CN^3) - \frac{G_N(N-N_t)\mid E\mid^2}{1+\epsilon\mid E\mid^2} \quad (5)$$

where $E(t) = E_1(t) + iE_2(t)$ is the complex electrical field and $\xi(t) = \xi_1(t) + i\xi_2(t)$ is the complex Gaussian white noise with zero average and correlation given by $< \xi(t)\xi^*(t') >= \delta(t-t')$ that represents the spontaneous emission noise, and where we have considered that $R_{sp}(N) = \beta BN^2$. These equations exactly correspond to our initial model because the application of the rules for the change of variables in the Ito's calculus [11] to $P = | E |^2 = E_1^2 + E_2^2$ and $\phi = \arctan(E_2/E_1)$ in Equations (4) and (5) gives Equations (1)–(3). Instabilities do not appear because P is not inside the square root factor that multiplies the noise term.

Up to now we have considered an equation for N that has not any noise term. Carrier noise can also be important for describing statistics in semiconductor laser dynamics [6]. These fluctuations can be taken into account if we substitute Equation (5) by

$$\frac{dN}{dt} = \frac{I(t)}{e} - (AN + BN^2 + CN^3) - \frac{G_N(N - N_t) \mid E \mid^2}{1 + \epsilon \mid E \mid^2} + \sqrt{2\left(AN + BN^2 + CN^3 + \frac{I(t)}{e}\right)}\xi_N - 2\sqrt{\beta B}N(E_1\xi_1 + E_2\xi_2) \quad (6)$$

where ξ_N is a real Gaussian white noise of zero average and correlation given by $< \xi_N(t)\xi_N(t') >= \delta(t - t')$ and statistically independent of $\xi(t)$ [6,10,37,40].

In this work we will numerically solve Equations (4) and (6) by using the Euler–Maruyama algorithm [3,39] with an integration time step of 0.001 ps. We will use the numerical values of the parameters that have been extracted for a discrete mode laser (DML) [35,37]. This device is a single longitudinal mode semiconductor laser emitting close to 1550 nm wavelength and $I_{th} = 14.14$ mA at a temperature of 25 °C. The values of the parameters are $G_N = 1.48 \times 10^4$ s^{-1}, $N_t = 1.93 \times 10^7$, $\epsilon = 7.73 \times 10^{-8}$, $\tau_p = 2.17$ ps, $\alpha = 3$, $\beta = 5.3 \times 10^{-6}$, $A = 2.8 \times 10^8$ s^{-1}, $B = 9.8$ s^{-1}, and $C = 3.84 \times 10^{-7}$ s^{-1} [35,37]. Simulation and experimental results have shown not only qualitative but also a remarkable quantitative agreement for a very wide range of gain-switching conditions [35,37,41].

3. Analysis of the Dynamics

We first analyze the dynamical evolution of relevant variables when the laser is modulated with $I_{on} = 30$ mA, $I_{off} = 7$ mA, and $T = 1$ ns. The laser is switched-off with a current close to $I_{th}/2$, for obtaining a significant effect of the spontaneous emission noise on the randomness of the phase. Figure 1a–c show the photon number, carrier number, and optical phase, respectively, as a function of time. We integrate the equations for consecutive bias current pulses in such a way that the initial conditions for one period correspond to the final values of the variables at the end of the previous period. Figure 1a shows P for three consecutive pulses. The laser is switched-on with I_{on} at $t = 1$ ns. After this time P begins to build-up from very small random values determined by the spontaneous emission noise events. After the emission of the pulse with the corresponding relaxation oscillations, P begins to decrease at $t = 1.5$ ns (when I_{off} is applied), reaching the small random values at which spontaneous emission noise dominates the device dynamics. N begins at $t = 1$ ns from a value well below the threshold carrier number, $N_{th} = N_t + 1/(G_N\tau_p) = 5.045 \times 10^7$, as it can be seen in Figure 1b. The characteristics relaxation oscillations of N associated to the pulse emission are followed by a monotonous decrease from $t = 1.5$ ns to 2 ns due to the value below threshold of I_{off}.

The optical phase is calculated at each integration step from E_1 and E_2 in such a way that it is a continuous function of t. The dynamical evolution of ϕ is shown in Figure 1c. When P is large (small) the noise term in Equation (2) is much smaller (larger) than the other term in that equation and ϕ mainly evolves in a deterministic (random) way. The deterministic decrease of ϕ is due to the value below threshold of the current when switching-off the laser: $G_N(N - N_t) - \frac{1}{\tau_p} < 0$ because $N < N_{th}$, and therefore ϕ decreases (see Equation (2)).

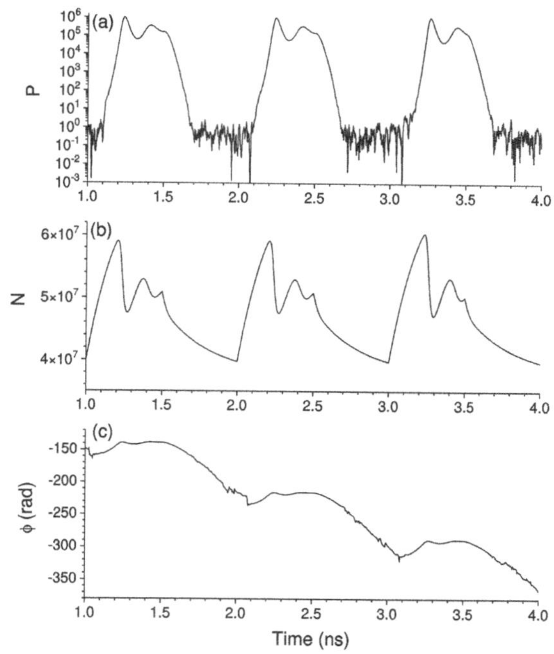

Figure 1. (**a**) Photon number, (**b**) carrier number, and (**c**) optical phase as a function of time for three consecutive pulses when $T = 1$ ns.

Visualization of different random trajectories and calculation of statistical moments of the phase, specially its standard deviation, $\sigma_\phi(t)$, have been usually done by overlaying them in a temporal window with a duration of a few periods [33–35]. For instance just one period is considered in references [34,35] to calculate the value of $\sigma_\phi(t) = \sqrt{<\phi^2>(t) - <\phi>^2(t)}$ with $0 \leq t \leq T$. To obtain well defined averages, $<\phi>(t)$ and $<\phi^2>(t)$, it is necessary to make a choice of the initial conditions at the beginning of each period because ϕ is an unbounded quantity, as shown in Figure 1. One choice is to take $P(0) =< P(0) >$, $N(0) =< N(0) >$, and $\phi(0) =< \phi(0) >$ [34], that is fixed initial conditions. A second choice is to take random initial conditions [35]. Photon and carrier numbers at $t = 0$ are those obtained at the end of the previous period, like in Figure 1. The change with respect to Figure 1 is related to the phase and it is based on the cyclic nature of angles: We consider that ϕ at the beginning of the period, $\phi(0)$, is that corresponding to ϕ at the end of the previous period, $\phi(T)$, but converted into the $[0, 2\pi)$ range, that is we consider that $\phi(0)$ is given by $\phi(T) - \text{int}(\frac{\phi(T)}{2\pi})2\pi$.

Figure 2 shows the temporal evolution of P, N and ϕ, plotted in a window of duration T, corresponding to the three consecutive pulses of Figure 1 and using the previous choice of random initial conditions. Figure 2a,c show that laser pulses that have a larger switch-on time, defined as the time at which P crosses a fixed level, have also a larger value of the maximum of N and P [9]. Figure 2b shows that ϕ takes values in a range of several multiples of 2π during one period. Figure 2b also shows, in a more clear way than in Figure 1, that the phase fluctuations are more important at the beginning and at the end of the pulse. Comparison between Figure 2a,b shows that pulses with a similar evolution of P can have a very different phase evolution (see black and red realizations). In the next section we will focus on the description of the temporal evolution of the phase statistics.

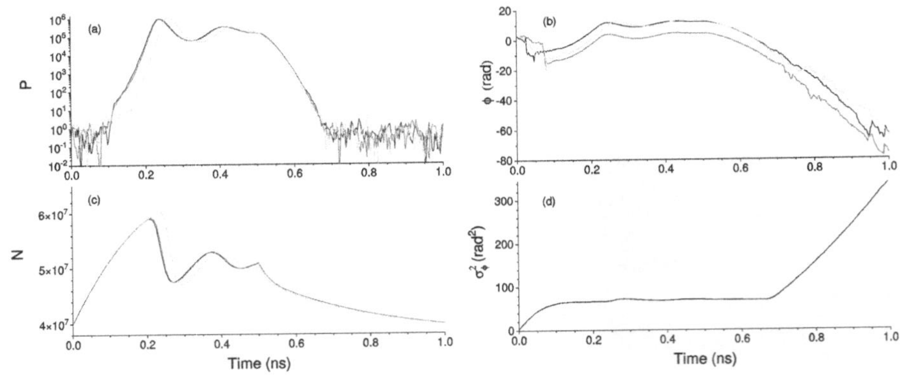

Figure 2. (**a**) Photon number, (**b**) phase, and (**c**) carrier number dynamical evolution for three different realizations are shown with black, red, and green lines in a temporal window of duration T. (**d**) Variance of the phase as a function of t. In this figure $T = 1$ ns and the three realizations are extracted from the time traces of Figure 1.

4. Analysis of the Phase Statistics

The dynamical evolution of the variance of the phase, σ_ϕ^2, is shown in Figure 2d for the case of random initial conditions and a temporal window of duration $T = 1$ ns. $\sigma_\phi^2(t)$ has been calculated by averaging over 2×10^4 temporal windows. $\sigma_\phi^2(0) > 0$ because of our choice of random initial conditions. Large increases of $\sigma_\phi^2(t)$ occur while P is small and dominated by the spontaneous emission noise, that is at the beginning and at the end of the period. While the evolutions of P and ϕ are deterministic and $I > I_{th}$ (0.15 ns < t < 0.5 ns) $\sigma_\phi^2(t)$ oscillates with the frequency of the relaxation oscillations around a value that increases linearly with time, similarly to what was observed by Henry [8]. These oscillations and the linear increase are barely seen in Figure 2d because of the vertical scale determined by the large values of the variance when the laser is switched-off. From 0.5 ns < t < 0.65 ns, while ϕ still has a deterministic evolution, there is a slight decrease of $\sigma_\phi^2(t)$. After that time, both ϕ and P become determined by the spontaneous emission noise. In this way the linear increase of $\sigma_\phi^2(t)$ with t, characteristic of phase diffusion, is observed until the end of the period, as it is seen in Figure 2d.

We now analyze the effect of carrier noise on the statistics of the phase. Figure 3 shows the probability density function (pdf) of ϕ at three different times when the carrier noise is considered (that is, integrating Equation (6)) and when it is neglected (considering instead Equation (5)). This figure has been obtained using the same conditions of Figure 2.

Figure 3 shows that the effect of carrier noise on the statistics of ϕ is very small. In fact, it has been shown that the consideration of noise in the carrier equation is not important during transient regimes [9,33], being only essential in the stationary regime for calculating quantities like the relative intensity noise [6]. Figure 3 also shows the Gaussian distributions of average and standard deviation given by the simulation with carrier noise. It is clear that the Gaussian distribution does not describe well the phase satististics, specially for short times ($t = 0.1$ and $t = 0.5$ ns). The Gaussian approximation becomes better at longer times ($t = 0.9$ ns).

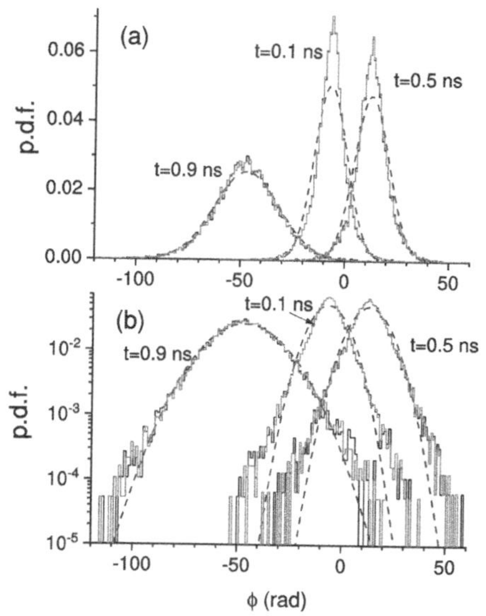

Figure 3. Probability density function of the phase at three different times in (**a**) linear, and (**b**) logarithmic vertical scale. Pdfs obtained with and without noise in the carrier number equation are plotted with red and black solid lines. Gaussian approximations are plotted with blue dashed lines.

A way of quantifying if the Gaussian distribution is suitable for describing the phase statistics is by calculating moments of ϕ of order higher than 2. Asymmetry and kurtosis coefficient of the simulated data are shown in Figure 4 as a function of time. Both coefficients must vanish if the distribution is Gaussian. Figure 4a shows with black lines the asymmetry, γ_r, and kurtosis, κ_r, coefficients obtained under the same conditions of Figure 2, that is with random initial conditions. While the phase distribution is symmetric ($\gamma_r \sim 0$), κ_r is significantly larger than zero. κ_r decreases fast until it develops a small peak close to the time at which the first relaxation oscillation appears. After that peak it reaches a plateau that finishes when P reaches the spontaneous emission noise level (around $t = 0.7$ ns). From that time ϕ diffuses and κ_r monotonously decreases reaching values that are closer to zero at the end of the period ($\kappa_r = 0.65$ at $t = 0.9$ ns). Figure 4b shows γ_r and κ_r when $T = 2$ ns. In this case ϕ has more time to diffuse when the laser is switched-off and then the Gaussian approximation is better at the end of the period ($\kappa_r = 0.14$ at $t = 2$ ns).

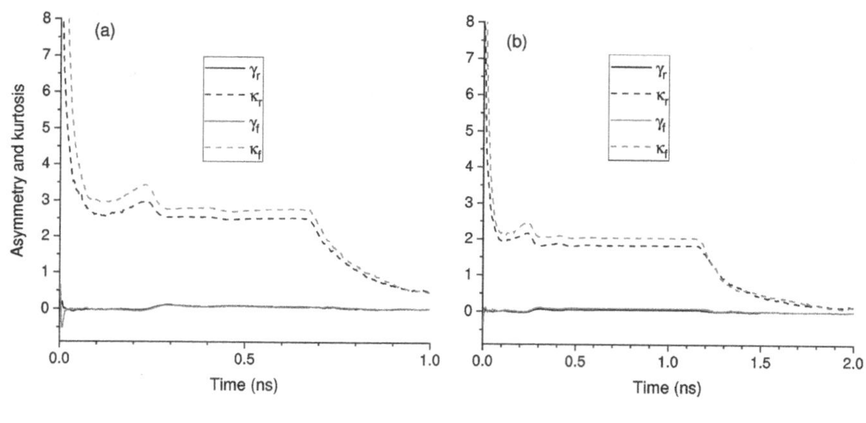

Figure 4. Asymmetry and kurtosis coefficients of the phase as a function of time for (**a**) $T = 1$ ns, and (**b**) $T = 2$ ns. Asymmetry and kurtosis coefficients are plotted with solid and dashed lines, respectively. Results for random and fixed initial conditions are plotted with black and red lines, respectively.

The reason why ϕ is not Gaussian can be understood by plotting the pdf of ϕ at $t = 0$. Figure 5 shows that distribution for the case of $T = 1$ ns. The distribution corresponds to a uniform random variable in $[0, 2\pi)$. This is because of the way random initial conditions are chosen: Doing the operation $\phi(0) = \phi(T) - \text{int}(\frac{\phi(T)}{2\pi})2\pi$ from a broad nearly Gaussian distribution for $\phi(T)$ makes $\phi(0)$ a uniform random variable, $\text{U}(0, 2\pi)$. The kurtosis of $\text{U}(0, 2\pi)$ is $354/5 \sim 70.8$. Diffusion of ϕ at the beginning of the period (see Figure 2) makes κ_r to decrease quickly, but not enough for becoming strictly Gaussian, even at the end of the period.

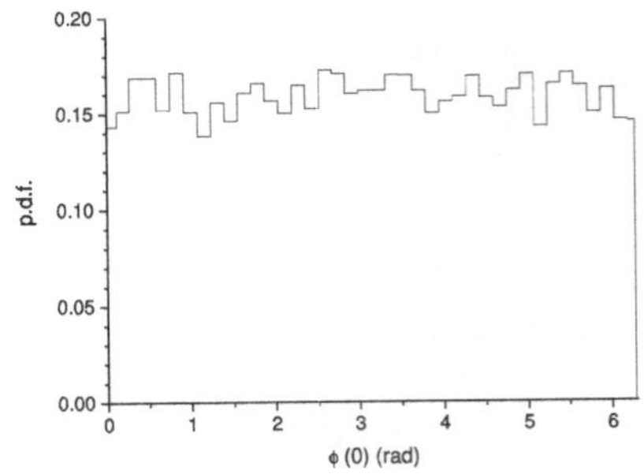

Figure 5. Probability density function of the initial value of the phase for $T = 1$ ns.

Of course these results depend on the way initial conditions are chosen. Another way of choosing these values is by considering fixed initial values for $E(0)$, and $N(0)$. Figure 4 shows, with red lines, asymmetry and kurtosis coefficients, γ_f and κ_f, when fixed initial conditions are used. We choose these values in the following way. We first integrate Equations (4) and (6) from arbitrary initial conditions corresponding to below threshold operation in order to find the averaged $< P(t) >, < N(t) >$, and $< \phi(t) >$ for $0 \leq t \leq T$. Then we choose $N(0) =< N(T) >$, and $E(0) = \sqrt{< P(T) >}(\cos < \phi(T) > + i \sin < \phi(T) >)$. This election is similar to that considered in [34]. Figure 4 shows that the evolution of γ_f and κ_f is very similar to that of γ_r and κ_r, respectively. $\kappa_f > \kappa_r$ because the initial delta-like distribution of $\phi(0)$ produce larger values of the kurtosis. These differences decrease with t, specially when spontaneous emission dominates the phase evolution: In Figure 4a,b when $t > 0.7$ ns ($t > 1.2$ ns).

The choice of initial conditions also impacts on the values of the standard deviation as a function of t. In Figure 6a the dynamical evolution of σ_ϕ for both, random and fixed initial conditions, is shown when $T = 1$ ns. Again both standard deviations have similar trends but the value for random initial conditions is larger than that obtained for the fixed ones. This is due to the non-zero value of $\sigma_\phi(0)$ obtained with the uniform distribution of $\phi(0)$ in contrast to the zero value obtained for fixed initial conditions. Relative differences between both quantities enhance if the speed of QRNG increases as it can be seen in Figure 6b where results obtained for $T = 0.4$ ns have been plotted. For instance, σ_ϕ at 0.2 ns is around 20 % smaller for the case of fixed initial conditions.

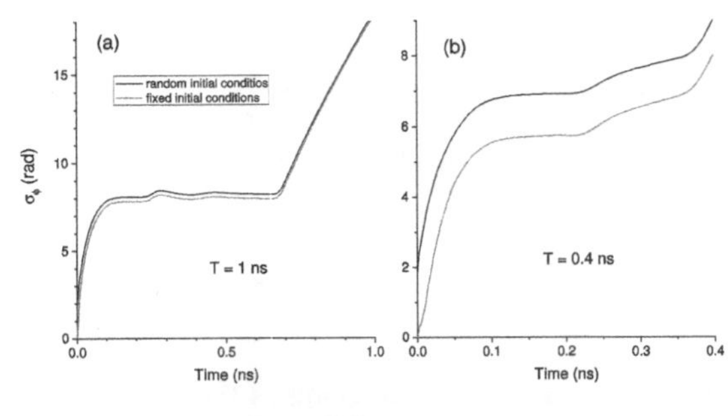

Figure 6. Standard deviation of the phase as a function of time for (**a**) T =1 ns, and (**b**) T =0.4 ns. Results for random and fixed initial conditions are plotted with black and red lines, respectively.

The dependence of the phase statistics on the way initial conditions are chosen suggests that averages must be done in a different way in order to lose the memory of those initial conditions. We have been considering averages performed in a temporal window with the same duration than the period of the current, T. From now on we will consider longer temporal windows for calculating statistical averages. Figure 7 illustrates the situation found when averages are calculated over a window of duration $2T$. Random initial conditions are considered such that $\phi(0) = \phi(2T) - \mathrm{int}(\frac{\phi(2T)}{2\pi})2\pi$. Averages have been done over 2×10^4 $2T$-windows, where $T = 1$ ns, in order to compare with situations illustrated in previous figures. Figure 7a shows the averaged phase vs t. The drift towards decreasing values of the phase is similar to that shown in Figure 1c. Standard deviation and variance of the phase are shown in Figure 7b,c, respectively. $< \phi(t) >$, $\sigma_\phi(t)$ and $\sigma^2_\phi(t)$ during the second half of the $2T$-window are basically replicas of what was found in the first half. The continuity of ϕ along the $2T$-window makes that $\sigma_\phi(t)$ and $\sigma^2_\phi(t)$ monotonously increase. However the situation is different when considering the kurtosis coefficient as Figure 7d shows. In this case, during the second half of the window κ_r keeps on decreasing towards the zero value. This means that the distribution of the phase keeps on approaching to the Gaussian shape. In fact $\kappa_r = 0.22$ when $t = 2$ ns.

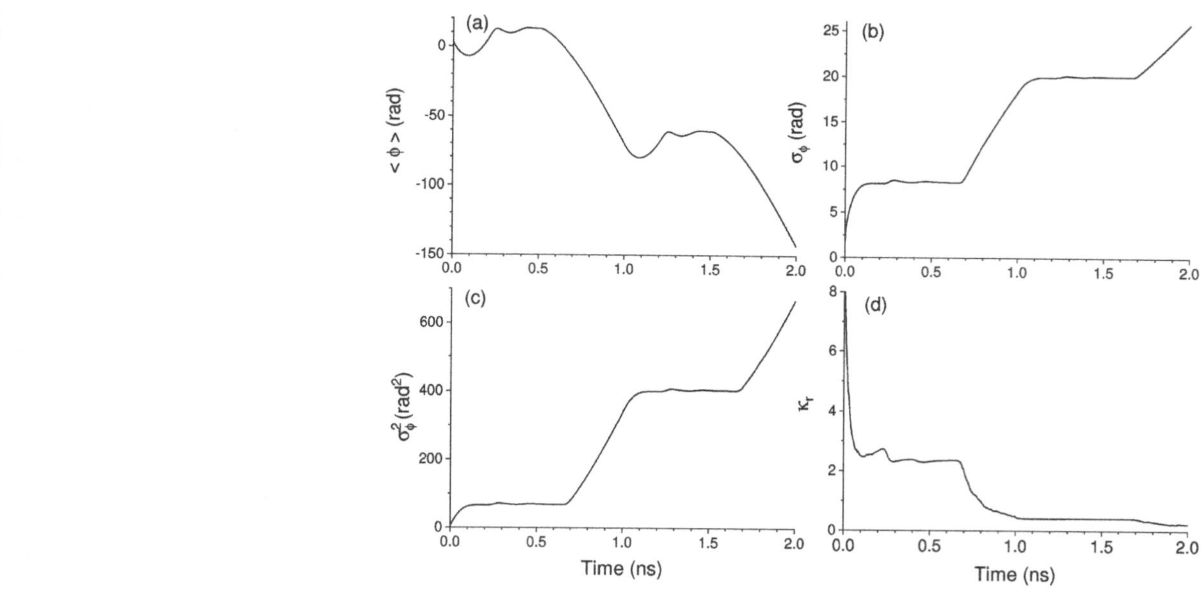

Figure 7. (**a**) Average, (**b**) standard deviation, (**c**) variance, and (**d**) kurtosis coefficient of the phase as a function of time for a two-period window with $T = 1$ ns.

That approach can be illustrated by plotting the phase pdf at two different times. Figure 8 shows those distributions at times $t = 0$ and $t = 1.1$ ns. The phase at $t = 0$ is a $\mathrm{U}(0, 2\pi)$ random variable, similarly to Figure 5. The phase at $t = 1.1$ ns is approximately Gaussian as it can be seen when comparing with the Gaussian of average and standard deviation obtained from simulations. The kurtosis coefficient in Figure 8b is 0.4. Figure 8b can also be compared with the pdf obtained at $t = 0.1$ ns in Figure 3b because both distributions correspond to 0.1 ns after switching-on the bias current. The pdf in Figure 3b is not Gaussian while the pdf in Figure 8c is approximately Gaussian. This indicates that in order to have a phase distributed as a Gaussian it is necessary to calculate and average the phase in windows with durations of several modulation periods. In this way the memory of the initial conditions and their distribution is lost.

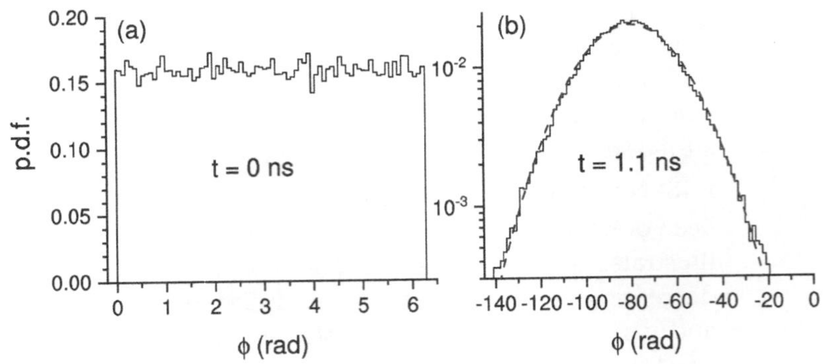

Figure 8. Probability density function of the phase at (**a**) $t = 0$ ns, and (**b**) $t = 1.1$ ns for a two-period window with $T = 1$ ns. The Gaussian approximation is plotted with a blue dashed line.

5. Discussion and Summary

In our study we have considered two types of initial conditions, corresponding to deterministic and random values of the variables. Fixed initial conditions have been considered because they have been used in previous studies of QRNG. They are not the best choice for simulation of these systems because the spontaneous emission noise, that is always present in the system, causes fluctuations in the variables of the system at all times. These include the times at which each period begins, and so initial conditions must be also random, as it is also expected in an experimental realization of the system. We have chosen these random initial values by calculating the phase angle in the $[0, 2\pi)$ range that corresponds to the final value in the previous averaging window. Note that the conversion to the $[0, 2\pi)$ range is necessary if a calculation of well defined statistical moments of the phase is required. If no conversion is done, not even $< \phi(t) >$ could be calculated because ϕ decreases in each averaging window in a magnitude of more than several 2π, as illustrated for instance in Figure 1c.

Deterministic initial conditions and phase averages over windows of T-duration have been recently used for describing the phase statistics [34]. While these conditions can give an approximation to the phase distribution and their statistical moments, our results show that it is necessary to consider averages over windows of several T-duration and random initial conditions for obtaining Gaussian statistics for the phase at the end of the averaging period.

We now briefly discuss the effect of two laser parameters, the non-linear gain and the Auger coefficients, on the standard deviation of the phase. The number of relaxation oscillation peaks increases when the non-linear gain coefficient decreases. The standard deviation of the phase at the end of the modulation period oscillates when changing I_{on} [35]. The number of these oscillations is directly related to the number of relaxation oscillation peaks that are excited. In this way, the main effect of having a small nonlinear gain coefficient is to observe more oscillations of the standard deviation of the phase as a function of I_{on}. The effect of the Auger coefficient is also important for describing the standard deviation of the phase. In fact we have shown that the Auger term must be considered in the carrier recombination term for achieving good agreement between experiments and theory [36].

Summarizing, we have theoretically analyzed the phase diffusion in gain-switched semiconductor lasers by performing numerical simulations of the corresponding stochastic rate equations. We have focused on the calculation of the temporal dependence of the statistical moments and distribution of the phase. We have considered several types of initial conditions for the phase. By using the temporal dependence of the kurtosis coefficient we have shown that the phase pdf becomes Gaussian only after the memory of the statistical distribution of the initial conditions is lost. We show that under the typical gain-switching with square-wave modulation used in QRNGs, the time it takes for the phase to become

Gaussian is in the ns scale. We finally compared the variance of the phase obtained with random and fixed initial conditions to show that their differences are more important as the modulation speed is increased. This is precisely the situation in which faster generation bit rates are achieved when using QRNGs based on gain-switched laser diodes.

Abbreviations

The following abbreviations are used in this manuscript:

QRNG	Quantum random number generation
DML	Discrete mode laser
PDF	Probability density function
ASE	Amplified spontaneous emission
RNG	Random number generation
LED	Light emitting diode

References

1. Lax, M. Quantum noise. IV. Quantum theory of noise sources. *Phys. Rev.* **1966**, *145*, 110. [CrossRef]
2. Lax, M.; Louisell, W. Quantum noise. XII. Density-operator treatment of field and population fluctuations. *Phys. Rev.* **1969**, *185*, 568. [CrossRef]
3. Risken, H. Fokker–Planck equation. In *The Fokker–Planck Equation*; Springer: Berlin/Heidelberg, Germany, 1996.
4. Henry, C.H.; Kazarinov, R.F. Quantum noise in photonics. *Rev. Mod. Phys.* **1996**, *68*, 801. [CrossRef]
5. Arecchi, F.; Degiorgio, V.; Querzola, B. Time-dependent statistical properties of the laser radiation. *Phys. Rev. Lett.* **1967**, *19*, 1168. [CrossRef]
6. Coldren, L.A.; Corzine, S.W.; Mashanovitch, M.L. *Diode Lasers and Photonic Integrated Circuits*; John Wiley & Sons: Hoboken, NJ, USA, 2012; Volume 218.
7. Agrawal, G.P.; Dutta, N.K. *Semiconductor Lasers*; Springer: Berlin/Heidelberg, Germany, 2013.
8. Henry, C. Phase noise in semiconductor lasers. *J. Light. Technol.* **1986**, *4*, 298–311. [CrossRef]
9. Balle, S.; Colet, P.; San Miguel, M. Statistics for the transient response of single-mode semiconductor laser gain switching. *Phys. Rev. A* **1991**, *43*, 498. [CrossRef]
10. Balle, S.; De Pasquale, F.; Abraham, N.; San Miguel, M. Statistics of the transient frequency modulation in the switch-on of a single-mode semiconductor laser. *Phys. Rev. A* **1992**, *45*, 1955. [CrossRef] [PubMed]
11. Gardiner, C.W. *Handbook of Stochastic Methods*; Springer: Berlin/Heidelberg, Germany, 1985; Volume 3.
12. Stipčević, M.; Koç, Ç.K. True random number generators. In *Open Problems in Mathematics and Computational Science*; Springer: Berlin/Heidelberg, Germany, 2014; pp. 275–315.
13. Herrero-Collantes, M.; Garcia-Escartin, J.C. Quantum random number generators. *Rev. Mod. Phys.* **2017**, *89*, 015004. [CrossRef]
14. Sciamanna, M.; Shore, K.A. Physics and applications of laser diode chaos. *Nat. Photonics* **2015**, *9*, 151–162. [CrossRef]
15. Jennewein, T.; Achleitner, U.; Weihs, G.; Weinfurter, H.; Zeilinger, A. A fast and compact quantum random number generator. *Rev. Sci. Instrum.* **2000**, *71*, 1675–1680. [CrossRef]
16. Stipčević, M.; Rogina, B.M. Quantum random number generator based on photonic emission in semiconductors. *Rev. Sci. Instrum.* **2007**, *78*, 045104. [CrossRef]
17. Fürst, H.; Weier, H.; Nauerth, S.; Marangon, D.G.; Kurtsiefer, C.; Weinfurter, H. High speed optical quantum random number generation. *Opt. Express* **2010**, *18*, 13029–13037. [CrossRef] [PubMed]
18. Wei, W.; Guo, H. Bias-free true random-number generator. *Opt. Lett.* **2009**, *34*, 1876–1878. [CrossRef]
19. Durt, T.; Belmonte, C.; Lamoureux, L.P.; Panajotov, K.; Van den Berghe, F.; Thienpont, H. Fast quantum-optical random-number generators. *Phys. Rev. A* **2013**, *87*, 022339. [CrossRef]
20. Shen, Y.; Tian, L.; Zou, H. Practical quantum random number generator based on measuring the shot noise of vacuum states. *Phys. Rev. A* **2010**, *81*, 063814. [CrossRef]
21. Argyris, A.; Pikasis, E.; Deligiannidis, S.; Syvridis, D. Sub-Tb/s physical random bit generators based on direct detection of amplified spontaneous emission signals. *J. Light. Technol.* **2012**, *30*, 1329–1334. [CrossRef]

22. Guo, Y.; Cai, Q.; Li, P.; Jia, Z.; Xu, B.; Zhang, Q.; Zhang, Y.; Zhang, R.; Gao, Z.; Shore, K.A.; et al. 40 Gb/s quantum random number generation based on optically sampled amplified spontaneous emission. *APL Photonics* **2021**, *6*, 066105. [CrossRef]
23. Guo, H.; Tang, W.; Liu, Y.; Wei, W. Truly random number generation based on measurement of phase noise of a laser. *Phys. Rev. E* **2010**, *81*, 051137. [CrossRef]
24. Qi, B.; Chi, Y.M.; Lo, H.K.; Qian, L. High-speed quantum random number generation by measuring phase noise of a single-mode laser. *Opt. Lett.* **2010**, *35*, 312–314. [CrossRef]
25. Xu, F.; Qi, B.; Ma, X.; Xu, H.; Zheng, H.; Lo, H.K. Ultrafast quantum random number generation based on quantum phase fluctuations. *Opt. Express* **2012**, *20*, 12366–12377. [CrossRef]
26. Jofre, M.; Curty, M.; Steinlechner, F.; Anzolin, G.; Torres, J.; Mitchell, M.; Pruneri, V. True random numbers from amplified quantum vacuum. *Opt. Express* **2011**, *19*, 20665–20672. [CrossRef] [PubMed]
27. Abellán, C.; Amaya, W.; Jofre, M.; Curty, M.; Acín, A.; Capmany, J.; Pruneri, V.; Mitchell, M. Ultra-fast quantum randomness generation by accelerated phase diffusion in a pulsed laser diode. *Opt. Express* **2014**, *22*, 1645–1654. [CrossRef] [PubMed]
28. Yuan, Z.; Lucamarini, M.; Dynes, J.; Fröhlich, B.; Plews, A.; Shields, A. Robust random number generation using steady-state emission of gain-switched laser diodes. *Appl. Phys. Lett.* **2014**, *104*, 261112. [CrossRef]
29. Marangon, D.G.; Plews, A.; Lucamarini, M.; Dynes, J.F.; Sharpe, A.W.; Yuan, Z.; Shields, A.J. Long-term test of a fast and compact quantum random number generator. *J. Light. Technol.* **2018**, *36*, 3778–3784. [CrossRef]
30. Abellan, C.; Amaya, W.; Domenech, D.; Muñoz, P.; Capmany, J.; Longhi, S.; Mitchell, M.W.; Pruneri, V. Quantum entropy source on an InP photonic integrated circuit for random number generation. *Optica* **2016**, *3*, 989–994. [CrossRef]
31. Shakhovoy, R.; Sharoglazova, V.; Udaltsov, A.; Duplinskiy, A.; Kurochkin, V.; Kurochkin, Y. Influence of Chirp, Jitter, and Relaxation Oscillations on Probabilistic Properties of Laser Pulse Interference. *IEEE J. Quantum Electron.* **2021**, *57*, 1–7. [CrossRef]
32. Nakata, K.; Tomita, A.; Fujiwara, M.; Yoshino, K.i.; Tajima, A.; Okamoto, A.; Ogawa, K. Intensity fluctuation of a gain-switched semiconductor laser for quantum key distribution systems. *Opt. Express* **2017**, *25*, 622–634. [CrossRef]
33. Septriani, B.; de Vries, O.; Steinlechner, F.; Gräfe, M. Parametric study of the phase diffusion process in a gain-switched semiconductor laser for randomness assessment in quantum random number generator. *AIP Adv.* **2020**, *10*, 055022. [CrossRef]
34. Shakhovoy, R.; Tumachek, A.; Andronova, N.; Mironov, Y.; Kurochkin, Y. Phase randomness in a gain-switched semiconductor laser: Stochastic differential equation analysis. *arXiv* **2020**, arXiv:2011.10401.
35. Quirce, A.; Valle, A. Spontaneous emission rate in gain-switched laser diodes for quantum random number generation. Preprint.
36. Quirce, A.; Valle, A. Phase diffusion in gain-switched semiconductor lasers for quantum random number generation. Preprint.
37. Rosado, A.; Pérez-Serrano, A.; Tijero, J.M.G.; Valle, A.; Pesquera, L.; Esquivias, I. Numerical and experimental analysis of optical frequency comb generation in gain-switched semiconductor lasers. *IEEE J. Quantum Electron.* **2019**, *55*, 1–12. [CrossRef]
38. Schunk, N.; Petermann, K. Noise analysis of injection-locked semiconductor injection lasers. *IEEE J. Quantum Electron.* **1986**, *22*, 642–650. [CrossRef]
39. Kloeden, P.E.; Platen, E. Stochastic differential equations. In *Numerical Solution of Stochastic Differential Equations*; Springer: Berlin/Heidelberg, Germany, 1992.
40. McDaniel, A.; Mahalov, A. Stochastic differential equation model for spontaneous emission and carrier noise in semiconductor lasers. *IEEE J. Quantum Electron.* **2018**, *54*, 1–6. [CrossRef]
41. Quirce, A.; Rosado, A.; Díez, J.; Valle, A.; Pérez-Serrano, A.; Tijero, J.M.G.; Pesquera, L.; Esquivias, I. Nonlinear dynamics induced by optical injection in optical frequency combs generated by gain-switching of laser diodes. *IEEE Photonics J.* **2020**, *12*, 1–14. [CrossRef]

Organic Diode Laser Dynamics: Rate-Equation Model, Reabsorption, Validation and Threshold Predictions

Abstract: We present and analyze a simple model based on six rate equations for an electrically pumped organic diode laser. The model applies to organic host-guest systems and includes Stoke-shifted reabsorption in a self-consistent manner. With the validated model for the Alq3:DCM host-guest system, we predict the threshold for short-pulse laser operation. We predict laser operation characterized by damped relaxation oscillations in the GHz regime and several orders of magnitude linewidth narrowing. Prospect for CW steady-state laser operation is discussed.

Keywords: optoelectronics; OLED; laser; organic laser diode; laser dynamics

---

1. Introduction

An Organic Diode Laser (ODL) is the lasing manifestation of an Organic Light Emitting Diode (OLED). It represents a promising class of new lasers with foreseen applications in spectroscopy, sensing, environmental monitoring, optical communication, short haul data transfer [1,2]. Since Heeger's demonstration of "plastic" conductivity in 1977 [3], organic semiconductor technology has made a huge step forward. With relatively simple, economic, and environmentally friendly production processes, and virtually unlimited availability of amorphous organic semiconductors, organic optoelectronics has become a large research field for various device types such as organic photovoltaic cells (OPV), organic transistors (OFETs), and OLEDs [4–7]. Developments in OLEDs have resulted in successful applications including lighting or display technologies such as screens for TV and mobile phones, but they have so far been underused in optical transmission systems in comparison to their inorganic counterparts, namely the conventional light emitting diode devices (LEDs).

The ODL will open a new era in the field of lasing. Firstly, because solid-state organic materials, contrary to their III-V counterparts, cover continuously the whole visible spectrum as well as part of the IR and UV spectrum. Secondly, they can be deposited more easily on almost any substrate with less energy consumption for the manufacturing process than conventional epitaxially-grown III-V materials [8]. Thirdly, this new device combines properties from dye-lasers and III-V diode lasers and as such will open new perspectives and potential applications. Fourthly, organic electronics is a low-carbon industry, unlike the III-V industry.

Regarding perspectives, organic materials exhibit dependence of the refractive index on the carrier density different from conventional III-V semiconductors, which is largely due to the specific mobility of disordered organic semiconductors [9,10]. Therefore, new and interesting dynamical behavior will occur, especially when the laser is submitted to different types of external perturbations, such as optical injection and feedback [11]. Potential applications and new possibilities will be facilitated by the ease of deposition of organic heterostructures on a large variety of substrates including silicon, silica, glass, as well as flexible substrates, and by the availability of an almost unlimited library of electroluminescent organic materials [12].

Until now, solid-state organic lasers have been realized by optical pumping of OLEDs provided with an integrated optical cavity [13–15]. Lasing based on electrical injection appears to be much more difficult, because of gain quenching due to triplet accumulation and absorption from the metallic anode [13]. We will summarize our rate-equation theory for an electrically injected ODL [16], extend the theory to include a detailed treatment of the self-consistent reabsorption, and present simulation results for operation below and above laser threshold.

2. Characterization of a Laser OLED

Organic Diode Lasers (ODLs) are organic hetero structures with an integrated optical cavity to enhance the interaction time of photons with the active molecules before they are emitted, thus enabling a sufficiently high level of stimulated emission to generate laser light. A schematic of the layer structure of an organic hetero structure is given in Figure 1a. An important difference with conventional III-V semiconductor devices is that the charges are not just electrons and holes, but rather electron-like (i.e., negatively charged) polarons and hole-like (positively charged) polarons. These polarons are special organic molecules brought in excited states under influence of an applied voltage and created in the regions indicated electron injection layer (EIL) and hole transport layer (HTL), respectively, indicated in Figure 1b. The polarons have an effective mobility based on their diffusion by hopping of the excitation from one molecule to the next. In the emitting layer (EL, see Figure 1b) both type of polarons will be present allowing them to recombine forming excitons, with 25% chance of a singlet exciton and 75% chance of a triplet. Only the singlet excitons can decay optically to the ground state, whereas the decay of the triplets to ground state is optically forbidden. The schematic energy level diagram corresponding to Figure 1a is depicted in Figure 1c. The hole-blocking role of TPBi can be explained with the HOMO energy difference between TPBi (HBL) and Alq3 (EL) being 0.5 eV whereas it is only 0.2 eV for the corresponding LUMO.

The layer structure is integrated with a horizontal cavity consisting of a second-order Bragg grating sandwiched between two first-order Bragg gratings, such that the photons are emitted downward due to the second-order grating. This configuration is sketched in Figure 1e, with a top-view photograph of the Bragg gratings in Figure 1d. The blue arrow starting from Figure 1b points to one of the various organic heterostructure units in Figure 1e that provide the optical gain in the cavity formed by the grating structure.

The optical gain is provided by the singlet excitons in the emitting layer. When a singlet decays radiatively, the molecule is left in the ground state and the exciton will disappear. There are several other decay channels for the singlets, that is, intersystem crossing (ISC), singlet-singlet annihilation (SSA), singlet-triplet annihilation (STA), and singlet-polaron annihilation (SPA). These decay processes will be explained briefly here; they are extensively discussed in [17]. ISC is a spin-flip induced intra-molecular process in which the singlet decays to the triplet on the same molecule, i.e., a loss of 1 singlet and at the same time a gain of 1 triplet. The other annihilation processes are of bi-molecular nature. In case of SSA, the interaction of two singlet excitons yields one ground state molecule plus one exciton with 25% chance of a singlet and 75% chance of a triplet, i.e., on average a net loss of 7/4 singlet and a net gain of 3/4 triplet [17]. In case of STA, the interaction of one singlet exciton and one triplet exciton leads to annihilation of the singlet exciton, which has decayed to the ground state, i.e., a net loss of 1 singlet. The interaction between a polaron and a singlet exciton in case of SPA leads to annihilation of the singlet, both for hole-like and electron-like polarons.

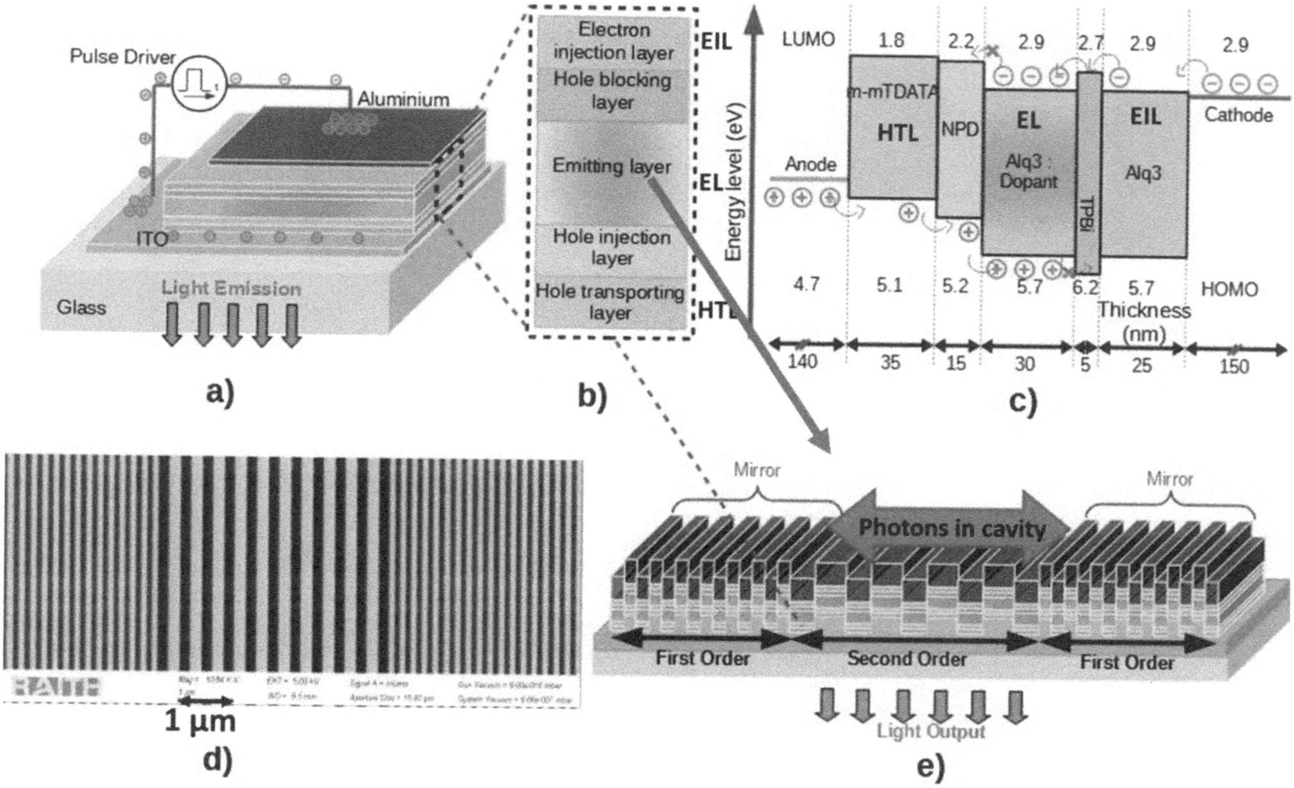

Figure 1. Schematics and direction of light emission of (**a**) the layer structure indicated in (**b**) with (**c**) the energy level diagram. In (**d**) a top view photograph is shown of the second-order grating sandwiched between the two first-order gratings and (**e**) presents a sketch of the OLD structure indicating the confined light between the mirrors formed by the first-order gratings.

In view of STA, proper accounting of the triplet population is crucial for the calculation of the optical gain (or: will be decisive for the available optical gain). Triplets are annihilated in the bi-molecular processes of triplet-triplet annihilation (TTA) and triplet-polaron annihilation (TPA). Like SSA, in TTA the interaction of two triplet excitons yields one ground state molecule plus one exciton with 25% chance of a singlet and 75% chance of a triplet. Hence, TTA leads, on average, to a net loss of 5/4 triplet and a net gain of 1/4 singlet. TPA leads to a loss of the triplet, just as SPA for the singlet does. In the rate equations that will be presented in Section 3, each of the above described processes corresponds to a term with corresponding rate coefficient.

3. Rate-Equation Model for the ODL

In the model we assume that the hole-type and electron-type polarons that participate in charge transfer across the organic semiconductor layers recombine in the emitting layer to form singlet and triplet excitons. We consider the situation where the emitting layer is composed of host molecules (the matrix), doped with a few percent of guest molecules (the dopant), and where the excitonic states are quickly transferred from the host molecules to the singlet and triplet excitons of the dopant molecules by Förster transfer and, to a lesser extent, by Dexter transfer, respectively. With a host-guest system like Alq3:DCM [18], the host (Alq3) singlets have their optical transition in the green part of the spectrum (~530 nm), whereas the dopant (DCM) singlets provide both spontaneous and stimulated emission in the red spectrum (~620 nm).

The model equations read

$$\frac{d}{dt}N_P = \frac{J(t)P_0}{ed} - \gamma N_P^2; \tag{1}$$

$$\frac{d}{dt}N_S = \frac{1}{4}\gamma N_P^2 + \frac{1}{4}\kappa_{TT}N_T^2 - (\kappa_{FRET}P_{0D} + \kappa_S + \kappa_{ISC})N_S - \left(\frac{7}{4}\kappa_{SS}N_S + \kappa_{SP}N_P + \kappa_{ST}N_T\right)N_S; \tag{2}$$

$$\frac{d}{dt}N_T = \frac{3}{4}\gamma N_P^2 + \kappa_{ISC}N_S + \frac{3}{4}\kappa_{SS}N_S^2 - (\kappa_{DEXT}P_{0D} + \kappa_T + \kappa_{TP}N_P)N_T - \frac{5}{4}\kappa_{TT}N_T^2; \tag{3}$$

$$\frac{d}{dt}N_{SD} = \kappa_{FRET}P_{0D}N_S + \frac{1}{4}\kappa_{TTD}N_{TD}^2 - (\kappa_{SD} + \kappa_{ISCD})N_{SD} -$$

$$\left(\frac{7}{4}\kappa_{SSD}N_{SD} + \kappa_{SPD}N_P + \kappa_{STD}N_{TD}\right)N_{SD} - \xi_E M(E,E,CAV)(N_{SD} - WN_{0D})P_{HO}; \tag{4}$$

$$\frac{d}{dt}N_{TD} = \kappa_{DEXT}P_{0D}N_T + \kappa_{ISCD}N_{SD} + \frac{3}{4}K_{SSD}N_{SD}^2 - \kappa_{TD}N_{TD} - \frac{5}{4}\kappa_{TTD}N_{TD}^2 - \kappa_{TPD}N_{TD}N_P; \tag{5}$$

$$\frac{d}{dt}P_{HO} = \beta_{sp}\kappa_{SD}N_{SD} + \{\Gamma\xi_E M(E,E,CAV)(N_{SD} - WN_{0D}) - \kappa_{CAV}\}P_{HO}; \tag{6}$$

$$N_0 = N_{HOST} - 2N_P - N_S - N_T; N_{0D} = N_{DOP} - N_{SD} - N_{TD};$$

$$N_{DOP} = CN_{MOL}; N_{HOST} = (1-C)N_{MOL}; \tag{7}$$

$$P_0 = \frac{N_0}{N_{MOL}}; P_{0D} = \frac{N_{0D}}{N_{DOP}}. \tag{8}$$

These equations are valid in the emitting layer and the variables are: N_P the polaron density, N_S the density of singlet excitons, N_T the density of triplet excitons, N_0 the density of ground-state molecules, all in the host; N_{SD}, N_{TD} and N_{0D} the respective dopant singlet, triplet and ground-state population densities. P_{HO} is the photon density, $J(t)$ the current density, N_{MOL} is the molecular density, N_{DOP} and N_{HOST} the respective densities of dopant and host molecules. P_0 and P_{0D} are the respective probabilities that a host or dopant molecule is in the ground state. Finally, W represents the overlap between the dopant absorption spectrum $S_A(\lambda)$ and the emission spectrum $S_E(\lambda)$,

$$W \equiv \frac{\xi_A M(A,E,CAV)}{\xi_E M(E,E,CAV)} \equiv \frac{\xi_A \int d\lambda S_A(\lambda)S_E(\lambda)S_{CAV}(\lambda)}{\xi_E \int d\lambda S_E^2(\lambda)S_{CAV}(\lambda)}, \tag{9}$$

where ξ_X are the coefficients for emission ($X = E$) and absorption ($X = A$) of the dopant and the normalization should be such that $\int d\lambda S_{CAV}(\lambda) = 1$ and $S_X(\lambda_X) = 1$, with λ_X the wavelength for which S_X is maximal ($X = E,A$). Note that W in (9) also depends on the cavity mode wavelength λ_{CAV}.

The derivation of (9) and the definition of M are given in Appendix A. W accounts for the fraction of dopant ground-state molecules that participate in the absorption of the emitted light. Note that $W = 1$ for identical spectra and $\xi_A = \xi_E$. The various parameters in (1) to (8) are listed in Table 1 together with their values. More about W will be discussed in Section 3.1.

Before we proceed with a brief discussion of the processes described by Equations (1)–(8), two remarks should be made. The first remark concerns the light emission by the host singlet excitons (green in case of Alq3). As we will see in Section 4, the build-up of N_S remains relatively small, compared to N_{SD}. Moreover, no resonating structure is considered for the green light. Nevertheless, the host singlets will decay under spontaneous emission of green light. This photonic interaction is not considered in the rate equations.

As the second remark, note that the emission spectrum of the organic dopant emitter (DCM) is Stoke-shifted to the red by 160 nm from its absorption spectrum [19]. This implies that W will depend on the shift between the emission and absorption spectra as well as their respective widths. We estimate, using (A9) and (A10) (see Appendix A), that in the weak micro-cavity limit with $\kappa_{CAV} = 1.0 \times 10^{14}\ s^{-1}, Q \sim 18$, we estimate $W \approx 0.019$, but as κ_{CAV} decreases with increasing cavity quality factor and the threshold for lasing is approached, the emitted spectrum will narrow, implying smaller values for W. Therefore, W is a dynamic quantity, and this will be studied in more detail in Section 3.1.8.

Table 1. Model Parameters (Alq3:DCM).

Symbol	Name	Value	Ref.
S	OLED active area	10^{-4} cm^2	
d	OLED active layer thickness	30 nm	
γ	Langevin recombination rate	6.2×10^{-12} cm^3 s^{-1} to 2.0×10^{-9} cm^3 s^{-1}	[20,21]
N_{MOL}	Molecular density	2.1×10^{21} cm^{-3}	
C	Dopant concentration	2%	
κ_{FRET}	Förster transfer rate	1.15×10^{10} s^{-1}	[19,22,23]
κ_{DEXT}	Dexter transfer rate	1.0×10^{10} s^{-1} to 5.0×10^{15} s^{-1}	
κ_S	Host singlet-exciton decay rate	8.0×10^{7} s^{-1}	[24,25]
κ_{SD}	Dopant singlet-exciton decay rate	1.0×10^{9} s^{-1}	[24]
κ_T	Host triplet decay rate	6.5×10^{2} s^{-1} to 4.0×10^{4} s^{-1}	[24,26]
κ_{TD}	Dopant triplet decay rate	6.6×10^{2} s^{-1}	[26]
κ_{ISC}	Host inter-system crossing rate	2.2×10^{4} s^{-1} to 1.0×10^{7} s^{-1}	[17,27]
κ_{ISCD}	Dopant inter-system crossing rate	2.2×10^{4} s^{-1} to 1.0×10^{7} s^{-1}	[17,27]
κ_{SS}	Host singlet-singlet annihilation (SSA) rate	3.5×10^{-12} cm^3 s^{-1}	[24]
κ_{SSD}	Dopant singlet-singlet annihilation (SSA) rate	9.6×10^{-13} cm^3 s^{-1}	[24]
κ_{SP}	Host singlet-polaron annihilation (SPA) rate	3.0×10^{-10} s^{-1}	[24]
κ_{SPD}	Dopant singlet-polaron annihilation (SPA) rate	3.0×10^{-10} cm^3 s^{-1}	[24]
κ_{TP}	Host triplet-polaron annihilation (TPA) rate	2.8×10^{-13} cm^3 s^{-1}	[24]
κ_{TPD}	Dopant triplet-polaron annihilation (TPA) rate	5.6×10^{-13} cm^3 s^{-1}	[27]
κ_{ST}	Host singlet-triplet annihilation (STA) rate	1.9×10^{-10} cm^3 s^{-1}	[17,24]
κ_{STD}	Dopant singlet-triplet annihilation (STA) rate	1.9×10^{-10} cm^3 s^{-1}	[28,29]
κ_{TT}	Host triplet-triplet annihilation (TTA) rate	2.2×10^{-12} cm^3 s^{-1}	[24]
κ_{TTD}	Dopant triplet-triplet annihilation (TTA) rate	2.4×10^{-15} cm^3 s^{-1}	[27]
Γ	Confinement factor	0.29	
ξ_E	Dopant stimulated emission gain coefficient	1.4×10^{-5} cm^3 s^{-1}	[17,20]
ξ_A	Dopant absorption coefficient	1.4×10^{-5} cm^3 s^{-1}	
κ_{CAV}	Cavity photon decay rate	1–300 $\times 10^{12}$ s^{-1}	
β_{SP}	Spontaneous emission factor	<0.15	

3.1. *Brief Discussion of Equations (1)–(8)*

3.1.1. The Polaron Recombination

Polarons appear in two manifestations, positively charged hole-like polarons (density $N_P{}^+$) and negatively charged electron-like polarons (density $N_P{}^-$), where in view of assumed charge neutrality both populations are equal, $N_P{}^+ = N_P{}^-$. Moreover, each neutral polaron pair recombines to form one exciton together with one neutral molecule, which occurs at the Langevin-recombination rate γ [17,30]. This recombination process drives the electrical current and leads to the sink term in (1). Since γ is related to the polaron mobilities $\mu_h \wedge \mu_e$ as $\gamma = \frac{e}{\epsilon}(\mu_h + \mu_e)$, and since according to the Poole-Frenkel model the mobilities show an exponential dependence on the square root of the electric field F, we expect the value of γ to increase substantially with increasing applied diode voltage. In ref. [20] the zero-field value $\gamma = 6.2 \times 10^{-12}$ cm^3 s^{-1} is evaluated.

3.1.2. Host Singlet Excitons

The first term on the right-hand side (r.h.s.) of (2) is a source for the singlet excitons originating from the above-mentioned polaron recombination term. The factor 1/4 is due to the randomly injected spin statistics. The second term is a source term arising from triplet-triplet annihilation with generation rate κ_{TT} [24]. All other terms are sink terms. The first sink term describes the Förster Resonance Energy Transfer (FRET) of singlet excitons from host to dopant molecules with transfer rate κ_{FRET}. The probability P_{0D} accounts for the potential depopulation of the dopant-ground state that would limit the energy transfer.

The second sink term describes the decay of the singlet exciton due to both radiative and non-radiative processes. The third sink term accounts for the inter-system crossing (ISC), a non-radiative mechanism, i.e., a spin-flip-induced intra-molecular energy transfer from singlet to triplet with a decay rate κ_{ISC}. The last sink terms in (2) describe the depopulation of the host singlet density with different annihilation terms: singlet-singlet annihilation (SSA) with decay rate κ_{SS} [24], singlet-polaron annihilation (SPA) with decay rate κ_{SP} [24], and singlet-triplet annihilation (STA) with decay rate κ_{ST} [17,24].

3.1.3. Host Triplet Excitons

Rate Equation (3) describes the variation of host triplet excitons. The first three terms in the r.h.s. are sources. The first is a contribution arising from the polaron recombination. With a 3/4 factor resulting from the spin statistics, this source term, when added to the first singlet source term in (2), matches the first sink term for the polaron recombination in (1). The second term describes the increase of N_T due to ISC in the same way as it decreases N_S in (2). The third term corresponds to the decay of the triplet excitons with rate κ_T [24,26]. The fourth and fifth terms correspond respectively to triplet-triplet annihilation (TTA) [24], and triplet-polaron annihilation (TPA) [24].

3.1.4. Dopant Singlet Excitons

The dynamics of the dopant-singlet density N_{SD} is described by (4). The first term on the r.h.s. is the source because of the Förster energy transfer [19,22,23]. This term matches the corresponding sink term in (2). The second term is a relatively small and indirect source term due to the dopant triplet-triplet annihilation (TTA) leading to generation of singlets at rate κ_{TTD} [27]. Except for the last term on the r.h.s., all other terms are the corresponding counterparts of terms in (2). In the first sink term, the dopant singlets decay radiatively at rate κ_{SD}. For the Alq3-DCM host-guest system we have taken the value $\kappa_{SD} = 1.0 \times 10^9\ \mathrm{s}^{-1}$ [24]. The last term describes the dopant singlet interaction with the photons due to stimulated emission with differential gain coefficient ξ.

3.1.5. Dopant Triplet Excitons

Rate Equation (5) describes the dopant triplet density N_{TD} variations. The first term matches the corresponding Dexter transfer term in (3). The second term is the source resulting from the ISC matching the corresponding fourth term in (4). The third term represents the decay of the dopant triplet density at rate κ_{TD} [26] by de-excitation, while other terms correspond to the absorption processes TTA (κ_{TTD}) and TPA (κ_{TPD}) [27].

3.1.6. Photons and Linewidth

Rate Equation (6) accounts for the dynamics of the photon density P_{HO}. The first term on the r.h.s. gives the spontaneous-emission contribution arising from the radiative recombination of the dopant singlets N_{SD} at the rate κ_{SD} where the spontaneous-emission factor β_{sp} is the fraction of emitted photons within the lasing mode. The second term gives the net-amplification rate due to stimulated-emission

$$A_{STIM} \equiv \Gamma \xi_E M(E,E,CAV)(N_{SD} - WN_{0D}) - \kappa_{CAV}, \tag{10}$$

which will be large and negative so long the device operates below the lasing threshold but will climb up to a value close to zero if lasing is to be reached. In (10), $(N_{SD} - WN_{0D})$ is the effective inversion. Γ is the confinement factor introduced to consider the fact that only the part of the photons inside the emitting layer is amplified. The last term on the r.h.s. accounts for the photon losses out of the cavity, with decay rate $\kappa_{CAV} = 1/\tau_{CAV}$, where

τ_{CAV} is the cavity photon lifetime. The dopant singlet density, for which the photon net loss, $WN_{0D} + \frac{\kappa_{CAV}}{\Gamma\xi_E M(E,E,CAV)}$, is precisely compensated, defines the threshold for lasing, i.e.,

$$N_{SD|thr} = WN_{0D} + \frac{\kappa_{CAV}}{\Gamma\xi_E M(E,E,CAV)}, \tag{11}$$

As we will see in Section 4, laser operation is characterized by the clamping of the dopant singlet density at a value very close to the value defined in (11) at the same time the net-amplification rate (11) is clamping near zero.

The frequency linewidth $\Delta\nu$ of the emitted light can be related to the effective photon cavity decay rate (see the last term in (6))

$$\kappa_{CAV,eff} = \kappa_{CAV} + \Gamma\xi_E M(E,E,CAV)(WN_{0D} - N_{SD}) \tag{12}$$

as [31]

$$\Delta\nu = \frac{\kappa_{CAV,eff}}{2\pi}, \tag{13}$$

valid as long the system is quasi cw and no linewidth enhancement due to amplitude-phase coupling occurs. The cavity width Δ_{CAV} that should be substituted in $M(E,E,CAV)$ and W (see (A7), (A8), and (A11)), is related to the linewidth (13) as $\Delta_{CAV} = \frac{\lambda_E^2}{2\pi c}\Delta\nu$, with c the vacuum light velocity.

3.1.7. Cavity Quality Factor

The outcoupling, diffraction, and absorption of the light in the cavity define a relationship between the cavity photon lifetime τ_{cav} and the corresponding quality factor Q which reads:

$$Q = \omega_0\tau_{CAV}, \tag{14}$$

where ω_0 is the resonance (angular) frequency of the cavity mode. The cavity photon decay rate κ_{CAV} can be expressed in the quality factor Q and the resonance wavelength in vacuum λ_{CAV} as

$$\kappa_{CAV} = \frac{1}{\tau_{CAV}} = \frac{2\pi c}{n\lambda_0 Q}, \tag{15}$$

where n is the refractive index. At 620 nm wavelength, a typical value for an OLED undergoing a parasitic weak microcavity is $Q \approx 6$, corresponding to a cavity decay rate of $\kappa_{CAV} \approx 3.0 \times 10^{14}\ \text{s}^{-1}$. In a DFB-type laser cavity, a reasonable value for the quality factor $Q \approx 1800$ is achievable, and this corresponds to a cavity decay rate $\kappa_{CAV} \approx 1.0 \times 10^{12}\ \text{s}^{-1}$.

3.1.8. (Re)Absorption Factor W

Despite the Stoke shift, the absorption spectrum $S_A(\lambda)$ and the emission spectrum $S_E(\lambda)$ of the emitted light by the dopant show some overlap, which induces a residual reabsorption of the photons emitted in the cavity by the dopant singlet excitons N_{SD}. With W representing the spectral overlap, the reabsorption rate per unit photon density equals $\Gamma\xi_E M(E,E,CAV)WN_{0D}$. Note that the reabsorption of photons yields a source term for the dopant singlet population in (4). In the bad-cavity limit, the broad cavity spectrum S_{CAV} and the absorption spectrum S_A maximally overlap, hence re-absorption is maximal. When approaching the lasing threshold, the effective cavity spectrum narrows, and W will assume its smallest value.

In Appendix A an expression for W is derived in terms of integrals of intersecting spectra. For a model with Gaussian absorption and emission spectra, this expression can be written as (see (A8) and (A12))

$$W \equiv \frac{\xi_A M(A,E,CAV)}{\xi_E M(E,E,CAV)} = \frac{\xi_A C(A,E)\Delta_0(A,E)V(\lambda_{CAV} - \lambda_0(A,E);\Delta_0(A,E),\Delta_{CAV})}{\xi_E C(E,E)\Delta_0(E,E)V(\lambda_{CAV} - \lambda_0(E,E);\Delta_0(E,E),\Delta_{CAV})}, \tag{16}$$

where V is the Voigt function [32], and definitions for $C(X,Y), \Delta_0(X,Y), \lambda_0(X,Y)$, and $\Delta_0(X,Y)$ are given in ((A13)–(A15)). Note that $W = 1$ in case of identical spectra ($\lambda_A = \lambda_E$, $\xi_A = \xi_E$ and $\Delta_A = \Delta_E$), irrespective of λ_{CAV} and Δ_{CAV}. Taking $\lambda_A = 460$ nm, $\Delta_A = 50$ nm, $\lambda_E = \lambda_{CAV} = 620$ nm and $\Delta_E = \Delta_{CAV} = 50$ nm, we find $W = 0.035$.

4. Simulations

4.1. Below Laser Threshold

We will first present results for an OLED without a special optical cavity, that is, for which $\kappa_{CAV} = 1 \times 10^{14}$ s^{-1}, corresponding to $Q \cong 18$. The current density of ~500 A/cm^2, switched on at time 0, is applied during 300 ns and the parameters are as in Table 1, except for some values mentioned in the caption of Figure 2. In Figure 2a, apart from the current density, time evolutions are seen for the ground-state probabilities P_0 for a host molecule, P_{0D} for a dopant molecule, the spectral overlap W and the linewidth $\Delta\upsilon$ of the light emitted by the dopant single excitons. W and $\Delta\upsilon$ remain nearly constant at respective values 5×10^{-2} and 1.6×10^{13} s^{-1}. Due to the formation of excitons, the fraction of dopant molecules in the ground state falls to zero in ~100 ns, when the total number of dopant excitons approaches the total number of dopant molecules. This is demonstrated in Figure 2b, where it is seen that after ~100 ns the dopant triplet density N_{TD} is already at the level of the total dopant density of 4.2×10^{19} cm^{-3}. We recall that the light is emitted by the dopant singlet density N_{SD} and the photon density P_{HO} is seen to exhibit the same time development, apart from a proportionality constant. This is indicative for spontaneous emission. An important feature visible in Figure 2b is the sharp decrease of N_{SD} after reaching its maximum ~7.5 ns after the current offset. The reason for this is the rapid increase of N_{TD} and the associated singlet-triplet absorption STA leading to the dominant contribution to the singlet decay rate (see (4)) $\kappa_{STD} N_{TD} \sim 7.6 \times 10^9$s^{-1}. Therefore, with the parameters as in Table 1, sufficient amount of gain to reach a laser threshold can only be expected in a small time interval below ~10 ns.

4.2. Validation of the Model for an OLED

To validate our model, we have confronted our simulation with an experimental analysis of an electrically pumped OLED without a special cavity. In this device, the organic hetero-structure itself defines a residual weak micro-cavity effect ($Q \sim 6$, $\kappa_{CAV} \sim 3.0 \times 10^{14}$ s^{-1} and reabsorption fraction $W \sim 8\%$. A 20 ns, 45 V pulse excitation voltage is applied to the OLED and the electrical injection current is measured and recorded together with the emitted light intensity. This measured current is taken as the source term in the polaron rate Equation (1). The exciton and photon densities are then calculated from the set of Equations (1)–(8) with model parameters from Table 1, except for the fitted parameters given in Table 2. The results are plotted in Figure 3, where Figure 3d shows the measured and simulated photon densities in one plot for comparison.

The values for γ and κ_{DEXT} in Table 2 are the result of detailed fitting of the shape of the simulated photon response to the measured data. Variations in κ_{DEXT} values mainly affect the leading edge and the maximum of the photon response. The black dashed curve in Figure 4d is the simulated photon density if the literature value $\gamma = 6.2 \times 10^{-12}$ cm^3 s^{-1} instead of the fitted value in Table 2 is taken. The nearly two orders of magnitude larger value for γ stems from the Poole-Frenkel effect for the mobility due to the internal electric field, induced by the high voltage of 45 V applied in the experiment (see Section 3.1.1). The validation of our model for a sub-threshold case is important, since it validates the gain behavior represented by the dopant singlet density N_{SD}. When the cavity quality factor is increased, the effective amplification rate by stimulated emission (see Equation (6)) will increase from large negative toward zero, without changing the underlying exciton dynamics, except N_{SD} near and above the threshold. In fact, here the last term in Equation (4) will become the dominant loss, leading to clamping of N_{SD} to the threshold value given in (10). Hence, we can use our validated model to predict laser operation behavior based on simulations where we increase the Q-factor of the OLED.

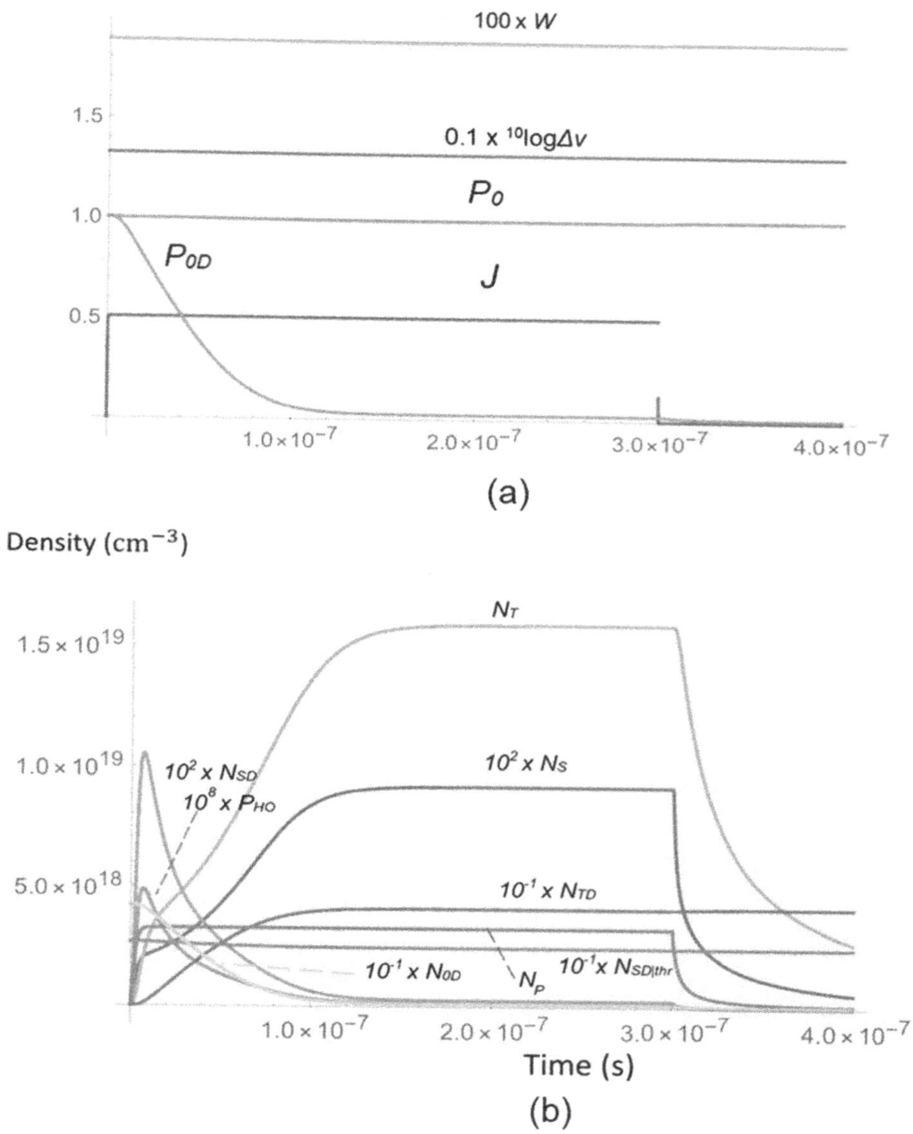

Figure 2. Time evolutions of various quantities for an OLED far below the threshold for lasing. The applied injection current has amplitude 0.5 kA/cm^2, and a duration of 300 ns. There is no special cavity arrangement assumed; $\kappa_{CAV} = 1 \times 10^{14}\ \text{s}^{-1}$, corresponding to $Q \cong 18$, $\gamma = 1 \times 10^{-10}\ \text{cm}^3\ \text{s}^{-1}$, $\beta_{sp} = 0.05$, $\kappa_{DEXT} = 2 \times 10^8\ \text{s}^{-1}$, $\kappa_T = 6.5 \times 10^2\ \text{s}^{-1}$, $\kappa_{ISC(D)} = 2.2 \times 10^4\ \text{s}^{-1}$. (**a**) Current density J in kA/cm^2, ground-state probabilities P_0 and P_{OD} respectively for host and dopant molecules and the linewidth Δv of the emitted light. (**b**) Polaron density N_P, exciton densities for host and dopant molecules and the photon density P_{HO}. The light blue curve is $10^{-1} \times N_{SD}|_{thr}$ (see (11); clearly, the maximum of N_{SD} (~ 10^{17} cm^{-3}) is far below the threshold value (~2.5×10^{19} cm^{-3}).

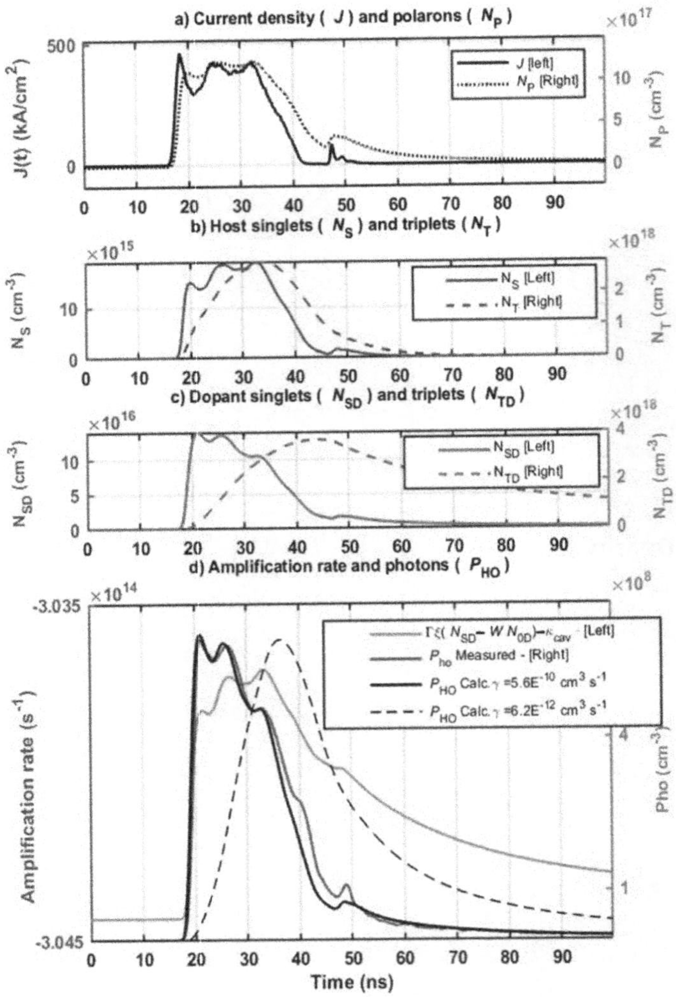

Figure 3. Comparison between measurements and simulation of the dynamical optical responses of an OLED under a peak current density of 0.462 kA/cm^2. In (**a**) the measured current and the simulated polaron density are depicted, in (**b**) the simulated host singlet and triplet densities and in (**c**) the dopant excitons. In (**d**) the measured (red curve) and simulated (black curve) photon densities (right scale) are shown together with the amplification rate (10) (blue curve; left scale). Parameters used in the simulation are given in Table 1, except for those given in Table 2. The dashed black curve in (**d**) is the simulated photon density for the zero-voltage value $\gamma = 6.2 \times 10^{-12}$ cm^3 s^{-1}.

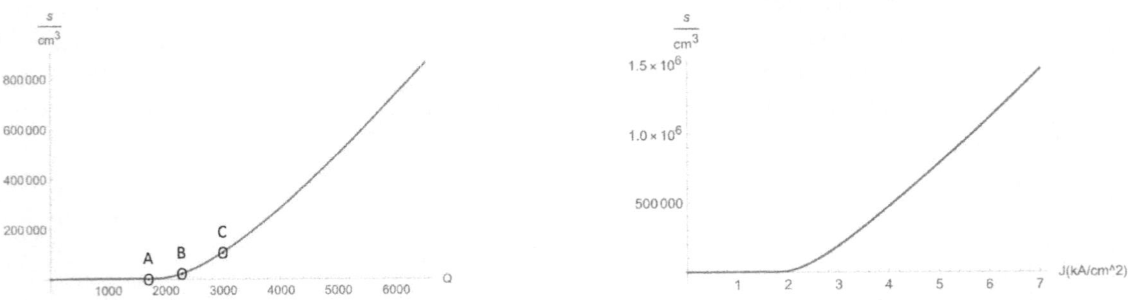

(a) (b)

Figure 4. Simulated time-integrated photon density versus (**a**) the cavity quality factor Q and (**b**) the applied electrical current density J. In (**a**) the pump current is 2.0 kA/cm^2 and the points A, B, C correspond to the cases of Figure 5a–c, respectively; in (**b**) the quality factor is 2000 ($\kappa_{CAV} \sim 9 \times 10^{11}$ s^{-1}). The duration of the pulse is 20 ns; the parameter values are for the validated case in Figure 3 and $\beta_{sp} = 3 \times 10^{-4}$.

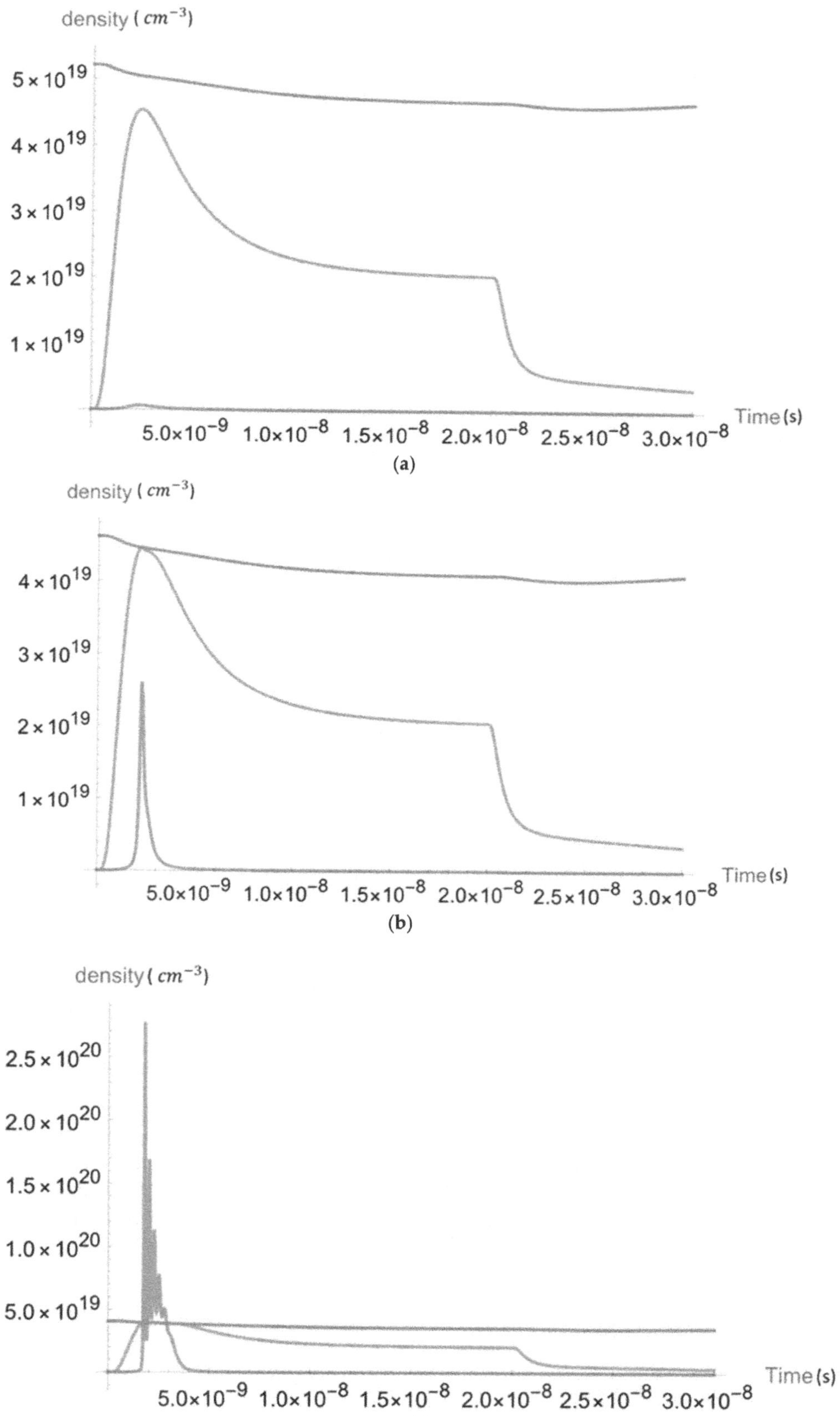

Figure 5. Time evolutions of the photon density $P_{HO}(\times 10^6)$ (red), singlet density $N_{SD}(\times 10^2)$ (green) and threshold density $N_{SD}|_{thr}(\times 10^2)$ (blue; see (11)) for (**a**) below threshold $Q = 1700$, (**b**) at threshold $Q = 2200$ and (**c**) above threshold $Q = 3000$. In each case the pump current density is 2 kA/cm^2 and applied from 0 to 20 ns while $\beta_{sp} = 3 \times 10^{-4}$.

Table 2. Model Parameters for Figure 3.

Symbol	Name	Value
S	OLED active area	1.5×10^{-4} cm^2
γ	Langevin recombination rate	5.6×10^{-10} cm^3 s^{-1}
κ_{DEXT}	Dexter transfer rate	2.0×10^8 s^{-1}
κ_S	Host singlet-exciton decay rate	8.3×10^7 s^{-1}
κ_T	Host triplet decay rate	6.5×10^2 s^{-1}
$\kappa_{ISC(D)}$	Dopant inter-system crossing rate	2.2×10^4 s^{-1}
κ_{SP}	Host SPA rate	1.0×10^{-11} s^{-1}
κ_{SPD}	Dopant SPA rate	3.0×10^{-10} cm^3 s^{-1}
κ_{TPD}	Dopant TPA rate	9.0×10^{-11} cm^3 s^{-1}
κ_{ST}	Host STA rate	2.5×10^{-10} cm^3 s^{-1}
κ_{STD}	Dopant STA rate	3.7×10^{-10} cm^3 s^{-1}
κ_{TTD}	Dopant TTA rate	8.0×10^{-12} cm^3 s^{-1}
Γ	Confinement factor	0.29
β_{sp}	Spontaneous emission factor	1.3×10^{-3}

4.3. Laser Predictions

So far, in the literature only one apparent though reserved and modest claim of observed lasing in an electrically pumped organic diode is given. It is reported in [33] and the OLED has the organic BsB-Cz as gain material in a configuration as sketched in Figure 1. Our model is validated for sub-threshold behavior but as we argued in the previous section, we can predict laser operation behavior by simulations with increasing Q-factor.

Figure 4a shows the simulated LQ-characteristic for the case of pump current $J = 2$ kA/cm^2. The pump pulse duration is 20 ns and the parameter values are for the validated case in Figure 3. A laser threshold is clearly seen at $Q_{TH} \sim 2200$ ($\kappa_{CAV} \sim 8.1 \times 10^{11}$ s^{-1}). In Figure 4b the integrated photon density versus the pump current is depicted for fixed quality factor $Q = 2000$ ($\kappa_{CAV} \sim 9.0 \times 10^{11}$ s^{-1}). This is the more usual LI-curve and shows the threshold at $J_{TH} \sim 2.2$ kA/cm^2. For the operation points labeled A, B, and C, the corresponding simulated photon density $P_{HO}(t)$, the singlet density $N_{SD}(t)$ and its threshold value N_{SDthr} are plotted in the respective Figure 5a–c. In Figure 5a the maximum of the singlet density (green curve) remains below the threshold value for lasing (blue curve); in Figure 5b the singlet top just touches the threshold value at $t \sim 3$ ns producing a short laser pulse during ~1 ns. In Figure 5c, the system is above threshold, the singlet density is clamped to its threshold value from $t = 1.6$ ns up to $t = 3.9$ ns and during this time interval the system emits stimulated emission. The photon density after the onset of lasing shows a damped oscillation with frequency 3.8 GHz.

Such a relaxation oscillation is well known to occur in conventional III-V semiconductor lasers and more generally in class-B lasers [11,34]. For the case of Figure 5c, the evolutions of the reabsorption fraction $W(t)$ and the instantaneous linewidth $\Delta\nu$ (see (13) of the emitted light are shown in Figure 6 together with the photon density oscillations. Note the 3 to 4 orders of magnitude linewidth reduction during the lasing phase.

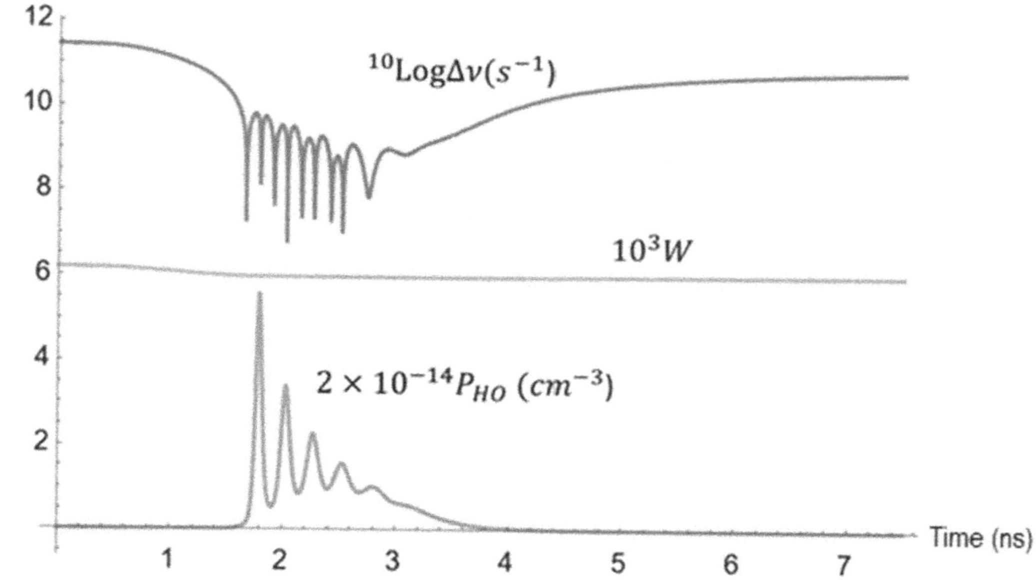

Figure 6. Time evolutions of the photon density (red), instantaneous linewidth (purple) and reabsorption factor (yellow) for the case of Figure 5c, i.e., $Q = 3000$ and $J = 2\ \text{kA/cm}^2$.

5. Discussion

With the formulation of the rate Equations (1)–(6) we have established the simplest possible model for an organic laser diode with a host-guest system that includes all known processes which underly the gain mechanism and the buildup of the photons in the cavity. The model applies to a host-guest system where the optical transitions of the host system can be disregarded and takes the reabsorption of the dopant into account in a self-consistent manner. The numerical simulations are for Alq3 as the host and a small volume fraction of 2% DCM as the dopant. Extension of the model to include light emitted from the host molecules is straightforward, and so is a different dopant fraction. The characteristic properties of the host and dopant molecules are reflected by the molecular parameter values, whereas the interaction of the gain with the optical field involves the optical-cavity parameters, such as the confinement and quality factor.

We have simulated the dynamics of the various molecular entities, i.e., the polarons and excitons responding to an electrical injection of 0.5 kA/cm^2 with a relatively long duration of 300 ns, which shows that a (quasi) steady state is reached after ~100 ns, which is characterized by a fully quenched gain. The reason is that the buildup of triplet excitons continues until all dopant molecules are used and no gain-providing singlets are present anymore. The bi-molecular interaction process of triplet-singlet absorption (TPA) starts to hamper the buildup of singlets already a few nanoseconds after the onset of the electrical injection. From this information it is concluded that if laser action with Alq3:DCM is to occur, it only will happen during a short time interval of a few nanoseconds.

The model has been validated by applying it to an experimental analysis of an electrically pumped OLED with weak residual micro-cavity effect defined by the organic heterostructure itself. The measured electrical current is taken as the source term in (1) and the emitted light intensity obtained with the model simulation is compared with the observed intensity. By fitting some of the parameters, good agreement was obtained (Figure 3d). Although in this case the emitted light is amplified spontaneous emission, we argue that based on the validation we can extrapolate to the case of laser emission. The argument is that most of the molecular dynamics remains unchanged; it is rather the cavity quality factor that will be different.

This leads to predictions for single-pulse laser operation during a few nanoseconds at most and accompanied by several orders of magnitude linewidth reduction and relaxation oscillations. For an applied current density amplitude of 2 kA/cm^2 the threshold can be reached for $Q \sim 2200$, and for $Q \sim 2000$ the threshold current $J \sim 2.2$ kA/cm^2. These are all feasible values for practical systems.

Finally, we can speculate on the feasibility of CW laser operation. Inspection of Figure 5c suggests that if the singlet density N_{SD} could be maintained at a larger value, it might be possible to extend the laser operation to a longer time interval. Apart from the singlet-exciton decay rates $\kappa_{S(D)}$, the parameters that influence the singlet decay most are the singlet-triplet absorption and to a lesser extent the singlet-polaron absorption rate. Indeed, assuming somewhat smaller values for $\kappa_{ST(D)}$ and $\kappa_{TP(D)}$ than in Table 2, we find CW laser operation with a linewidth of ~65 MHz in a simulation longer than 1000 ns electrical pumping for $J = 2.8$ kA/cm^2, $Q = 2000$ and $\kappa_{ST(D)} = 5 \times 10^{-11}$ cm^3 s^{-1}, $\kappa_{SP(D)} = 1.0 \times 10^{-11}$ cm^3 s^{-1}. For this case the laser threshold for CW operation is at ~ 0.8 kA/cm^2. We have indications that these parameter values apply to the organic material BsB-Cz [16,33]. A systematic investigation of CW laser operation in dependance of molecular parameters will be published separately.

Appendix A

To derive an expression for W, we decompose the total photon density in its cavity mode continuum. The photons in the active layer are distributed in wavelength according to the emission spectrum $S_E(\lambda)$ intersected with the normalized Lorentzian cavity profile $S_{CAV}(\lambda)$ with $\int d\lambda S_{CAV}(\lambda) = 1$. Hence

$$P_{HO}(\lambda) \equiv S_{CAV}(\lambda) S_E(\lambda) P_{HO}, \tag{A1}$$

$$P_{HO} = \int d\lambda P_{HO}(\lambda). \tag{A2}$$

The stimulated emission rate in the wavelength interval $d\lambda$ is

$$\dot{P}_{HO}(\lambda)|_{SE} d\lambda = \xi_E S_E(\lambda) N_{SD} P_{HO}(\lambda) d\lambda \tag{A3}$$

and the absorption is

$$\dot{P}_{HO}(\lambda)|_{ABS} d\lambda = \xi_A S_A(\lambda) N_{0D} P_{HO}(\lambda) d\lambda \tag{A4}$$

where $S_X(\lambda)$ is the emission respectively absorption spectrum for X=E, A, with

$$S_X(\lambda_X) = 1 \tag{A5}$$

and λ_X is the wavelength for which the corresponding spectrum assumes its maximum value.

In principle, the respective linearized gain and absorption coefficients, ξ_A and ξ_E, may be different. Subtracting (A4) from (A3), integrating over λ and using (A2), the net stimulated emission can be written as

$$\dot{P}_{HO|stim} = \xi_E P_{HO} M(E,E,CAV)(N_{SD} - \frac{\xi_A M(A,E,CAV)}{\xi_E M(E,E,CAV)} N_{OD}), \tag{A6}$$

with

$$M(X,Y,CAV) \equiv \int d\lambda S_X(\lambda) S_Y(\lambda) S_{CAV}(\lambda),\ (X,Y = A,E). \tag{A7}$$

Hence, comparing (A6) with (6), we obtain

$$W \equiv \frac{\xi_A M(A,E,CAV)}{\xi_E M(E,E,CAV)} = \frac{\xi_A \int d\lambda S_A(\lambda) S_E(\lambda) S_{CAV}(\lambda)}{\xi_E \int d\lambda S_E^2(\lambda) S_{CAV}(\lambda)}. \tag{A8}$$

For numerical purposes, the following explicit forms for the spectra are introduced:

$$S_A(\lambda) = e^{-\frac{(\lambda-\lambda_A)^2}{2\Delta_A^2}} \text{ (absorption spectrum approximated by a Gaussian);} \tag{A9}$$

$$S_E(\lambda) = e^{-\frac{(\lambda-\lambda_E)^2}{2\Delta_E^2}} \text{ (emission spectrum approximated by a Gaussian).} \tag{A10}$$

Then, with the normalized Lorentzian cavity spectrum

$$S_{CAV}(\lambda) = \frac{\Delta_{CAV}}{\pi((\lambda-\lambda_{CAV})^2 + \Delta_{CAV}^2)}, \tag{A11}$$

the spectral overlap integral $M(X,Y,CAV)$ can be expressed in the Voigt function V [32],

$$M(X,Y,CAV) = C(X,Y)\sqrt{2\pi}\Delta_0(X,Y)V(\lambda_{CAV} - \lambda_0(X,Y); \Delta_0(X,Y), \Delta_{CAV}), \tag{A12}$$

where V is a standard built-in function and

$$C(X,Y) \equiv e^{-\frac{(\lambda||X-\lambda_Y)^2}{2(\Delta_X^2+\Delta_Y^2)}}, \tag{A13}$$

$$\lambda_0(X,Y) \equiv \frac{\Delta_Y^2\lambda_X + \Delta_X^2\lambda_Y}{\Delta_Y^2 + \Delta_X^2}, \tag{A14}$$

$$\Delta_0(X,Y) \equiv \frac{\Delta_X\Delta_Y}{\sqrt{\Delta_X^2 + \Delta_Y^2}}. \tag{A15}$$

In case emission and absorption spectra are equal, i.e., $S_A(\lambda) = S_E(\lambda), \xi_A = \xi_E$ and $\lambda_A = \lambda_E$ we find $W = 1$. With $\lambda_A = 460$ nm, $\Delta_A = 50$ nm, $\lambda_{CAV} = \lambda_E = 620$ nm, $\Delta_E = \Delta_{CAV} = 50$ nm, we calculate $W = 0.035$. For $\Delta_{CAV} \to 0$, i.e., very narrow cavity line ($Q \to \infty$), we find $W \to S_A(\lambda_E) = 0.006$. With $\Delta_{CAV} = \lambda_E^2\kappa_{CAV}/(2\pi c)$ and $\kappa_{CAV} = 10^{14}\ \mathrm{s}^{-1}$, we find $\Delta_{CAV} = 20$ nm and $W = 0.019$.

References

1. Yang, Y.; Turnbull, G.A.; Samuel, D.W. Sensitive Explosive Vapor Detection with Polyfluorene Lasers. *Adv. Funct. Mater.* **2010**, *20*, 2093–2097. [CrossRef]
2. Yoshida, K.; Manousiadis, P.P.; Bian, R.; Chen, Z.; Murawski, C.; Gather, M.C.; Haas, H.; Turnbull, G.A.; Samuel, I.D.W. 245 MHz bandwidth organic light-emitting diodes used in a gigabit optical wireless data link. *Nat. Commun.* **2020**, *11*, 1–7. [CrossRef] [PubMed]
3. Chiang, C.K.; Fincher, C.R., Jr.; Park, Y.W.; Heeger, A.J.; Shirakawa, H.; Louis, E.J.; Gau, S.C.; MacDiarmid, A.G. Electrical Conductivity in Doped Polyacetylene. *Phys. Rev. Lett.* **1977**, *39*, 1098. [CrossRef]
4. Fitzner, R.; Mena-Osteritz, E.; Mishra, A.; Schulz, G.; Reinold, E.; Weil, M.; Körner, C.; Ziehlke, H.; Elschner, C.; Leo, K.; et al. Correlation of π-Conjugated Oligomer Structure with Film Morphology and Organic Solar Cell Performance. *J. Am. Chem. Soc.* **2012**, *134*, 11064–11067. [CrossRef] [PubMed]
5. Lee, K.-H.; Leem, D.-S.; Sul, S.; Park, K.-B.; Lim, S.-J.; Han, H.; Kim, K.-S.; Jin, Y.W.; Lee, S.; Park, S.Y. A high-performance green-sensitive organic photodiode comprising a bulk heterojunction of dimethyl-quinacridone and dicyanovinyl terthiophene. *J. Mater. Chem. C* **2013**, *1*, 2666. [CrossRef]
6. Wang, C.; Dong, H.; Hu, W.; Liu, Y.; Zhu, D. Semiconducting π-Conjugated Systems in Field-Effect Transistors: A Material Odyssey of Organic Electronics. *Chem. Rev.* **2011**, *112*, 2208. [CrossRef]
7. Han, T.-H.; Lee, Y.; Choi, M.-R.; Woo, S.; Bae, S.-H.; Hong, B.H.; Ahn, J.-H.; Lee, T.-W. Extremely efficient flexible organic light-emitting diodes with modified graphene anode. *Nat. Photon* **2012**, *6*, 105–110. [CrossRef]
8. Herrnsdorf, J.; Guilhabert, B.; Chen, Y.; Kanibolotsky, A.L.; Mackintosh, A.R.; Pethrick, R.A.; Skabara, P.J.; Gu, E.; Laurand, N.; Dawson, M.D. Flexible blue-emitting encapsulated organic semiconductor DFB laser. *Opt. Express* **2010**, *18*, 25535–25545. [CrossRef]
9. Torricelli, F.; Colalongo, L. Unified Mobility Model for Disordered Organic Semiconductors. *IEEE Electron. Device Lett.* **2009**, *30*, 1048–1050. [CrossRef]
10. Murgatroyd, P.N. Theory of space-charge-limited current enhanced by Frenkel effect. *J. Phys. D Appl. Phys.* **1970**, *3*, 151–156. [CrossRef]
11. Ohtsubo, J. *Semiconductor Lasers, Stability, Instability and Chaos*; Springer: Berlin, Germany, 2017. [CrossRef]
12. Segura, J.L. The chemistry of electroluminescent organic materials. *Acta Polym.* **1998**, *49*, 319–344. [CrossRef]
13. Chenais, S.; Forget, S. Recent Advances in Solid-State Organic Lasers. *Polym. Int.* **2011**, *61*, 390–406. [CrossRef]
14. Gozhyk, I. Polarization and Gain Phenomena in Dye-Doped Polymermicro-Lasers. Ph.D. Thesis, Laboratoire de Photonique Quantique et Moléculaire, Cachan, France, 2012.
15. Senevirathne, C.A.M.; Sandanayaka, A.S.D.; Karunathilaka, B.S.B.; Fujihara, T.; Bencheikh, F.; Qin, C.; Goushi, K.; Matsushima, T.; Adachi, C. Markedly Improved Performance of Optically Pumped Organic Lasers with Two-Dimensional Distributed-Feedback Gratings. *ACS Photonics Artic. ASAP* **2021**. [CrossRef]
16. Ouirimi, A.; Chamberlain Chime, A.; Loganathan, N.; Chakaroun, M.; Fischer, A.P.A.; Lenstra, D. Threshold estimation of an Organic Laser Diode using a rate-equation model validated experimentally with a microcavity OLED submitted to nanosecond electrical pulses. *Org. Electron.* **2021**, *97*, 106190. [CrossRef]
17. Gärtner, C. Organic Laser Diodes: Modelling and Simulation. Ph.D. Thesis, Universität Karlsruhe, Karlsruhe, Germany, 2008.
18. Chua, S.-L.; Zhen, B.; Lee, J.; Bravo-Abad, J.; Shapira, O.; Soljačić, M. Modeling of threshold and dynamics behavior of organic nanostructured lasers. *J. Mater. Chem. C* **2014**, *2*, 1463–1473. [CrossRef]
19. Shoustikov, A.A.; You, Y.; Thompson, M.E. Electroluminescence color tuning by dye doping in organic light-emitting diodes. *IEEE J. Select. Top. Quantum Electron.* **1998**, *4*, 3–13. [CrossRef]
20. Gärtner, C.; Karnutsch, C.; Pflumm, C.; Lemmer, U. Numerical Device Simulation of Double-Heterostructure Organic Laser Diodes Including Current-Induced Absorption Processes. *IEEE J. Quantum Electron.* **2007**, *43*, 1006–1017. [CrossRef]
21. Juhasz, P.; Nevrela, J.; Micjan, M.; Novota, M.; Uhrik, J.; Stuchlikova, L.; Jakabovic, J.; Harmatha, L.; Weis, M. Charge injection and transport properties of an organic light-emitting diode. *Beilstein J. Nanotechnol.* **2016**, *7*, 47–52. [CrossRef]
22. Tsutsumi, N.; Hinode, T. Tunable organic distributed feedback dye laser device excited through Förster mechanism. *Appl. Phys. B* **2017**, *123*, 93. [CrossRef]
23. Kozlov, V.G.; Bulovic, V.; Burrows, P.E.; Baldo, M.; Khalfin, V.B.; Parthasaraty, G.; Forrest, S.R.; You, Y.; Thompson, E. Study of lasing action based on Förster energy transfer in optically pumped organic semiconductor thin films. *J. Appl. Phys.* **2014**, *84*, 14. [CrossRef]
24. Kasemann, D.; Brückner, R.; Fröb, H.; Leo, K. Organic light emitting diodes under high currents explored by transient electroluminescence on the nanosecond scale. *Phys. Rev. B* **2011**, *84*, 115208. [CrossRef]
25. Mori, T.; Obata, K.; Miyachi, K.; Mizutani, T.; Kawakami, Y. Fluorescence Lifetime of Organic Thin Films Alternately Deposited with Diamine Derivative and Aluminum Quinoline. *Jpn. J. Appl. Phys.* **1997**, *36*, 7239. [CrossRef]
26. Lehnhardt, M.; Riedl, T.; Rabe, T.; Kowalsky, W. Room temperature lifetime of triplet excitons in fluorescent host/guest systems. *Org. Electron.* **2011**, *12*, 486–491. [CrossRef]
27. Zhang, S.; Song, J.; Kreouzis, T.; Gillin, W.P. Measurement of the intersystem crossing rate in aluminum tris(8-hydroxyquinoline) and its modulation by an applied magnetic field. *J. Appl. Phys.* **2009**, *106*, 043511. [CrossRef]

28. Giebink, N.C.; Forrest, S.R. Temporal response of optically pumped organic semiconductor lasers and its implication for reaching threshold under electrical excitation. *Phys. Rev. B* **2009**, *79*, 073302. [CrossRef]
29. Gispert, J.R. *Coordination Chemistry*; Wiley-VCH: Hoboken, NJ, USA, 2008; p. 483, ISBN 3-527-31802-X. [CrossRef]
30. Zeng, L.; Chamberlain Chime, A.; Chakaroun, M.; Bensmida, S.; Nkwawo, H.; Boudrioua, A.; Fischer, A.P.A. Electrical and Optical Impulse Response of High-Speed Micro-OLEDs Under UltraShort Pulse Excitation. *IEEE Trans. Electron. Devices* **2017**, *64*, 2942–2948. [CrossRef]
31. Coldren, L.A.; Corzine, S.; Mashanovitch, M. *Diode Lasers and Photonic Integrated Circuits*, 2nd ed.; Wiley: Hoboken, NJ, USA, 2012; p. 293, ISBN 978-0-470-48412-8.
32. Olver, F.W.J.; Lozier, D.W.; Boisvert, R.F.; Clark, C.W. *NIST Handbook of Mathematical Functions*; Cambridge University Press: New York, NY, USA, 2010; Chapter 7.19; ISBN 978-0-521-19225-5.
33. Sandanayaka, A.S.D.; Matsushima, T.; Bencheikh, F.; Terakawa, S.; Potscavage, W.J., Jr.; Qin, C.; Fujihara, T.; Goushi, K.; Ribierre, J.-C.; Adachi, C. Indication of current-injection lasing from an organic semiconductor. *Appl. Phys. Express* **2019**, *12*, 061010. [CrossRef]
34. Erneux, T.; Glorieux, P. *Laser Dynamics*; Cambridge University Press: Cambridge, UK, 2010; ISBN 978-0-521-83040-9.

The Overlap Factor Model of Spin-Polarised Coupled Lasers

Abstract: A general model for the dynamics of arrays of coupled spin-polarised lasers is derived. The general model is able to deal with waveguides of any geometry with any number of supported normal modes. A unique feature of the model is that it allows for independent polarisation of the pumping in each laser. The particular geometry is shown to be introduced via 'overlap factors', which are a generalisation of the optical confinement factor. These factors play an important role in determining the laser dynamics. The model is specialised to the case of a general double-guided structure, which is shown to reduce to both the spin flip model in a single cavity and the coupled mode model for a pair of guides in the appropriate limit. This is applied to the particular case of a circular-guide laser pair, which is analysed and simulated numerically. It is found that increasing the ellipticity of the pumping tends to reduce the region of instability in the plane of pumping strength versus guide separation.

Keywords: spin-VCSELs; laser arrays; laser dynamics; spin flip model; coupled lasers

---

1. Introduction

Detailed control of the polarization dynamics of vertical cavity surface emitting laser (VCSEL) arrays opens up many technological opportunities. In addition to being high-power sources, potential applications may lie in emerging areas, such as reservoir computing based on polarization dynamics [1] and secure key distribution based on chaos synchronization [2,3], as well as an enhanced understanding of anticipation in polarisation chaos synchronisation of coupled VCSELs [4] for secure communications.

Key to this development is the understanding of both polarisation dynamics in a given laser and the evanescent coupling to nearby elements in the laser array. The dynamics in a single laser are governed by the interplay between spin-polarised carriers and the circularly polarised components of the optical field. The spin flip model (SFM) [5] is now well-established as providing an excellent quantitative description of these effects. However, despite extensive developments of the SFM, it remains restricted to the description of single lasers.

The dynamics of coupled lasers are typically studied using a coupled mode model. However, this approach has the shortcoming that it does not allow for independent polarisation of the pump in each laser, whether this is optical or via spin-polarised injection current. Nor does it allow us to investigate the spatial variation in the optical polarisation. What is required to deal properly with this scenario is a supermode (also known as normal mode) model, in which one finds the optical solutions of the entire array. This also allows for the spatial distribution of the polarisation to be studied. Whilst supermode studies of laser arrays have been undertaken [6], to our knowledge none have incorporated the optical polarisation or carrier spin-dynamics . This current work is intended to remedy this deficit, in which we extend the SFM to apply to a general supermode model for laser arrays.

The basic SFM consists of four coupled rate equations (two for spin-polarised carriers and two for polarised field components) and it includes rates of carrier recombination, photon field decay, and electron spin relaxation (spin relaxation of holes is usually assumed to be instantaneous). The nonlinear dispersion that couples the carrier concentrations to the phases of the optical fields is described by the linewidth enhancement factor, and the field interactions due to nonlinear anisotropies are included via rates of birefringence and dichroism. For conventional VCSELs, the SFM has been applied to explain the experimental results of polarisation switching (PS) [7,8]. An "extended SFM" [9] that accounts for thermal effects and includes a realistic spectral dependence of the gain and the index of refraction of the QWs has been used [10] to explain the experimental results on elliptically polarised dynamical states that occur in the polarisation dynamics of VCSELs in the vicinity of one type of PS. For a more complete discussion of polarisation dynamics in VCSELs, the reader is referred to [11].

A further development of the extended SFM [12] includes a description of the spatial variation of the electromagnetic modes and the carrier densities. The variation in the longitudinal direction is dealt with by integration over the length of the VCSEL cavity, whilst the radial and azimuthal variation is described by accurate solutions of the wave equation. The model assumes a given functional dependence of the guiding mechanisms (built-in refractive index and thermal lensing) as well as the spatial dependence of the current density. The transverse mode behaviour of gain-guided, bottom, and top-emitter VCSELs were studied, and it was shown that the stronger the thermal lens, the stronger the tendency toward multimode operation, which indicates that high lateral uniformity of the temperature is required in order to maintain single mode operation in gain-guided VCSELs. Additionally, close-to-threshold numerical simulations showed that, depending on the current profile, thermal lensing strength and relative detuning, different transverse modes could be selected.

Another version of the extended SFM [13] includes a rate equation for the temperature of the active region, which takes decay to a fixed substrate temperature, Joule heating, and heating due to non-radiative recombination into account. The temperature dependence of the PS point is characterised in terms of various model parameters, such as the room-temperature gain-cavity offset, the substrate temperature, and the size of the active region.

The SFM has also been widely applied to describe the behaviour of spin-VCSELs whose output polarisation can be controlled by the injection of spin-polarised electrons using either electrical or optical pumping (for a review with more details, see [14]). In the latter case the polarisation of the optical pump is included [15] to reveal its effect on the output polarisation [16–18]. The SFM has also been used [19–21] to explain the experimental results on high-speed polarisation oscillations that result from competition between the spin-flip processes, dichroism, and birefringence.

From this brief summary of the SFM and its applications, it is clear that the structures studied have been limited to single lasers, either conventional electrically driven VCSELs or spin-VCSELs that may be pumped electrically or optically. In the present contribution, we seek to extend the range of application to include structures, where two or more evanescently-coupled lasers are arranged in parallel to form arrays with the possibility of different lasers having differing pumping polarisation. To the best of our knowledge, this configuration has not been analysed previously, although there is one report [22] of an experiment where optical pumping with orthogonally polarised beams was used to study the interaction between two VCSELs as a function of their separation. There is considerable literature on laser arrays because of their important practical applications as high-power sources (including, most recently, for three-dimensional (3D) sensing in smartphones [23]) and very sophisticated models of VCSEL arrays have been developed [24]. Arrays of coupled lasers are also of fundamental interest in view of the range of nonlinear dynamics that they can exhibit (see, for example, [25] and references cited therein). Although there is sometimes a need to stabilise the polarisation of such arrays of VCSELs, the possibility of manipulating the output polarisation of an array by means of independent pumping polarisations has not yet been considered. Moreover new potential applications of the theory developed here may lie in emerging areas, such as reservoir computing based on polarization dynamics [1] and secure key distribution based on chaos synchronization [2,3], as well as an enhanced understanding of anticipation in polarisation chaos synchronisation of coupled VCSELs [4] for secure communications. Hitherto, these topics have been studied by adding delayed optical injection terms into the SFM equations [26], whereas the present theory permits a more general analysis which is applicable to a much wider range of VCSEL array applications. Thus, these developments, together with the issue of how the array dynamics is affected by spin-polarised pumping, provide the motivation for the present study (the reader is also referred to complementary work reported in [27], which focusses on numerical results in terms of new dynamics and regions of bistability obtained with this theory).

In Section 2, we derive a model for guided mode lasers of general geometry with any number of real-index guides and any number of normal modes. An important aspect of this model is the introduction of the 'overlap factors', as discussed in detail in Section 3. These are calculated by integrating products of the spatial mode solutions of the Helmholtz equation over the active regions. As such, they represent a generalisation of the optical confinement factor. It is through these factors that the particular geometry of the waveguide is introduced and their effect on the laser dynamics can be quite significant, as indicated in Ref. [28] in comparison to the coupled mode model [29].

Familiarity with the overlap factors should give the necessary physical intuition into their properties and limiting behaviour that we frequently exploit in the derivation of the double-guided model in Section 4. Here, we specialise to the case of just two guides and only consider the lowest two normal modes. This model is still quite general in regards to the waveguide geometry that may be simulated, although it is particularly appropriate for the case of symmetric waveguides. In this paper, we look at two particular cases: equal slab guides and equal circular guides, both with real, stepped refractive index profiles, as shown schematically in Figure 1. The application to coupled VCSELs with circular guides is indicated in the schematic of Figure 2, omitting the Bragg mirrors, substrate, and other structural details. Note that, in Figures 1 and 2, a resonant cavity is assumed with propagation in the z-direction, i.e., normal to the plane of optical confinement. No further account is taken of the z-direction in what follows and the values of parameters appearing in the analysis are assumed to be averaged over the cavity length. For widely separated guides, we show, in Section 4.2, that the model reduces to both the SFM [5,15,30] and coupled mode model [29] in the appropriate limits.

Having established the mathematical model, we investigate the effect of varying the optical pump polarisation in each guide via numerical simulation in Section 6. A novel feature of this model is that it allows for us to examine the spatial variation of the circularly polarised components of the optical intensity and the optical ellipticity throughout the waveguide structure. Section 6.2 provides examples of this. In Section 6.3, we give some introductory examples of stability boundaries in the plane of total pump power and normalised guide separation. This illustrates how we can use this model to investigate the effect of independently varying the pump polarisation in each guide. More generally, we may also vary the overall pump power or adjust the relative sizes of each guide, thereby introducing an effective frequency detuning. Such investigations are deferred for future study.

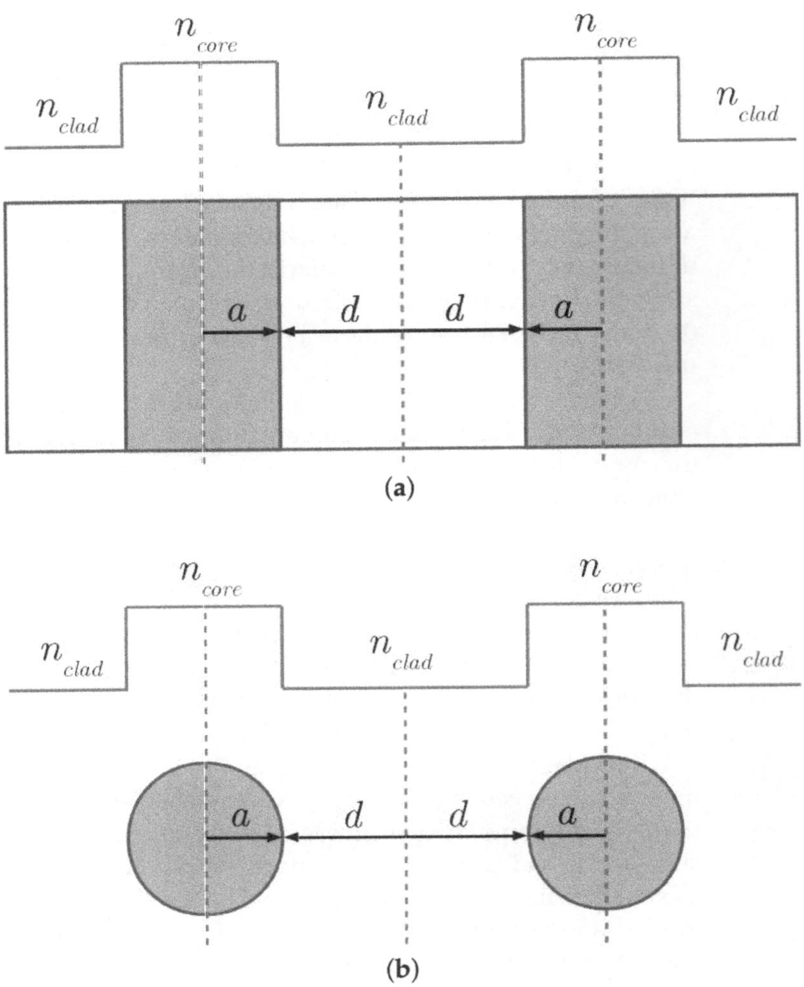

Figure 1. The refractive index profiles of the double guided structures used in this work. Here, the distance between the guides is given as $2d$, whilst a is used both for the half-width of a slab guide and the radius of a circular guide. Elsewhere in this work, we use $n_1 = n_{core}$ and $n_2 = n_{clad}$ for brevity. (**a**) Slab waveguide. (**b**) Circular waveguide.

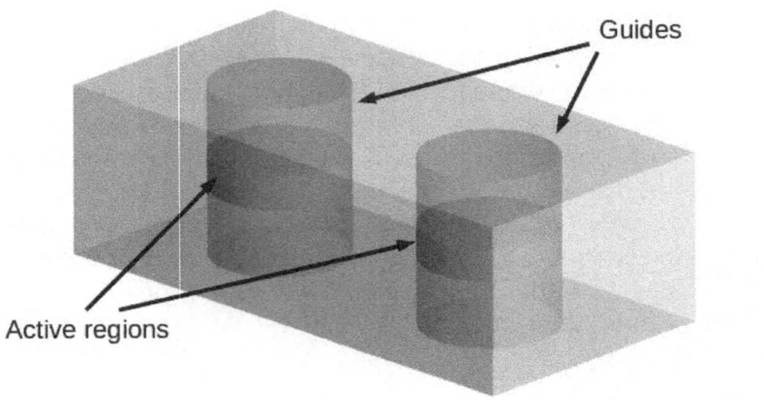

Figure 2. A three-dimensional (3D) schematic of two coupled circular waveguides encapsulating the essence of the application to a pair of vertical cavity surface emitting laser (VCSEL) cavities. Shown are the cylindrical waveguide regions incorporating the active areas. Pumping is assumed to be confined to these regions. Note that we have omitted the Bragg stack mirrors and substrate from this figure.

1.1. Summary Overview of the Paper

Briefly summarising the structure of the paper:

- Section 2: derivation of the general model for the dynamics of arrays of coupled, spin-polarised lasers;
- Section 3: discussion of the 'overlap factors', which incorporate the details of the geometry of the arrays;
- Section 4: specialisation to a double-guided structure and demonstration that the model reduces to both the spin-flip model and coupled mode model in the appropriate limit;
- Section 5: discussion of steady state solutions, giving exact and approximate expressions;
- Section 6: numerical solutions, illustrating spatial profiles for the optical ellipticity and stability boundaries; and,
- Section 7: conclusions.

2. The General Model

2.1. The Optical Rate Equations

In any waveguiding structure that is defined by a spatially-dependent relative permittivity $\epsilon(\mathbf{r})$, we will have optical mode solutions $\Phi_k(\mathbf{r})$ satisfying the Helmholtz equation

$$\left[\nabla^2 + \frac{\Omega^2 \epsilon(\mathbf{r})}{c^2}\right]\Phi_k(\mathbf{r}) = 0, \tag{1}$$

where Ω is a reference frequency that is taken to be the average of the modal frequencies and k is the transverse mode index. The $\Phi_k(\mathbf{r})$ are known as the normal modes or, sometimes, supermodes of the waveguide. For modes of a given order, the orthogonal polarisations have almost exactly the same spatial profile (having checked this numerically for cases of interest), and we shall assume this to be precisely true. Thus, each mode $\Phi_k(\mathbf{r})$ may be associated with two polarisations. Later, we shall explicitly formulate this in terms of left and right-circularly polarised light.

After cancelling a phase factor $e^{i\beta z}$, where β is the propagation constant along the cavity length (see Appendix A of Ref. [31]), the total optical field may then be written as a superposition of the normal modes, as

$$\mathbf{E}(x,y,t) = \sum_k \mathbf{E}_k(t)\Phi_k(x,y)e^{-i\nu_k t}, \tag{2}$$

where $\mathbf{E}_k(t)$ is a time dependent Jones vector that incorporates the polarisation and ν_k is the modal frequency, determined via a solution of the Helmholtz equation for the mode. Hereafter, we shall use $\mathbf{r} = x\mathbf{e}_x = y\mathbf{e}_y$ for brevity wherever we need to denote spatial coordinates, but it should be remembered that $\mathbf{r}$ is confined to the $x-y$ plane.

Starting from the general form of Maxwell's wave equation and applying the slowly varying envelope approximation (SVEA), as described in Appendix A of Ref. [31], we may obtain a set of optical rate equations for the complex amplitudes of the normal modes

$$\begin{aligned}\frac{\partial E_{k,\pm}}{\partial t} &= \left[i(\nu_k - \Omega) - \frac{1}{2\tau_p}\right]E_{k,\pm} - \left[\gamma_a + i\gamma_p\right]E_{k,\mp} \\ &+ \sum_{k'} \frac{c}{2n_g}(1+i\alpha)E_{k',\pm}e^{i\Delta\nu_{kk'}t}\sum_i \overline{g}_\pm^{(i)}\Gamma_{kk'}^{(i)}.\end{aligned} \tag{3}$$

Here, the $\pm$ subscripts denote the right ($+$) and left ($-$) circularly polarised components, τ_p is photon lifetime, c is the speed of light, n_g is the group refractive index, and α is the linewidth enhancement factor, as defined in terms of the change in the real and imaginary components of the electric susceptibility, $\Delta\chi'_\pm$ and $\Delta\chi''_\pm$, respectively, by $\alpha = -\Delta\chi'_\pm/\Delta\chi''_\pm$. Note that we have adopted this sign convention for consistency with the SFM model [5,15,30] and it is opposite to that used in Ref. [29]. Hence, the values of α that are used in this work take the opposite sign to that in the latter reference.

Each polarisation component is coupled to the other via the birefringence rate γ_p and dichroism rate γ_a. The time-dependent exponential factor involves the difference between modal frequencies $\Delta\nu_{kk'} = \nu_k - \nu_{k'}$.

The summation over i in the last term of (3) is over the optically confining guides. Here, we have defined optical overlap factors $\Gamma^{(i)}_{kk'}$ for the ith guide via

$$\overline{g}^{(i)}_{\pm}\Gamma^{(i)}_{kk'} \equiv \int_{(i)} g_{\pm}(\mathbf{r})\Phi_k(\mathbf{r})\Phi_{k'}(\mathbf{r})\,\mathrm{d}^2\mathbf{r},$$

where $\overline{g}^{(i)}_{\pm}$ is the average gain for each polarisation in guide (i) and the integral is over the ith guide. In practice, we take the gain to be spatially constant over a guide and zero outside it. Hence, in this paper, the overlap factors are simply defined by

$$\Gamma^{(i)}_{kk'} \equiv \int_{(i)} \Phi_k(\mathbf{r})\Phi_{k'}(\mathbf{r})\,\mathrm{d}^2\mathbf{r} \tag{4}$$

and we normalise the spatial profiles, so that

$$\int |\Phi_k(\mathbf{r})|^2\,\mathrm{d}^2\mathbf{r} = 1,$$

where the integral is over all space.

2.2. *The Carrier Rate Equations*

The rate equations for spin-polarised populations of carriers may be derived from the optical Bloch equations. The general result for the spatially dependent carrier concentrations $N_{\pm}(\mathbf{r},t)$ and circularly polarised optical fields $E_{\pm}(\mathbf{r},t)$, including spin relaxation, may be found to be given by

$$\frac{\partial N_{\pm}}{\partial t} = -\frac{N_{\pm}}{\tau_N} + \Lambda_{\pm} - \gamma_J(N_{\pm} - N_{\mp}) - \frac{c}{n_g}g_{\pm}(N_{\pm})|E_{\pm}|^2, \tag{5}$$

where τ_N is the carrier lifetime, $\Lambda_{\pm}$ is the pumping rate, and γ_J is the spin relaxation rate. The $\pm$ subscripts on N refer to spin up (+) and spin-down (−) carriers, which directly couple with right (+) and left (−) circularly polarised photons respectively. Here, we assume that all carrier pumping, whether that be optical or electrical, is confined to the active region, and, hence, the effects of lateral diffusion are neglected at this time.

Note that we take $|E_{\pm}(\mathbf{r},t)|^2 = S(\mathbf{r},t)$ to be the photon density and, hence, $E_{\pm}$ does not have dimensions of electric field in (5). This is unproblematic, since the optical rate Equations (3) may be multiplied by any arbitrary factor to match the dimensions of $E_{\pm}$ in (5) without changing the dynamics.

The earlier assumption that the gain is spatially constant over a given guide and zero between guides requires a similar assumption for the carrier concentrations. We shall assume a linear gain model of the form $g(N) = g_0(N - N_0)$, where N_0 is the transparency concentration, so, if $g(N)$ is a step function, $N \le N_0$ outside the active regions. With the pumping confined to the active regions and no spatial diffusion, we may take any optical loss in the cladding regions to have been absorbed into the cavity loss rate κ. Therefore, we may take the carrier concentration in this region to be exactly N_0.

Taking the spatial dependence to be in the $x-y$ plane only, we may then put

$$N_{\pm}(t,x,y) = \sum_i N^{(i)}_{\pm}(t)\xi^{(i)}(x,y), \tag{6}$$

where $N^{(i)}(t)$ is the time dependent carrier concentration in the ith guide and $\xi^{(i)}(x,y)$ is a step function. Because we may subtract N_0 from either side (which we do on normalisation), we may take this as effectively giving zero outside the active regions. Applying this assumption, we find that the rate equations for the spin-polarised concentrations in the (i)th guide are given by

$$\frac{\partial N_{\pm}^{(i)}}{\partial t} = -\frac{N_{\pm}^{(i)}}{\tau_N} + \Lambda_{\pm}^{(i)} - \gamma_J\left(N_{\pm}^{(i)} - N_{\mp}^{(i)}\right) - \frac{c}{n_g}\sum_{k,k'} E_{k,\pm}^{*} E_{k',\pm} \overline{g}_{\pm}^{(i)} \Gamma_{kk'}^{(i)} e^{i\Delta\nu_{kk'}t}, \quad (7)$$

where the (i) superscripts on a quantity label the values of that quantity in each guide. Appendix B of Ref. provide the details of the derivation [31].

In this study, we assume that $|\Delta\nu_{k,k'}| \ll \nu_k, \nu_{k'}$. This assumption is physically relevant, provided that the coupled waveguides are well separated (relative to a characteristic length). In that case, the modal frequency of the symmetric and anti-symmetric modes will become very similar. However, the assumption may not be so accurate for the frequency difference between different orders of transverse modes. Our main interest lies only in the symmetric and anti-symmetric versions of the lowest order mode of a two-guide structure, in which case the assumption is well-justified. Under this assumption, the non-autonomous Equations (3) and (7) may have their explicit time dependence removed and simplified to

$$\frac{\partial \tilde{E}_{k,\pm}}{\partial t} = \left[i(\nu_k - \Omega) - \frac{1}{2\tau_p}\right]\tilde{E}_{k,\pm} - \left[\gamma_a + i\gamma_p\right]\tilde{E}_{k,\mp} + \sum_{k'} \frac{c}{2n_g}(1 + i\alpha)\tilde{E}_{k',\pm} \sum_i \overline{g}_{\pm}^{(i)} \Gamma_{kk'}^{(i)}. \quad (8)$$

and

$$\frac{\partial N_{\pm}^{(i)}}{\partial t} = -\frac{N_{\pm}^{(i)}}{\tau_N} + \Lambda_{\pm}^{(i)} - \gamma_J\left(N_{\pm}^{(i)} - N_{\mp}^{(i)}\right) - \frac{c}{n_g}\sum_{k,k'} \tilde{E}_{k,\pm}^{*} \tilde{E}_{k',\pm} \overline{g}_{\pm}^{(i)} \Gamma_{kk'}^{(i)}, \quad (9)$$

where we have used the tilde notation to denote the transformed optical field. Equations (8) and (9) then represent the general model for any number of normal modes and any number of confining guides.

In the following, instead of the model (3) and (7), we will analyse (8) and (9). Nevertheless, the analysis of the latter equations will still be valid in recognising unstable solutions of the former ones with a critical eigenvalue λ that is much larger than $|\Delta\nu_{k,k'}|$. which is because, before the factor $\exp(i\Delta\nu_{k,k'}t)$, which is slowly varying, starts to have any effect in the system, the unstable solution will already show its instability.

Stable solutions of (8) and (9) will also correspond to stable solutions of the model (3) and (7) if the time frame is of order $\mathcal{O}(1/|\Delta\nu_{k,k'}|)$. In this way, we also conjecture that, if all the eigenvalues of a solution are far away from the imaginary axis, then the presence of the slowly varying phase $\exp(i\Delta\nu_{k,k'}t)$ should not change the eigenvalues much.

3. The Overlap Factors

3.1. Equal Guides

The overlap factors that are defined by (4) are calculated from the spatial modal solutions of the Helmholtz equation $\Phi(\mathbf{r})$. Details of the solutions used in this work are given in Appendix E of Ref. [31]. Firstly, we shall just consider equal guides with the same refractive index n_1 in the core regions and n_2 elsewhere, although our treatment of the rate equations in Section 4 is general enough to deal with twin guides of any geometry. Table 1 lists the parameters used in our example calculations.

Table 1. Waveguide parameters used in the solution of the Helmholtz equation.

Parameter	Value	Unit	Description
n_1	3.400971	-	Core refractive index
n_2	3.4	-	Cladding refractive index
a	4	μm	Half guide width/radius
λ_0	1.3	μm	Free-space wavelength

The guides may be characterised by a normalised decay constants u (in the core regions) and w (in the cladding regions)

$$u = a\sqrt{\left(\frac{n_1 \nu_k}{c}\right)^2 - \beta^2} \tag{10}$$

and

$$w = a\sqrt{\beta^2 - \left(\frac{n_2 \nu_k}{c}\right)^2}, \tag{11}$$

where a is either the half-width of a slab waveguide or the radius of a circular guide, as illustrated in Figure 1. These decay constants can also be combined into a conventional 'normalised frequency' v, which is defined as

$$v = \sqrt{u^2 + w^2}. \tag{12}$$

In practice, we shall take (10)–(12) to refer to the value for a single isolated guide. In a solitary slab guide, for values of $v < \pi/2$, only one guided TE mode is supported. For equal double slab guides with this same index profile, only two TE modes are supported. Therefore, we refer to such structures as being 'weakly-guiding'.

Because of the symmetry of equal guides, the lowest mode has even parity and the second lowest, odd parity. We refer to these as the 'symmetric' and 'anti-symmetric' modes, respectively, and then label them by suffixes s and a, respectively. Note that the anti-symmetric mode $\Phi_a(x)$ always goes through zero in between the guides, whereas the $\Phi_s(x)$ does not. This qualitative behaviour persists, even when we break the symmetry of the guides, so that we may still use s and a as labels, although they would then be distinguished by topology rather than geometric symmetry.

The overlap factors $\Gamma^{(i)}_{k'k}$ are found by integrating the products of the spatial modes over each guide, as in (4). For closely spaced guides, the products $\Phi_s^2(x)$, $\Phi_a^2(x)$ and the modulus $|\Phi_s(x)\Phi_a(x)|$ are noticeably different (see Figures 3 and 4 of Ref. [31]). Hence, we note that, in general,

$$\Gamma^{(i)}_{ss} \neq \Gamma^{(i)}_{aa} \neq \left|\Gamma^{(i)}_{sa}\right|.$$

However, in the case of equal guides, by symmetry, we do have

$$\Gamma^{(1)}_{ss} = \Gamma^{(2)}_{ss} \text{ and } \Gamma^{(1)}_{aa} = \Gamma^{(2)}_{aa} \quad \text{(equal guides)}.$$

Additionally, by symmetry, the integral of $|\Phi_s(x)\Phi_a(x)|$ will be the same in each guide, although the sign will be opposite, so

$$\Gamma^{(1)}_{sa} = -\Gamma^{(2)}_{sa}.$$

As the separation between the guides gets larger, the spatial profiles become like those of isolated guides. The difference between these profiles and those of an isolated guide are: (i) the normalisation—the integral of the squared modulus of the spatial modes will be half that of the optical confinement factor—and (ii) the sign of the anti-symmetric mode is flipped in one of the guides. In fact, as the separation tends to infinity, we may obtain the wave functions of the isolated guides by adding and subtracting the modes via

$$\Phi_1(x) = \lim_{d\to\infty} \frac{1}{\sqrt{2}}(\Phi_s(x) + \Phi_a(x)) \quad \text{(equal guides)}$$

and

$$\Phi_2(x) = \lim_{d\to\infty} \frac{1}{\sqrt{2}}(\Phi_s(x) - \Phi_a(x)) \quad \text{(equal guides)}.$$

This is the basis for the definition of the 'composite modes' in terms of the normal modes that are defined in (18) and (19) defined in the next section. We then have

$$\int_{(1)} |\Phi_1(x)|^2 dx = \int_{(2)} |\Phi_2(x)|^2 dx = \Gamma_S \quad \text{(equal guides)},$$

where Γ_S is the optical confinement factor of a single guide. In this limit, we also have

$$\lim_{d\to\infty} \Gamma_{ss}^{(i)} = \lim_{d\to\infty} \Gamma_{aa}^{(i)} = \lim_{d\to\infty} \left|\Gamma_{sa}^{(i)}\right| = \frac{\Gamma_S}{2} \quad \text{(equal guides)}.$$

Figure 3 shows the variation of the overlap factors for equal width slab guides with $v = \pi/2$ as a function of guide separation. The overlap factors are divided by Γ_S, and we clearly see the tendency of all values to $\Gamma_S/2$ at large separation. Only the factors for guide (1) are shown, since the values for guide (2) are the same, except for the change of sign on $\Gamma_{sa}^{(2)}$.

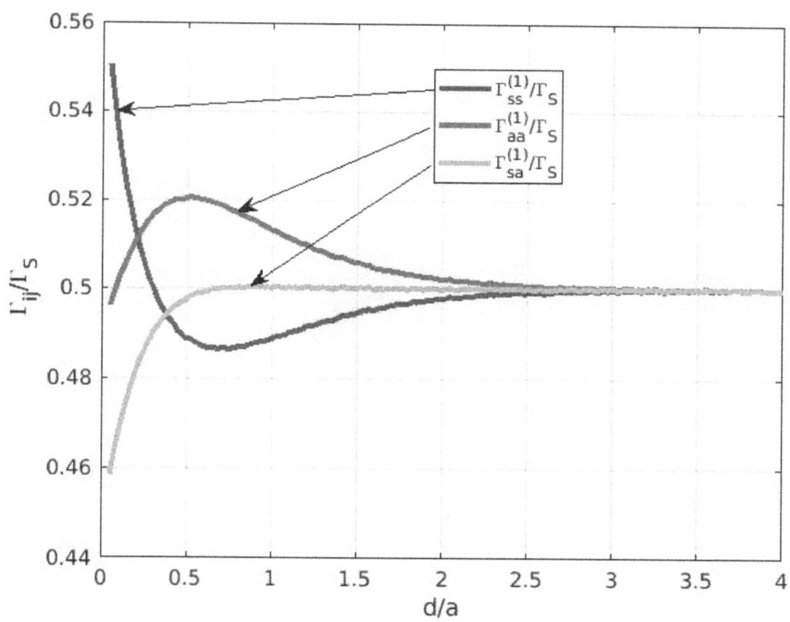

Figure 3. Plot of the overlap factors relative to Γ_S for a weakly-guiding ($v = \pi/2$) symmetric slab structure as a function of spatial separation between the guides. Here, $2d$ is the edge-to-edge separation between the guides and $2a$ is the guide width (8 μm in this case). Only the overlap factors for guide (1) are shown. The factors for guide (2) are the same, except that $\Gamma_{sa}^{(2)}$ is negative.

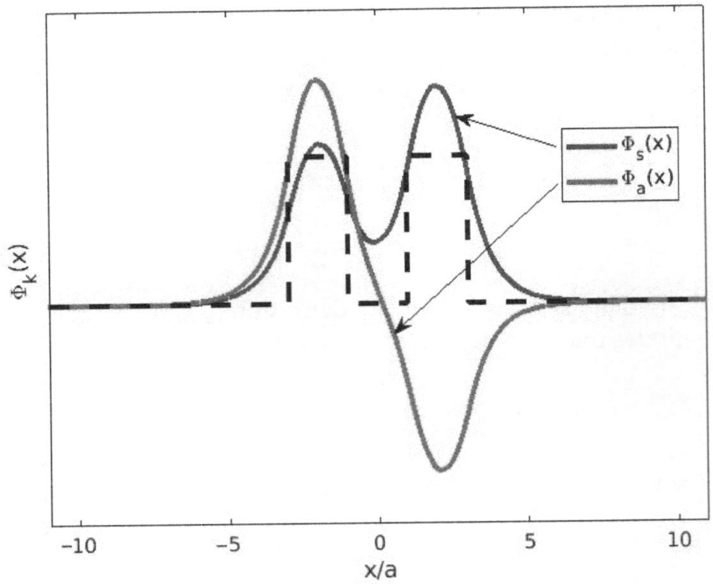

(a)

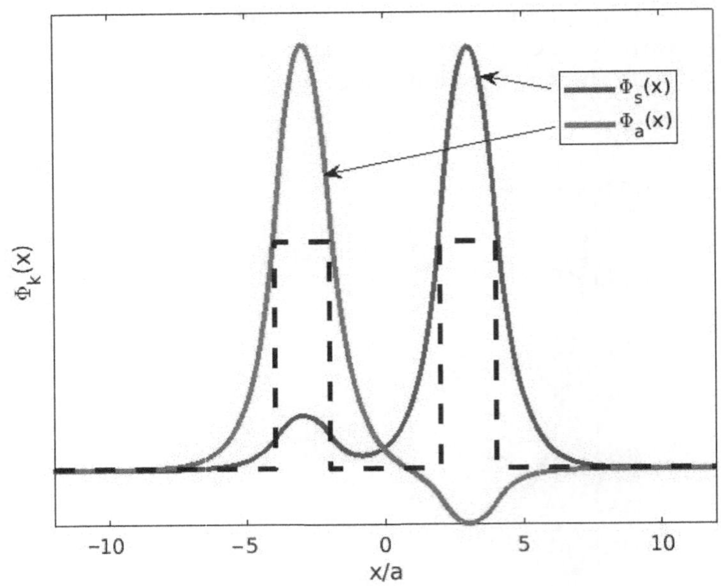

(b)

Figure 4. Spatial modes for slab guides with widths $w_1 = 7.9$ μm and $w_2 = 8.1$ μm (the wider guide is on the right). In this case, we define $a = (w_1 + w_2)/4$. The normalised frequency for an averaged guide is $v = \pi/2$. Note that, as the guide separation increases, the s mode becomes more greatly confined to the wider guide, whilst the a mode is confined to the narrower. (**a**) Spatial modes for unequal guides with $d/a = 1$. (**b**) Spatial modes for unequal guides with $d/a = 2$.

For such a weakly guiding structure, there is significant variation of the factors for $d/a < 2$. For more strongly guiding structures, $v > \pi/2$, this variation from $\Gamma_{sa}^{(1)}$ is greatly reduced.

3.2. Unequal Guides

For double-guided structures with unequal guiding regions, we lose the symmetric relations that were previously found. Figure 4 illustrates two examples with guide widths $w_1 = 7.9$ μm and $w_2 = 8.1$ μm, where the wider guide is on the right. In these cases, we have put $a = (w_1 + w_2)/4$, whilst the same refractive index difference has been used as for the equal width guides. It can be clearly seen that the component of each mode is different in each guide, and it is now the case that

$$\Gamma_{ss}^{(1)} \neq \Gamma_{ss}^{(2)} \text{ and } \Gamma_{aa}^{(1)} \neq \Gamma_{aa}^{(2)} \quad \text{(unequal guides)}.$$

We also note that each mode becomes more localised to a particular guide with increasing separation, with the s-labelled mode tending to the wider guide. This may be understood from basic waveguiding theory, since the s mode has the biggest effective index (or propagation constant) and the effective index (or propagation constant) of a mode of a single guide varies as the width. As the separation between the guides tends to infinity, each normal mode approaches the mode of a single isolated guide. Hence, labelling the narrow and wide guides (1) and (2), respectively,

$$\lim_{d\to\infty} \Gamma_{ss}^{(1)} = 0, \quad \lim_{d\to\infty} \Gamma_{ss}^{(2)} = \Gamma_2,$$

$$\lim_{d\to\infty} \Gamma_{aa}^{(1)} = \Gamma_1, \quad \lim_{d\to\infty} \Gamma_{aa}^{(2)} = 0$$

and

$$\lim_{d\to\infty} \Gamma_{sa}^{(1)} = \lim_{d\to\infty} \Gamma_{sa}^{(2)} = 0 \quad \text{(unequal guides)},$$

where Γ_1 and Γ_2 are the optical confinement factors of single guides of width w_1 and w_2.

Figure 5 illustrates these behaviours, which shows the variation of the overlap factors for a non-symmetric slab guide as a function of spatial separation. The calculations presented here use the same guide widths as for the modes that are shown in Figure 4. Note that, since Γ_S is the optical confinement factor of the averaged isolated guide, the limiting values of $\Gamma_{ss}^{(2)}/\Gamma_S$ and $\Gamma_{aa}^{(1)}/\Gamma_S$ are not exactly the same.

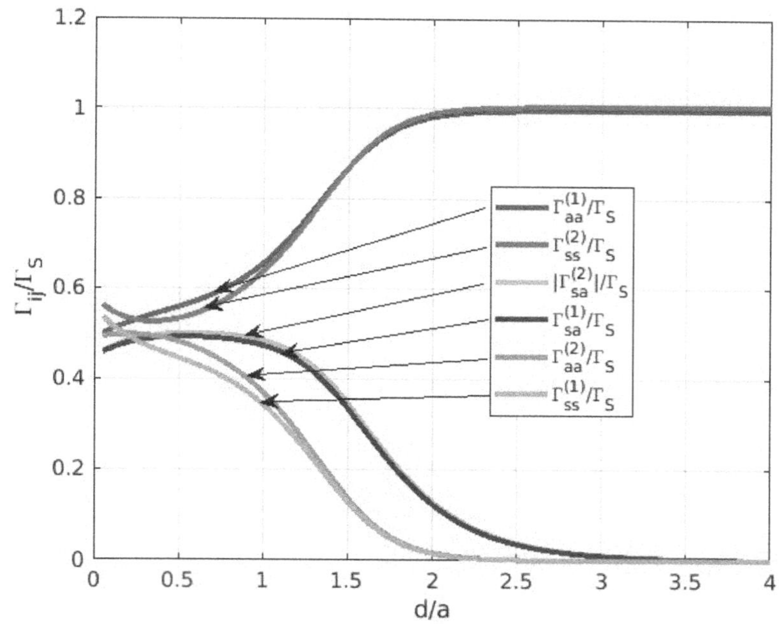

Figure 5. Plot of the overlap factors for a weakly-guiding ($v = \pi/2$) non-symmetric slab structure as a function of spatial separation between the guides. Here, $2d$ is the edge-to-edge separation between the guides and $2a = (w_1 + w_2)/2$ is the average guide width, using $w_1 = 7.9$ µm and $w_2 = 8.1$ µm, as in Figure 4. In contrast to the case of symmetric guides, in this case the s mode tends to occupy guide (2) (the wider guide) and the a mode occupies guide (1). As the separation increase, the corresponding overlap factors tend to the optical confinement factor, whilst all other factors tend to zero. Here, the modulus of $\Gamma_{sa}^{(2)}$ is shown, as this factor is negative.

3.3. Circular Guides

For circular guides, we used a commercial eigensolver to find solutions of the Helmholtz equation for symmetric structures at various guide separations. These solutions are for the two polarisation components of the lowest-order symmetric and antisymmetric mode. To interpolate between these results, we found that the overlap factors could be fitted very well by the following empirical formulae:

$$\Gamma_{ss}^{(i)}(d) = \frac{\Gamma_S}{2}\left(1 - e^{-a_1 d}\right) + \Gamma_1 e^{-b_1 d}, \tag{13}$$

$$\Gamma_{aa}^{(i)}(d) = \frac{\Gamma_S}{2}\left(1 + e^{-a_1 d}\right) - \Gamma_2 e^{-b_1 d} \tag{14}$$

and

$$\left|\Gamma_{sa}^{(i)}(d)\right| = \frac{\Gamma_S}{2}\left(1 + e^{-a_2 d}\right) - \Gamma_3 e^{-b_2 d}. \tag{15}$$

Table 2 lists the parameters used. Here, Γ_S is again the optical confinement factor associated with the lowest mode of a single isolated guide. For circular guides, this is HE11 (LP01) mode.

In Section 4.2.2, we discuss the reduction of the normal mode model to the coupled mode model. It is found that the coupling coefficient μ is given by the difference in normal mode frequencies $\mu = (\nu_s - \nu_a)/2$. Using the calculated values of these frequencies, the coupling coefficient for equal circular guides is found to be well-approximated by the Ogawa [32] expression

$$\mu \propto \frac{1}{d^{1/2}} \exp\left(-\frac{2wd}{a}\right),$$

where w is given by (11). The constant of proportionality may be found by fitting this to the calculated value of $(\nu_s - \nu_a)/2$ at $d/a = 1$.

Table 2. Parameters used in fitting functions for circular waveguides. Here, the cladding refractive index is $n_2 = 3.4$ and the core refractive index is $n_1 = n_2 + \Delta n$.

Parameter	$\Delta n = 0.000971$	Unit
a_1	0.200	μm^{-1}
a_2	0.399	μm^{-1}
b_1	0.247	μm^{-1}
b_2	0.441	μm^{-1}
Γ_S	0.5766	-
Γ_1	0.346	-
Γ_2	0.300	-
Γ_3	0.3156	-

4. Double-Guided Structure

4.1. Real Form of the Rate Equations

4.1.1. Optical Rate Equations

In this paper, we confine our attention to structures that only involve two weakly-confining guides supporting only two guided modes. The treatment that we shall follow in this section will be valid for the general case of unequal guides, although it is of particular use for the symmetric case, which we focus on in this paper.

For a double-guided structure, we may put reference frequency to $\Omega = (\nu_s + \nu_a)/2$. The optical rate equations of (8) may now be written

$$\frac{\partial \tilde{E}_{k,\pm}}{\partial t} = \left[\mathrm{i}\frac{\nu_k - \nu_{k'}}{2} - \frac{1}{2\tau_p}\right]\tilde{E}_{k,\pm} + \frac{c}{2n_g}(1 + \mathrm{i}\alpha)\tilde{\Sigma}_{k,\pm} - \left[\gamma_a + \mathrm{i}\gamma_p\right]\tilde{E}_{k,\mp}, \tag{16}$$

where $k = s, a$ for the symmetric and anti-symmetric modes, respectively and

$$\tilde{\Sigma}_{k,\pm} = \sum_{k'} \tilde{E}_{k',\pm} \sum_i \bar{g}_{\pm}^{(i)} \Gamma_{kk'}^{(i)}, \tag{17}$$

where $k' = a, s$. These are then the equations for the evolution of the normal modes. However, solutions in terms of the normal modes do not lend themselves well to physical intuition. When thinking of optical guides in close proximity, it is more natural to think of the optical intensity in each guide. To this end, it is convenient to introduce new optical field variables, as defined by

$$E_{1,\pm} = \frac{1}{\sqrt{2}}(\tilde{E}_{s,\pm} + \tilde{E}_{a,\pm}) \tag{18}$$

and

$$E_{2,\pm} = \frac{1}{\sqrt{2}}(\tilde{E}_{s,\pm} - \tilde{E}_{a,\pm}). \tag{19}$$

The motivation for this is that the squared modulus of these 'composite' modes becomes the optical intensity in each guide at infinite separation, which greatly aids in the visualisation of the laser dynamics.

In the following sections, we write all of the equations in their real form. Although these are often quite ungainly, there are practical advantages to rendering them in this way. In particular, any stability analysis of these equations must start with writing the equations out in real form anyway. Moreover, the real form is generally better suited, if not required, for numerical calculations, whether this be for dynamical simulation via the Runge–Kutta method or finding the steady state solutions using a nonlinear solver.

Defining the $\phi_{21\pm\pm}$ as the phase difference between $E_{2\pm}$ and $E_{1\pm}$ and ϕ_{kk+-} as the phase difference between E_{k+} and E_{k-}, in Appendix C of Ref. [31] we find that the optical rate equations may be written in real form as

$$\begin{aligned}\frac{\partial |E_{1,\pm}|}{\partial t} &= \left[-\frac{1}{2\tau_p} + \frac{c\Gamma_S}{2n_g} G_{12\pm}\right] |E_{1,\pm}| \\ &+ \left[\frac{c\Gamma_S}{2n_g} \Delta G_{\pm} (\cos(\phi_{21\pm\pm}) - \alpha \sin(\phi_{21\pm\pm})) - \mu \sin(\phi_{21\pm\pm})\right] |E_{2,\pm}| \\ &- \left[\gamma_a \cos(\phi_{11+-}) + \gamma_p \sin(\phi_{11+-})\right] |E_{1,\mp}|\end{aligned} \tag{20}$$

$$\begin{aligned}\frac{\partial |E_{2,\pm}|}{\partial t} &= \left[-\frac{1}{2\tau_p} + \frac{c\Gamma_S}{2n_g} G_{21\pm}\right] |E_{2,\pm}| \\ &+ \left[\frac{c\Gamma_S}{2n_g} \Delta G_{\pm} (\cos(\phi_{21\pm\pm}) + \alpha \sin(\phi_{21\pm\pm})) + \mu \sin(\phi_{21\pm\pm})\right] |E_{1,\pm}| \\ &- \left[\gamma_a \cos(\phi_{22+-}) + \gamma_p \sin(\phi_{22+-})\right] |E_{2,\mp}|\end{aligned} \tag{21}$$

$$\begin{aligned}\frac{\partial \phi_{21\pm\pm}}{\partial t} &= \frac{c\Gamma_S}{2n_g} \alpha (G_{21\pm} - G_{12\pm}) + \mu \cos(\phi_{21\pm\pm}) \left(\frac{|E_{1,\pm}|}{|E_{2,\pm}|} - \frac{|E_{2,\pm}|}{|E_{1,\pm}|}\right) \\ &+ \frac{c\Gamma_S}{2n_g} \Delta G_{\pm} \left[\alpha \cos(\phi_{21\pm\pm}) \left(\frac{|E_{1,\pm}|}{|E_{2,\pm}|} - \frac{|E_{2,\pm}|}{|E_{1,\pm}|}\right) - \sin(\phi_{21\pm\pm}) \left(\frac{|E_{1,\pm}|}{|E_{2,\pm}|} + \frac{|E_{2,\pm}|}{|E_{1,\pm}|}\right)\right] \\ &+ \gamma_p \left[\cos(\phi_{11+-}) \frac{|E_{1,\mp}|}{|E_{1,\pm}|} - \cos(\phi_{22+-}) \frac{|E_{2,\mp}|}{|E_{2,\pm}|}\right] \\ &\mp \gamma_a \left[\sin(\phi_{11+-}) \frac{|E_{1,\mp}|}{|E_{1,\pm}|} - \sin(\phi_{22+-}) \frac{|E_{2,\mp}|}{|E_{2,\pm}|}\right].\end{aligned} \tag{22}$$

and

$$\frac{\partial \phi_{11+-}}{\partial t} = \frac{c\Gamma_S}{2n_g}\alpha(G_{12+} - G_{12-}) + \mu\left(\cos(\phi_{21++})\frac{|E_{2,+}|}{|E_{1,+}|} - \cos(\phi_{21--})\frac{|E_{2,-}|}{|E_{1,-}|}\right)$$
$$+ \frac{c\Gamma_S}{2n_g}\left[\Delta G_+(\alpha\cos(\phi_{21++}) + \sin(\phi_{21++}))\frac{|E_{2,+}|}{|E_{1,+}|}\right.$$
$$\left. - \Delta G_-(\alpha\cos(\phi_{21--}) + \sin(\phi_{21--}))\frac{|E_{2,-}|}{|E_{1,-}|}\right]$$
$$+ \gamma_a \sin(\phi_{11+-})\left(\frac{|E_{1,+}|}{|E_{1,-}|} + \frac{|E_{1,-}|}{|E_{1,+}|}\right) + \gamma_p \cos(\phi_{11+-})\left(\frac{|E_{1,+}|}{|E_{1,-}|} - \frac{|E_{1,-}|}{|E_{1,+}|}\right) \quad (23)$$

Note that, since $\phi_{22+-} = \phi_{21++} - \phi_{21--} + \phi_{11+-}$, we do not need an equation for this last phase variable.

In these equations, we have used

$$\mu = \frac{\nu_S - \nu_a}{2}, \quad (24)$$

whilst the gain terms are defined by

$$G_{12\pm} = \frac{\Gamma_+^{(1)}\overline{g}_\pm^{(1)} + \Gamma_+^{(2)}\overline{g}_\pm^{(2)}}{\Gamma_S}, \quad (25)$$

$$G_{21\pm} = \frac{\Gamma_-^{(1)}\overline{g}_\pm^{(1)} + \Gamma_-^{(2)}\overline{g}_\pm^{(2)}}{\Gamma_S} \quad (26)$$

and

$$\Delta G_\pm = \frac{\Delta\Gamma^{(1)}\overline{g}_\pm^{(1)} + \Delta\Gamma^{(2)}\overline{g}_\pm^{(2)}}{\Gamma_S}. \quad (27)$$

The Γ terms that are introduced above are further defined in terms of the optical overlap factors via

$$\Gamma_\pm^{(i)} = \frac{\Gamma_{ss}^{(i)} + \Gamma_{aa}^{(i)} \pm 2\Gamma_{sa}^{(i)}}{2} \quad (28)$$

and

$$\Delta\Gamma^{(i)} = \frac{\Gamma_{ss}^{(i)} - \Gamma_{aa}^{(i)}}{2}. \quad (29)$$

It is worth noting a general limiting behaviour as the separation between guides tends to infinity that $\Gamma_{ss}^{(i)} \to \Gamma_{aa}^{(i)} \equiv \Gamma^{(i)}$ and $\Gamma_{sa}^{(i)} \to 0$, where $\Gamma^{(i)}$ is half the optical confinement factor in an isolated guide. Hence, in this limit (recalling that the separation between the guides is $2d$),

$$\lim_{d\to\infty} \Gamma_\pm^{(i)} \equiv \Gamma_\infty^{(i)} \text{ and } \lim_{d\to\infty} \Delta\Gamma^{(i)} = 0. \quad (30)$$

For equal guides, $2\Gamma_\infty^{(i)} = \Gamma_S$, the optical confinement factor of an isolated guide. For unequal guides, we may take Γ_S to be an average of the optical confinement factors. This does not undermine the generality of (20)–(23), since, in all cases, the factor of Γ_S in the denominator of the gain terms cancels with the factor multiplying it. The inclusion of Γ_S here is one of convenience to elucidate the limiting behaviour of the rate equations, as discussed later in Section 4.2.

4.1.2. Carrier Rate Equations

The carrier rate equations are straight-forward to render in the notation for the double guided structure. From (9), these are

$$\frac{\partial N_{\pm}^{(i)}}{\partial t} = -\frac{N_{\pm}^{(i)}}{\tau_N} + \Lambda_{\pm}^{(i)} - \frac{c}{n_g}\overline{g}_{\pm}^{(i)} I_{\pm}^{(i)} - \gamma_J\left(N_{\pm}^{(i)} - N_{\mp}^{(i)}\right), \tag{31}$$

where

$$I_{\pm}^{(i)} = \sum_{k,k'} \Gamma_{kk'}^{(i)} \tilde{E}_{k,\pm}^{*}(t)\tilde{E}_{k',\pm}(t). \tag{32}$$

Using $\Gamma_{sa}^{(i)} = \Gamma_{as}^{(i)}$, we can expand (32) in terms of $E_{1,\pm}$ and $E_{2,\pm}$, as

$$\begin{aligned} I_{\pm}^{(i)} &= \frac{\Gamma_{ss}^{(i)} + 2\Gamma_{sa}^{(i)} + \Gamma_{aa}^{(i)}}{2}|E_{1,\pm}|^2 \\ &+ \left(\Gamma_{ss}^{(i)} - \Gamma_{aa}^{(i)}\right)|E_{1,\pm}||E_{2,\pm}|\cos(\phi_{21\pm\pm}) \\ &+ \frac{\Gamma_{ss}^{(i)} - 2\Gamma_{sa}^{(i)} + \Gamma_{aa}^{(i)}}{2}|E_{2,\pm}|^2. \end{aligned}$$

Expressing this in terms of (28) and (29), we have

$$\begin{aligned} I_{\pm}^{(i)} &= \Gamma_{+}^{(i)}|E_{1,\pm}|^2 \\ &+ 2\Delta\Gamma^{(i)}|E_{1,\pm}||E_{2,\pm}|\cos(\phi_{21\pm\pm}) + \Gamma_{-}^{(i)}|E_{2,\pm}|^2. \end{aligned}$$

4.1.3. Normalised Rate Equations

Upon normalisation (described in detail in Appendix D of Ref. [31]), Equation (31) may be re-written as

$$\frac{\partial M_{\pm}^{(i)}}{\partial t} = \gamma\left[\eta_{\pm}^{(i)} - \left(1 + \mathcal{I}_{\pm}^{(i)}\right)M_{\pm}^{(i)}\right] - \gamma_J\left(M_{\pm}^{(i)} - M_{\mp}^{(i)}\right), \tag{33}$$

where the $M_{\pm}^{(i)}$ are the normalised carrier densities and

$$\begin{aligned} \mathcal{I}_{\pm}^{(i)} &= \frac{\Gamma_{+}^{(i)}}{\Gamma_S}|A_{1,\pm}|^2 + 2\frac{\Delta\Gamma^{(i)}}{\Gamma_S}|A_{1,\pm}||A_{2,\pm}|\cos(\phi_{21\pm\pm}) \\ &+ \frac{\Gamma_{-}^{(i)}}{\Gamma_S}|A_{2,\pm}|^2. \end{aligned} \tag{34}$$

Here, we have used assumed a linear gain model, having put

$$\overline{g} = g_0(N - N_0), \tag{35}$$

where g_0 is the differential gain and N_0 is the transparency concentration. Hence, in the steady state, we find the threshold carrier concentration N_S for the single guide to be given by

$$N_S - N_0 = \frac{n_g}{\Gamma_S g_0 c \tau_p}. \tag{36}$$

The normalised carrier concentrations appearing in (33) are then of the general form

$$M = \frac{N - N_0}{N_S - N_0}.$$

Hence, M will be zero at transparency and unity at threshold.

Further defining

$$M_{12\pm} = \frac{\Gamma_{+}^{(1)} M_{\pm}^{(1)} + \Gamma_{+}^{(2)} M_{\pm}^{(2)}}{\Gamma_S}, \tag{37}$$

$$M_{21\pm} = \frac{\Gamma_{-}^{(1)} M_{\pm}^{(1)} + \Gamma_{-}^{(2)} M_{\pm}^{(2)}}{\Gamma_S} \tag{38}$$

and

$$\Delta M_{\pm} = \frac{\Delta\Gamma^{(1)} M_{\pm}^{(1)} + \Delta\Gamma^{(2)} M_{\pm}^{(2)}}{\Gamma_S}, \tag{39}$$

and the cavity decay rate $\kappa = 1/(2\tau_p)$, the normalised optical rate equations are

$$\begin{aligned}\frac{\partial|A_{1,\pm}|}{\partial t} &= \kappa(M_{12\pm} - 1)|A_{1,\pm}| \\ &+ [\kappa\Delta M_{\pm}(\cos(\phi_{21\pm\pm}) - \alpha\sin(\phi_{21\pm\pm})) - \mu\sin(\phi_{21\pm\pm})]|A_{2,\pm}| \\ &- [\gamma_a\cos(\phi_{11+-}) \pm \gamma_p\sin(\phi_{11+-})]|A_{1,\mp}|,\end{aligned} \tag{40}$$

$$\begin{aligned}\frac{\partial|A_{2,\pm}|}{\partial t} &= \kappa(M_{21\pm} - 1)|A_{2,\pm}| \\ &+ [\kappa\Delta M_{\pm}(\cos(\phi_{21\pm\pm}) + \alpha\sin(\phi_{21\pm\pm})) + \mu\sin(\phi_{21\pm\pm})]|A_{1,\pm}| \\ &- [\gamma_a\cos(\phi_{22+-}) \pm \gamma_p\sin(\phi_{22+-})]|A_{2,\mp}|,\end{aligned} \tag{41}$$

$$\begin{aligned}\frac{\partial\phi_{21\pm\pm}}{\partial t} &= \kappa\alpha(M_{21\pm} - M_{12\pm}) + \mu\cos(\phi_{21\pm\pm})\left(\frac{|A_{1,\pm}|}{|A_{2,\pm}|} - \frac{|A_{2,\pm}|}{|A_{1,\pm}|}\right) \\ &+ \kappa\Delta M_{\pm}\left[\alpha\cos(\phi_{21\pm\pm})\left(\frac{|A_{1,\pm}|}{|A_{2,\pm}|} - \frac{|A_{2,\pm}|}{|A_{1,\pm}|}\right) - \sin(\phi_{21\pm\pm})\left(\frac{|A_{1,\pm}|}{|A_{2,\pm}|} + \frac{|A_{2,\pm}|}{|A_{1,\pm}|}\right)\right] \\ &+ \gamma_p\left[\cos(\phi_{11+-})\frac{|A_{1,\mp}|}{|A_{1,\pm}|} - \cos(\phi_{22+-})\frac{|A_{2,\mp}|}{|A_{2,\pm}|}\right] \\ &\mp \gamma_a\left[\sin(\phi_{11+-})\frac{|A_{1,\mp}|}{|A_{1,\pm}|} - \sin(\phi_{22+-})\frac{|A_{2,\mp}|}{|A_{2,\pm}|}\right],\end{aligned} \tag{42}$$

and

$$\begin{aligned}\frac{\partial\phi_{11+-}}{\partial t} &= \kappa\alpha(M_{12+} - M_{12-}) + \mu\left(\cos(\phi_{21++})\frac{|A_{2,+}|}{|A_{1,+}|} - \cos(\phi_{21--})\frac{|A_{2,-}|}{|A_{1,-}|}\right) \\ &+ \kappa\Delta M_{+}(\alpha\cos(\phi_{21++}) + \sin(\phi_{21++}))\frac{|A_{2,+}|}{|A_{1,+}|} \\ &- \kappa\Delta M_{-}(\alpha\cos(\phi_{21--}) + \sin(\phi_{21--}))\frac{|A_{2,-}|}{|A_{1,-}|} \\ &+ \gamma_a\sin(\phi_{11+-})\left(\frac{|A_{1,+}|}{|A_{1,-}|} + \frac{|A_{1,-}|}{|A_{1,+}|}\right) + \gamma_p\cos(\phi_{11+-})\left(\frac{|A_{1,+}|}{|A_{1,-}|} - \frac{|A_{1,-}|}{|A_{1,+}|}\right).\end{aligned} \tag{43}$$

Here, the amplitudes A are normalised according to

$$|A|^2 = \frac{\Gamma_S g_0 c \tau_N}{n_g}|\tilde{E}|^2.$$

Equations (33), (34), and (40)–(43) constitute the dynamical model of double-guided structure. This may be applied to any waveguide geometry that is restricted to two guides and the two lowest confined modes.

4.2. Limiting Behaviour for Widely Separated Guides

4.2.1. Reduction to the Spin-Flip Model (SFM)

In this section, we assume equal guides and, so, employ the results for the identities and limiting behaviour of the overlap factors that are found in Section 3.1. Because $d \to \infty$, the factors that are defined earlier in (28) and (29) become

$$\Gamma_{+}^{(1)} = \Gamma_S,\ \Gamma_{-}^{(1)} = 0,\ \Gamma_{+}^{(2)} = 0,\ \Gamma_{-}^{(2)} = \Gamma_S$$

and

$$\Delta\Gamma^{(i)} = 0.$$

The carrier terms that are defined in (37)–(39) then become

$$M_{12\pm} = M_{\pm}^{(1)},\ M_{21\pm} = M_{\pm}^{(2)} \text{ and } \Delta M_{\pm} = 0 \tag{44}$$

and the expression for the optical intensity of (34) appearing in the carrier rate equations becomes

$$\mathcal{I}_{\pm}^{(i)} = |A_{i,\pm}|^2, \tag{45}$$

for $i = 1, 2$. Additionally, note that the term given earlier in (24) as $\mu = (\nu_s - \nu_a)/2$ approaches zero as the guide separation approaches infinity and the frequencies of the normal modes become equal.

Hence, Equations (40)–(43) for the normalised optical rate equations reduce to

$$\begin{aligned}\frac{\partial|A_{i,\pm}|}{\partial t} &= \kappa\left(M_{\pm}^{(i)} - 1\right)|A_{i,\pm}| \\ &\quad - \left(\gamma_a \cos(\phi_{ii+-}) \pm \gamma_p \sin(\phi_{ii+-})\right)|A_{i,\mp}|,\end{aligned} \tag{46}$$

for $i = 1, 2$,

$$\begin{aligned}\frac{\partial\phi_{21\pm\pm}}{\partial t} &= \left[\kappa\alpha M_{\pm}^{(2)} - \left(\gamma_p \cos(\phi_{22+-}) \mp \gamma_a \sin(\phi_{22+-})\right)\frac{|A_{2,\mp}|}{|A_{2,\pm}|}\right] \\ &\quad - \left[\kappa\alpha M_{\pm}^{(1)} - \left(\gamma_p \cos(\phi_{11+-}) \mp \gamma_a \sin(\phi_{11+-})\right)\frac{|A_{1,\mp}|}{|A_{1,\pm}|}\right]\end{aligned}$$

and

$$\begin{aligned}\frac{\partial\phi_{11+-}}{\partial t} &= \kappa\alpha\left(M_{+}^{(1)} - M_{-}^{(1)}\right) + \gamma_a \sin(\phi_{11+-})\left(\frac{|A_{1,+}|}{|A_{1,-}|} + \frac{|A_{1,-}|}{|A_{1,+}|}\right) \\ &\quad + \gamma_p \cos(\phi_{11+-})\left(\frac{|A_{1,+}|}{|A_{1,-}|} - \frac{|A_{1,-}|}{|A_{1,+}|}\right)\end{aligned} \tag{47}$$

Recalling that $\phi_{22+-} = \phi_{21++} - \phi_{21--} + \phi_{11+-}$, we find that

$$\begin{aligned}\frac{\partial\phi_{22+-}}{\partial t} &= \frac{\partial\phi_{21++}}{\partial t} - \frac{\partial\phi_{21++}}{\partial t} + \frac{\partial\phi_{11+-}}{\partial t}, \\ &= \kappa\alpha\left(M_{+}^{(2)} - M_{-}^{(2)}\right) + \gamma_a \sin(\phi_{22+-})\left(\frac{|A_{2,+}|}{|A_{2,-}|} + \frac{|A_{2,-}|}{|A_{2,+}|}\right) \\ &\quad + \gamma_p \cos(\phi_{22+-})\left(\frac{|A_{2,+}|}{|A_{2,-}|} - \frac{|A_{2,-}|}{|A_{2,+}|}\right),\end{aligned} \tag{48}$$

Meanwhile, (33) for the carrier rate equations becomes

$$\frac{\partial M_{\pm}^{(i)}}{\partial t} = \gamma\left[\eta_{\pm}^{(i)} - \left(1+|A_{i,\pm}|^2\right)M_{\pm}^{(i)}\right] - \gamma_J\left(M_{\pm}^{(i)} - M_{\mp}^{(i)}\right), \tag{49}$$

The guides are now completely uncoupled and we have two sets of equivalent equations for each. Defining new variables $N = (M_{+}^{(i)} + M_{-}^{(i)})/2$, $m = (M_{+}^{(i)} - M_{-}^{(i)})/2$, $|A_{\pm}| = |A_{i,\pm}|/\sqrt{2}$ and $\phi = \phi_{ii+-}$ for each guide, we may re-write (46)–(49) as

$$\frac{\partial|A_{\pm}|}{\partial t} = \kappa(N \pm m - 1)|A_{\pm}| - (\gamma_a\cos\phi \pm \gamma_p\sin\phi)|A_{\mp}|, \tag{50}$$

$$\frac{\partial\phi}{\partial t} = 2\kappa\alpha m + \gamma_a\sin\phi\left(\frac{|A_+|}{|A_-|} + \frac{|A_-|}{|A_+|}\right) + \gamma_p\cos\phi\left(\frac{|A_+|}{|A_-|} - \frac{|A_-|}{|A_+|}\right), \tag{51}$$

$$\frac{\partial N}{\partial t} = \gamma\left[\eta - \left(1+|A_+|^2+|A_-|^2\right)N - \left(|A_+|^2-|A_-|^2\right)m\right] \tag{52}$$

and

$$\frac{\partial m}{\partial t} = \gamma\left[P\eta - \left(|A_+|^2-|A_-|^2\right)N - \left(|A_+|^2+|A_-|^2\right)m\right] - \gamma_s m. \tag{53}$$

Here, $\eta = (\eta_+ + \eta_-)/2$ (dropping the (i) superscripts), we have defined an effective spin relaxation rate $\gamma_s = \gamma + 2\gamma_J$ and P is the pump ellipticity defined by

$$P = \frac{\eta_+ - \eta_-}{\eta_+ + \eta_-}. \tag{54}$$

Equations (50)–(53) are the real form of the spin-flip model (SFM) [5,15,30]. Note that Gahl et al. [15] have the factor of $(1 + i\alpha)$ multiplying the cavity loss term in the complex rate equations, which we take to be unphysical, so there will be a discrepancy between their expressions and those above.

4.2.2. Reduction to the Coupled Mode Model (CMM)

We may consider an alternative scenario in which we retain the coupling term μ between the guides, but let the overlap factors take their limiting values as the guide separation tends to infinity. Additionally, we may remove the coupling between the spin polarised components by setting $\gamma_J = \gamma_a = \gamma_p = 0$. In this case, Equations (40)–(43) reduce to

$$\frac{\partial|A_{1,\pm}|}{\partial t} = \kappa\left(M_{\pm}^{(1)} - 1\right)|A_{1,\pm}| - \mu\sin(\phi_{21\pm\pm})|A_{2,\pm}|, \tag{55}$$

$$\frac{\partial|A_{2,\pm}|}{\partial t} = \kappa\left(M_{\pm}^{(2)} - 1\right)|A_{2,\pm}| + \mu\sin(\phi_{21\pm\pm})|A_{1,\pm}|, \tag{56}$$

$$\frac{\partial \phi_{21++}}{\partial t} = \kappa\alpha\left(M_+^{(2)} - M_+^{(1)}\right) + \mu\cos(\phi_{21++})\left(\frac{|A_{1,+}|}{|A_{2,+}|} - \frac{|A_{2,+}|}{|A_{1,+}|}\right). \tag{57}$$

and

$$\begin{aligned}\frac{\partial \phi_{21--}}{\partial t} &= \frac{\partial \phi_{21++}}{\partial t} + \frac{\partial \phi_{11+-}}{\partial t} - \frac{\partial \phi_{22+-}}{\partial t}, \\ &= \kappa\alpha\left(M_-^{(2)} - M_-^{(1)}\right) \\ &+ \mu\cos(\phi_{21--})\left(\frac{|A_{1,-}|}{|A_{2,-}|} - \frac{|A_{2,-}|}{|A_{1,-}|}\right),\end{aligned} \tag{58}$$

whilst the carrier rate equations become

$$\frac{\partial M_\pm^{(i)}}{\partial t} = \gamma\left[\eta_\pm^{(i)} - \left(1 + |A_{i,\pm}|^2\right)M_\pm^{(i)}\right]. \tag{59}$$

These give us two independent sets of equations for the polarisation components, each of which reproduces the coupled mode model of Ref. [29] with real coupling coefficient μ and no frequency detuning (although with a difference in sign on the α factor, due to opposite sign definitions). It is of particular note that the coupling coefficient μ, given in terms of the difference between the mode frequencies in (24), is consistent with the analysis of Marom et al. [33], who found the same relation between the coupled mode and normal mode models.

5. Steady State Solutions

In general, analytical steady state solutions of the double-guided structure are not obtainable. However, exact expressions and very good approximations are both available in certain limiting cases. In this section, we continue to consider only the case of symmetric guides. Only results are given here; for additional details of the derivations, see Ref. [31].

5.1. Effect of Spin Relaxation

In the steady state, the carrier rate equations of (33) yield

$$M_\pm^{(i)} = \frac{\left(I_\mp^{(i)} + 1 + \gamma_J/\gamma\right)\eta_\pm^{(i)} + (\gamma_J/\gamma)\eta_\mp^{(i)}}{\left(I_+^{(i)} + 1\right)\left(I_-^{(i)} + 1\right) + (\gamma_J/\gamma)\left(I_+^{(i)} + I_-^{(i)} + 2\right)}. \tag{60}$$

Alternatively, we may make the intensities the subject, giving

$$I_\pm^{(i)} = \frac{\eta_\pm^{(i)}}{M_\pm^{(i)}} - 1 - \frac{\gamma_J}{\gamma}\left(1 - \frac{M_\mp^{(i)}}{M_\pm^{(i)}}\right). \tag{61}$$

Close to threshold, we may take the optical intensities in (60) to be negligible and put these to zero, giving

$$M_{th\pm}^{(i)} \approx \frac{(1+\gamma_J/\gamma)\eta_{th\pm}^{(i)} + (\gamma_J/\gamma)\eta_{th\mp}^{(i)}}{1+2(\gamma_J/\gamma)}, \quad (\gamma_J \gg \gamma). \tag{62}$$

Note that, in general, the different polarisation components of the intensity will not go to zero at the same overall pumping rate $\eta^{(i)} = \eta_+^{(i)} + \eta_-^{(i)}$, so (62) is not exact. However, with a large spin relaxation rate $\gamma_J \gg \gamma$, Equation (62) is found to be a good approximation (in practice, $\gamma_J > 10\gamma$).

5.2. Equal Pumping

A useful simplification to make is to assume equal pumping in each guide. That is, the total pump power and pump ellipticity are both the same $\eta_{\pm}^{(1)} = \eta_{\pm}^{(2)}$. Under the equal pumping assumption, Equation (42) reduces to

$$\frac{\partial\phi_{21\pm\pm}}{\partial t} = -2\kappa\Delta M_{\pm}\sin(\phi_{21\pm\pm}). \tag{63}$$

In the steady state, this is satisfied for $\phi_{21\pm\pm} = 0, \pi$. On reduction to the coupled mode model [29], these are referred to as the 'in-phase' and 'out-of-phase' solutions, respectively.

Because $|A_{1,\pm}| = |A_{2,\pm}|$, Equations (40) and (41) may be written as

$$\begin{aligned}\frac{\partial|A_{i,\pm}|}{\partial t} &= \kappa(M_{ij\pm} - 1)|A_{i,\pm}| + (-1)^n\kappa\Delta M_{\pm}|A_{i,\pm}| \\ &\quad - \left[\gamma_a\cos(\phi_{ii+-}) \pm \gamma_p\sin(\phi_{ii+-})\right]|A_{i,\mp}|,\end{aligned} \tag{64}$$

where $n = 0, 1$ for the in-phase and out-of-phase solutions, respectively, and $i = 1, 2$ for each guide.

We may simplify (64) even further by assuming that there is no direct optical coupling between the polarisation components. That is, the dichroism and birefringence rates are both zero, $\gamma_a = \gamma_p = 0$ (note that these components may still be indirectly coupled via the spin-polarised carrier concentrations).

With this simplification, we find

$$M_{\pm}^{(i)} = \frac{\Gamma_S}{2\Gamma_{ss}^{(i)}}, \quad (\phi_{21\pm\pm} = 0; \gamma_a = \gamma_p = 0) \tag{65}$$

for the in-phase solution and

$$M_{\pm}^{(i)} = \frac{\Gamma_S}{2\Gamma_{aa}^{(i)}}, \quad (\phi_{21\pm\pm} = \pi; \gamma_a = \gamma_p = 0) \tag{66}$$

for the out-of-phase solution (at infinite separation, $\Gamma_{ss}^{(i)}, \Gamma_{aa}^{(i)} \to \Gamma_S/2$ and we would have $M_{\pm}^{(i)} = 1$ in both cases). Note that these solutions do not depend on carrier spin in any way; hence, we also have $M_{\pm}^{(i)} = M_{\mp}^{(i)}$.

Using Equations (34) and (61), we find

$$|A_{i,\pm}| = \sqrt{\eta_{\pm}^{(i)} - \frac{\Gamma_S}{2\Gamma_{ss}^{(i)}}}, \quad (\phi_{21\pm\pm} = 0; \gamma_a = \gamma_p = 0) \tag{67}$$

and

$$|A_{i,\pm}| = \sqrt{\eta_{\pm}^{(i)} - \frac{\Gamma_S}{2\Gamma_{aa}^{(i)}}}, \quad (\phi_{21\pm\pm} = \pi; \gamma_a = \gamma_p = 0). \tag{68}$$

With these results for $|A_{i,\pm}|$, $M_{\pm}^{(i)}$ and $\phi_{21\pm\pm}$, we see that nothing depends on ϕ_{11+-} (moreover, it may also be shown that $\partial\phi_{ii+-}/\partial t = 0$ follows without assuming it to be so). Hence, in this simplified case, we have found exact, analytic steady state solutions for all of the variables (whilst ϕ_{11+-} may be set arbitrarily).

If we now allow for γ_a and γ_p to be non-zero, we may obtain the condition

$$\tan\phi_{ii+-} = \left(\frac{\alpha\gamma_a - \gamma_p}{\alpha\gamma_p + \gamma_a}\right)\varepsilon^{(i)}, \tag{69}$$

where we have defined the modal optical ellipticity in the (i)th guide via

$$\varepsilon^{(i)} = \frac{|A_{i,+}|^2 - |A_{i,-}|^2}{|A_{i,+}|^2 + |A_{i,-}|^2}. \tag{70}$$

Equation (69) agrees with the result of taking the ratio of Equations (15) and (16) in Adams et al. [18], in the case where there is no misalignment of birefringence and dichroism.

We describe (70) as the 'modal' ellipticity, since it is terms of the composite mode amplitudes. Although this is defined for each guide, there is spatial dependence beyond this. Later, in Section 6.2, we shall define a spatially varying ellipticity, hence the reason for the specific nomenclature here. Note that, since $\tan(\phi_{ii+-}) = \tan(m\pi + \phi_{ii+-})$, where m is an integer, we have two possible solutions for ϕ_{ii+-} of ϕ_0 and $\phi_0 + \pi$, where ϕ_0 is the solution of the arctangent of Equation (69) for $\phi_0 \in [-\pi/2, \pi/2]$. Applying this consideration, we find

$$\begin{aligned} M_{\pm}^{(i)} = \frac{\Gamma_S}{2\Gamma_{kk}^{(i)}} \bigg[1 + \frac{(-1)^m}{\kappa} (\gamma_a \cos(\phi_{ii+-}) \\ \pm \gamma_p \sin(\phi_{ii+-})) \frac{|A_{i,\mp}|}{|A_{i,\pm}|} \bigg], \end{aligned} \tag{71}$$

where $m = 0, 1$. In the reduction to the spin-flip model, these solutions are also referred to as being 'in-phase' ($m = 0$) and 'out-of-phase' ($m = 1$). However, we will not use this terminology here, in order to avoid confusion with the previously defined meaning of these terms in the context of the coupled mode model.

In this case, no closed form expression for the optical amplitudes can be found. However, the results of numerical simulation show that using (67) and (68) in conjunction with (69) provides a very good approximation for the pump ellipticities of $|P^{(i)}| < \sim 0.8$.

6. Results and Discussion

In this paper, we mainly focus on the role of pump ellipticity. Further investigations into bistability and polarisation switching based on the theoretical model that was developed here were carried in [27], which may be viewed in conjunction with the present paper.

6.1. Numerical Solutions of the Rate Equations

For general solutions of the model, we employ a combination of numerical methods. For time series simulations of (33) and (40)–(43), we use an adaptive Runge–Kutta method of orders 4 and 5 [34,35]. This is very useful for finding both stable steady state solutions and simulating the temporal dynamics in regions of instability. For finding the unstable steady state solutions and computing the Jacobian, we use a nonlinear solver implementing a trust-region dogleg algorithm [36] based on the interior-reflective Newton method [37,38]. Where steady state solutions exist, this latter method is much faster than time series simulation, facilitating efficient routines for tracing stability boundaries and analysing the Jacobian eigenvalues to establish the nature of bifurcations.

In this section, we give the results for weakly guided structures with $v = \pi/2$. Table 1 lists the waveguide parameters, whilst Table 3 depicts the specific laser parameters used.

Table 3. Laser parameters used in numerical simulations. Note that we tabulate the effective spin relaxation rate $\gamma_s = \gamma + 2\gamma_J$ to aid direct comparison with the spin-flip model (SFM) model.

Parameter	Value	Unit	Description
α	-2		Linewidth enhancement
κ	70	ns^{-1}	Cavity loss rate
γ	1	ns^{-1}	Carrier loss rate
γ_a	0.1	ns^{-1}	Dichroism rate
γ_p	2	ns^{-1}	Birefringence rate
γ_s	100	ns^{-1}	Effective spin relaxation rate
N_0	1.1×10^{18}	cm^{-3}	Transparency density
g_0	1.1×10^{-15}	cm^2	Differential gain
n_g	3.4		Group refractive index

6.2. Spatial Profiles

The spatial profiles of the intensity and output ellipticity presented in this section were calculated directly from the numerical solutions of the Helmholtz equation for circular guides (i.e., no empirical interpolation was used). The spatially-dependent output ellipticity is defined as

$$\varepsilon(x,y) = \frac{\mathcal{I}_+(x,y) - \mathcal{I}_-(x,y)}{\mathcal{I}_+(x,y) + \mathcal{I}_-(x,y)}, \tag{72}$$

where the circularly-polarised intensities are given in terms of the normal mode amplitudes $A_{k,\pm}$, by

$$\mathcal{I}_\pm(x,y) = |A_{s,\pm}\Phi_s(x,y) + A_{a,\pm}\Phi_a(x,y)|^2. \tag{73}$$

The normal mode amplitudes $A_{k,\pm}$ are reconstructed from the composite mode solutions, as described in Appendix C of Ref. [31]. In Figure 6, we show the intensity results with pump ellipticities of $P^{(1)} = 0$ and $P^{(2)} = 1$ and a total normalised pump power in each guide of $\eta^{(i)} = \eta_+^{(i)} + \eta_-^{(i)} = 100$ for circular guides of radius $a = 4$ µm and an edge-to-edge separation that is given by $d/a = 1$ with an optical wavelength $\lambda_0 = 1.3$ µm.

Here, the mid-line joining both of the guides is in the y-direction and guide (2) is in the positive y half of the plane (to the left in the diagrams). We note a residual component of left-circularly-polarised light in guide (2). At this pumping power, this is largely due to spatial coupling between the guides.

Figure 7 shows the corresponding output ellipticity for these intensities. We note a dip in between the guides where the ellipticity goes to -1, even though the pump polarisation is $P^{(2)} = 1$ in guide (2) and $P^{(1)} = 0$ in guide (1). The reason for this can be seen in Figure 8, which plots the modal amplitudes against the pump ellipticity in guide (2). Across the whole range, the optical polarisation is dominated by the anti-symmetric mode, which follows the ellipticity of $P^{(2)}$. However, at $y = 0$, the anti-symmetric mode goes through zero, so the only contribution at this point comes from the much smaller symmetric component, for which the left-circularly polarised amplitude is slightly larger. Hence we see this dramatic dip. However, note that the optical intensity for both components is very small in this region.

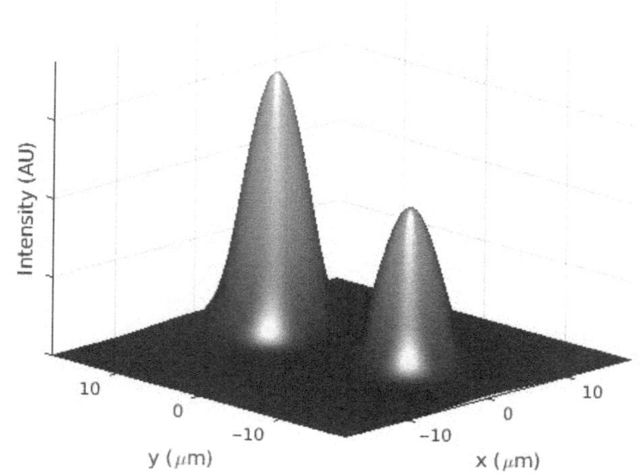

(a)

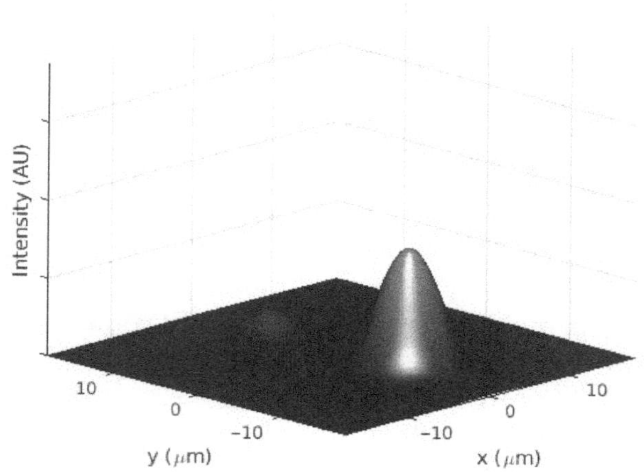

(b)

Figure 6. The spatial variation of the circularly-polarised components of the intensity, defined by (73) for pump ellipticities of $P^{(1)} = 0$ and $P^{(2)} = 1$ with a total normalised pump power in each guide of $\eta^{(i)} = \eta_+^{(i)} + \eta_-^{(i)} = 100$. Guide (2) is to the left in the diagrams (in the positive y direction). **(a)** Right-circularly polarised intensity $\mathcal{I}_+(x, y)$. **(b)** Left-circularly polarised intensity $\mathcal{I}_-(x, y)$.

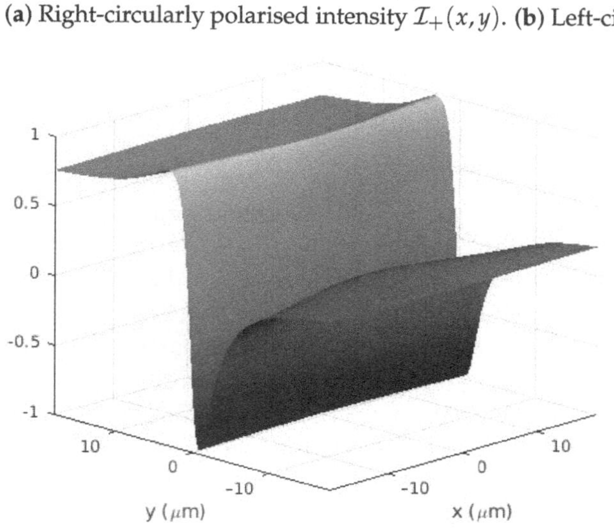

Figure 7. The spatial variation of the optical ellipticity, as defined by (72) for pump ellipticities of $P^{(1)} = 0$ and $P^{(2)} = 1$ with a total normalised pump power in each guide of $\eta^{(i)} = \eta_+^{(i)} + \eta_-^{(i)} = 100$. Guide (2) is to the left in the diagrams (in the positive y direction).

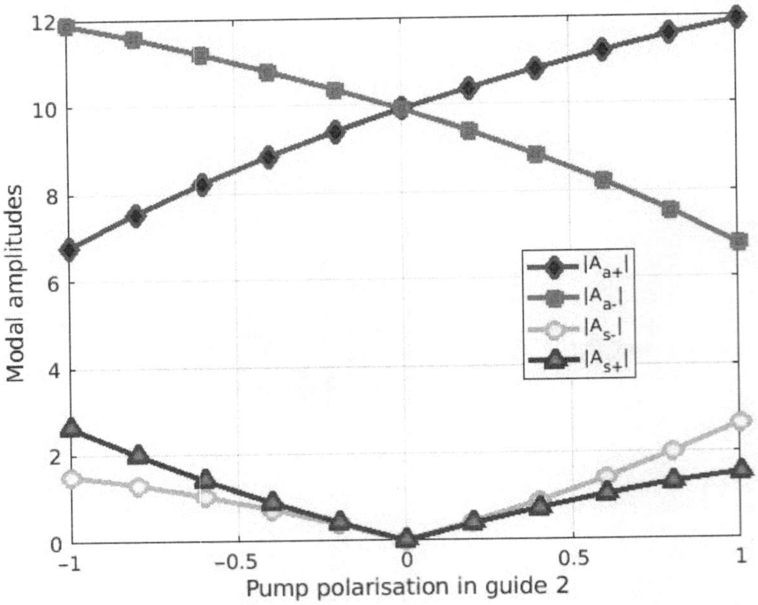

Figure 8. Variation of the normal mode amplitudes $|A_{k,\pm}|$ against $P^{(2)}$, the pump ellipticity in guide (2), for $P^{(1)} = 0$ and $\eta^{(i)} = \eta_+^{(i)} + \eta_-^{(i)} = 100$. Note that the polarisation is dominated by the anti-symmetric mode.

6.3. Stability Boundaries

An initial comparison of the normal mode model to the coupled mode model [29] (neglecting polarisation) has highlighted the importance of the overlap factors on the laser dynamics [28]. This is more significant for the weakly-guided structures that are designed to support only the lowest optical modes. Here we extend this initial investigation to explore the effects of including both optical and carrier-spin polarisation. In particular, we consider the stability of the laser dynamics as a function of the ratio of the optical pump power to the threshold pump against the normalised guide separation. In all cases, we take the total pump power in both guides to be equal, which is $\eta^{(1)} = \eta_+^{(1)} + \eta_-^{(1)} = \eta^{(2)} = \eta_+^{(2)} + \eta_-^{(2)}$, and so we may drop the guide index.

Following Ref. [29] in the case of zero pump ellipticity, we may define the pump-to-pump threshold ratio in terms of $Q = (\eta_+ + \eta_-)/2$ (the factor of two accounts for the normalisation method used) and consider the threshold pump power Λ_{th} at infinite guide separation. This gives us the relation

$$\frac{\Lambda}{\Lambda_{th}} = \frac{Q + C_Q}{1 + C_Q}, \tag{74}$$

where $C_Q = g_0 N_0 / g_{th}$ and the threshold gain is $g_{th} = 2\kappa n_g/(c\Gamma_S)$. The values of the differential gain g_0, transparency density N_0 and other laser parameters are given in Table 3. With the optical confinement factor Γ_S for a circular guide with refractive index step $\Delta n = 0.000971$ listed in Table 2, this gives a value of $C_Q = 43.9$.

For non-zero pump polarisation, we may expect the threshold pump to be affected. We retain the definition $Q = (\eta_+ + \eta_-)/2$ and, again, consider the behaviours at infinite separation. With no pump polarisation, the threshold pump would be $Q = 1$. In the same limiting conditions, our model is equivalent to the spin-flip model.

Following the same analysis that led to the steady state threshold density expression (71) and noting that, when the laser just turns on $Q_{th} = N_{th} = (M_+^{(i)} + M_-^{(i)})/2$, we find that

$$Q_{th} = 1 \pm \frac{1}{\kappa\sqrt{1-\varepsilon^2}}(\gamma_a \cos\phi_0 - \varepsilon\gamma_p \sin\phi_0), \tag{75}$$

where ε is the modal ellipticity in either guide, as given by (70) and ϕ_0 is the solution of (69) for $\phi_0 = \phi_{ii+-} \in [-\pi/2, \pi/2]$. Note that we have taken the limiting condition $\Gamma_S/(2\Gamma_{kk}^{(i)}) \to 1$.

Using the parameters shown in Table 3, we find, via numerical simulation, that, even for $P^{(i)} = 1$, the modal ellipticity does not exceed 0.6, and $Q_{th} - 1$ is of order 0.001. Hence, we may safely neglect the effect of the pump ellipticity on (74).

In the $\Lambda/\Lambda_{th} - d/a$ plane, we find a Hopf bifurcation separating the stable steady state solutions in the upper right half of the plane from unstable solutions in the lower left. Regions of stability also appear at low pump powers and small separation, again being separated by a Hopf bifurcation. In both of the regions, these stable solutions are the out-of-phase solutions, which have predominantly anti-symmetric modal components.

Solutions for slab waveguides neglecting the polarisation have been initially reported in Ref. [28] and compared to the coupled mode model results. It was found that, generally, the overlap factors tended to push the boundaries up in the direction of both increasing power and increasing separation, thus enlarging the regions of instability.

In this work, we focus on the effect of optical pump ellipticity and the simulation results described here are for weakly-guiding ($\Delta n = 0.000971$, $v = \pi/2$) circular guides. Starting with equal pump ellipticities $P^{(1)} = P^{(2)}$ in Figure 9. Here, we have plotted curves from $P^{(1)} = 0$ to $P^{(1)} = 1$ in steps of 0.2. The instability region is steadily reduced as $P^{(1)} = P^{(2)}$ is increased. On the other hand, the boundary of the lower stability region remains insensitive to these changes.

In Figure 10, the ellipticity in guide (1) is kept fixed at $P^{(1)} = 0$, whilst $P^{(2)}$ is varied from 0 to 1 in steps of 0.2. Here, we see the same qualitative behaviour, with the stability moving towards the origin with increasing $P^{(2)}$, although not to the same extent as in Figure 9. Note that the lower stability region has not been plotted in this case.

Figure 11 effectively shows the continuation of this set of results (but starting with $P^{(1)} = 1$ and $P^{(2)} = 0$) and increasing $P^{(2)}$ in steps of 0.2 to 1, ending with $P^{(1)} = P^{(2)} = 1$, as in Figure 9. Finally, Figure 12 shows the effect of putting $P^{(1)} = -P^{(2)}$ and increasing $P^{(1)}$ from 0 to 1. In this case, the final curve with $P^{(1)} = 1$ and $P^{(2)} = -1$ does not diminish the instability region to quite the same extent as $P^{(1)} = 1$ and $P^{(2)} = 1$, although we still see the same qualitative reduction of the region with increased pump ellipticity.

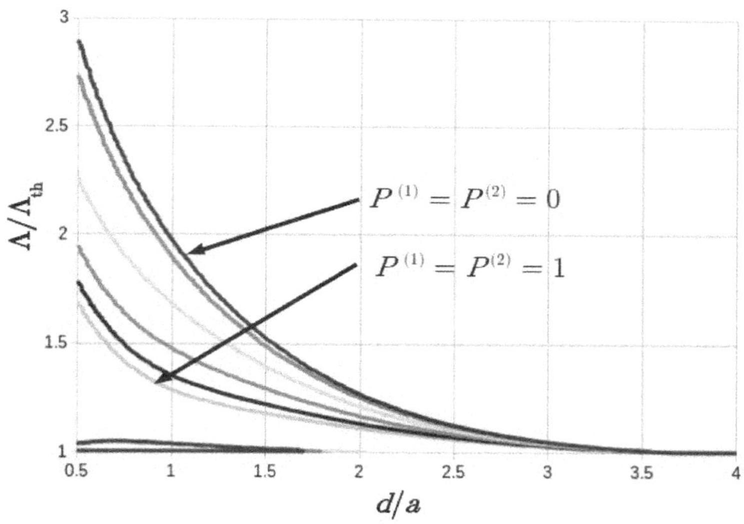

Figure 9. Stability boundaries for equal pump ellipticities. The calculated curves are for $P^{(1)} = P^{(2)} \in \{0, 0.2, 0.4, 0.6, 0.8, 1\}$, with $P^{(1)} = 0$ at the top and $P^{(1)} = 1$ at the bottom.

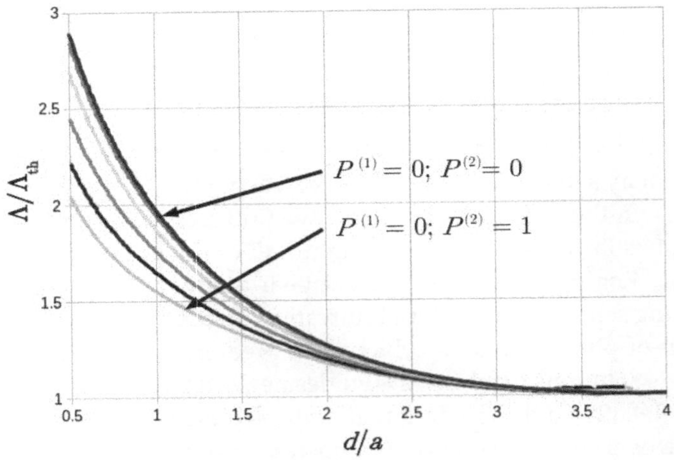

Figure 10. Stability boundary for pump ellipticity $P^{(1)} = 0$ in guide (1) and varying $P^{(2)}$ in guide (2) from 0 to 1 in steps on 0.2.

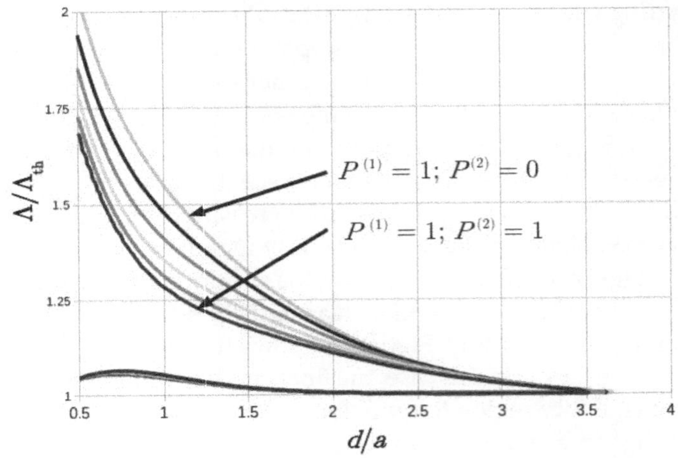

Figure 11. Stability boundary for pump ellipticity $P^{(1)} = 1$ in guide (1) and varying $P^{(2)}$ in guide (2) from 0 to 1 in steps on 0.2.

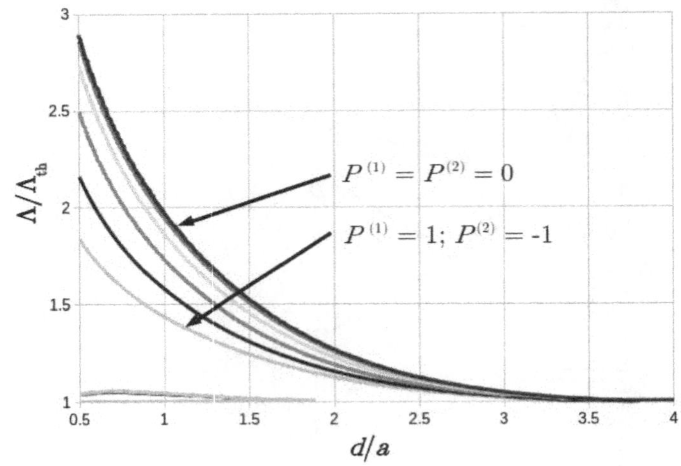

Figure 12. Stability boundaries for $P^{(1)} = -P^{(2)}$, varying $P^{(1)}$ from from 0 to 1 in steps on 0.2.

7. Conclusions

We have derived a set of general rate equations for the laser dynamics in a waveguide of arbitrary geometry supporting any number of guiding regions and normal modes.

The details of the geometry are encoded into the 'overlap factors' (generalisations of the optical confinement factor), which may then have a significant effect on the laser dynamics.

We have focused on the particular case of a double-guided structure in the case of just two supported normal modes and derived a set of real rate equations in terms of 'composite modes'. This treatment is particularly useful for the consideration of symmetric guides (with reflection symmetry) and it can be shown to reduce to both the spin flip model and the coupled mode model in the appropriate limiting case.

Assuming symmetric guides, we have found both exact and approximate analytical expressions for steady state solutions in certain simplified cases. In particular, we have looked at the case of equal power pumping in both guides and investigated the spatial solutions for the circularly polarised intensities (and optical ellipticity) and stability maps for different pump ellipticities. We have found that, in general, increasing the pump ellipticity reduces the region of instability in the pump power versus the normalised separation plane.

The effect of high birefringence has been explored in subsequent work [27] based on the model that was developed here, revealing new dynamics and regions of bistability. It has been shown that optical switching of the polarisation states of the lasers may be controlled through the optical pump and that, under certain conditions, the polarisation of one laser may be switched by controlling the intensity and polarisation in the other.

Although the equations that we have derived are general enough to deal with non-symmetric guides, where there is a large difference between guides in double guided structure, it may be more appropriate to use the general solutions, since the behaviour of the overlap factors tends to diverge from the symmetric case quite rapidly. We have not investigated the resulting dynamics that are associated with this in this paper and leave this matter to be addressed in more detail in future work.

Abbreviations

The following abbreviations are used in this manuscript:

CMM	Coupled mode model
PS	Polarisation switching
SFM	Spin flip model
SVEA	Slowly varying envelope approximation
VCSEL	Vertical cavity surface emitting laser

References

1. Guo, X.X.; Xiang, S.Y.; Zhang, Y.H.; Lin, L.; Wen, A.J.; Hao, Y. Four-channels reservoir computing based on polarization dynamics in mutually coupled VCSELs system. *Opt. Express* **2019**, *27*, 23293–23306. [CrossRef]
2. Jiang, N.; Xue, C.; Liu, D.; Lv, Y.; Qiu, K. Secure key distribution based on chaos synchronization of VCSELs subject to symmetric random-polarization optical injection. *Opt. Lett.* **2017**, *42*, 1055–1058. [CrossRef] [PubMed]
3. Zhang, H.; Guo, X.X.; Xiang, S.Y. Key distribution based on unidirectional injection of vertical cavity surface emitting laser system. *Acta Phys. Sin.* **2018**, *67*, 204202.

4. Jayaprasath, E.; Hou, Y.S.; Wu, Z.M.; Xia, G.Q. Anticipation in the Polarization Chaos Synchronization of Uni-Directionally Coupled Vertical-Cavity Surface-Emitting Lasers With Polarization-Preserved Optical Injection. *IEEE Access* **2018**, *6*, 58482–58490. [CrossRef]
5. San Miguel, M.; Feng, Q.; Moloney, J.V. Light-polarization dynamics in surface-emitting semiconductor lasers. *Phys. Rev. A* **1995**, *52*, 1728–1739. [CrossRef] [PubMed]
6. Erzgräber, H.; Wieczorek, S.; Krauskopf, B. Dynamics of two laterally coupled semiconductor lasers: Strong-and weak-coupling theory. *Phys. Rev. E* **2008**, *78*, 066201. [CrossRef]
7. Martin-Regalado, J.; Prati, F.; San Miguel, M.; Abraham, N. Polarization properties of vertical-cavity surface-emitting lasers. *IEEE J. Quantum Electron.* **1997**, *33*, 765–783. [CrossRef]
8. Travagnin, M.; van Exter, M.; Van Doorn, A.J.; Woerdman, J. Role of optical anisotropies in the polarization properties of surface-emitting semiconductor lasers. *Phys. Rev. A* **1996**, *54*, 1647–1660; Erratum in **1997**, *55*, 4641. [CrossRef]
9. Balle, S.; Tolkachova, E.; San Miguel, M.; Tredicce, J.R.; Martin-Regalado, J.; Gahl, A. Mechanisms of polarization switching in single-transverse-mode vertical-cavity surface-emitting lasers: Thermal shift and nonlinear semiconductor dynamics. *Opt. Lett.* **1999**, *24*, 1121–1123. [CrossRef]
10. Sondermann, M.; Ackemann, T.; Balle, S.; Mulet, J.; Panajotov, K. Experimental and theoretical investigations on elliptically polarized dynamical transition states in the polarization switching of vertical-cavity surface-emitting lasers. *Opt. Commun.* **2004**, *235*, 421–434. [CrossRef]
11. Panajotov, K.; Pratl, F. Polarization dynamics of VCSELs. In *VCSELs: Fundamentals, Technology and Applications of Vertical-Cavity Surface-Emitting Lasers*; Springer Series in Optical Sciences, Chapter 6; Michalzik, R., Ed.; Springer: Berlin/Heidelberg, Germany, 2012; Volume 166.
12. Mulet, J.; Balle, S. Spatio-temporal modeling of the optical properties of VCSELs in the presence of polarization effects. *IEEE J. Quantum Electron.* **2002**, *38*, 291–305. [CrossRef]
13. Masoller, C.; Torre, M. Modeling thermal effects and polarization competition in vertical-cavity surface-emitting lasers. *Opt. Express* **2008**, *16*, 21282–21296. [CrossRef]
14. Gerhardt, N.C.; Hofmann, M.R. Spin-controlled vertical-cavity surface-emitting lasers. *Adv. Opt. Technol.* **2012**, *2012*, 268949. [CrossRef]
15. Gahl, A.; Balle, S.; Miguel, M.S. Polarization dynamics of optically pumped VCSELs. *IEEE J. Quantum Electron.* **1999**, *35*, 342–351. [CrossRef]
16. Gerhardt, N.; Hovel, S.; Hofmann, M.; Yang, J.; Reuter, D.; Wieck, A. Enhancement of spin information with vertical cavity surface emitting lasers. *Electron. Lett.* **2006**, *42*, 88–89. [CrossRef]
17. Adams, M.J.; Alexandropoulos, D. Parametric analysis of spin-polarized VCSELs. *IEEE J. Quantum Electron.* **2009**, *45*, 744–749. [CrossRef]
18. Adams, M.; Li, N.; Cemlyn, B.; Susanto, H.; Henning, I. Algebraic expressions for the polarisation response of spin-VCSELs. *Semicond. Sci. Technol.* **2018**, *33*, 064002. [CrossRef]
19. Li, M.; Jähme, H.; Soldat, H.; Gerhardt, N.; Hofmann, M.; Ackemann, T. Birefringence controlled room-temperature picosecond spin dynamics close to the threshold of vertical-cavity surface-emitting laser devices. *Appl. Phys. Lett.* **2010**, *97*, 191114. [CrossRef]
20. Lindemann, M.; Pusch, T.; Michalzik, R.; Gerhardt, N.C.; Hofmann, M.R. Frequency tuning of polarization oscillations: Toward high-speed spin-lasers. *Appl. Phys. Lett.* **2016**, *108*, 042404. [CrossRef]
21. Torre, M.S.; Susanto, H.; Li, N.; Schires, K.; Salvide, M.; Henning, I.; Adams, M.; Hurtado, A. High frequency continuous birefringence-induced oscillations in spin-polarized vertical-cavity surface-emitting lasers. *Opt. Lett.* **2017**, *42*, 1628–1631. [CrossRef]
22. Hendriks, R.; Van Exter, M.; Woerdman, J.; Van der Poel, C. Phase coupling of two optically pumped vertical-cavity surface-emitting lasers. *Appl. Phys. Lett.* **1996**, *69*, 869–871. [CrossRef]
23. Ebeling, K.J.; Michalzik, R.; Moench, H. Vertical-cavity surface-emitting laser technology applications with focus on sensors and three-dimensional imaging. *Japan. J. Appl. Phys.* **2018**, *57*, 08PA02. [CrossRef]
24. Czyszanowski, T.; Sarzała, R.P.; Dems, M.; Walczak, J.; Wasiak, M.; Nakwaski, W.; Iakovlev, V.; Volet, N.; Kapon, E. Spatial-mode discrimination in guided and antiguided arrays of long-wavelength VCSELs. *IEEE J. Select. Top. Quantum Electron.* **2013**, *19*, 1702010. [CrossRef]
25. Blackbeard, N.; Wieczorek, S.; Erzgräber, H.; Dutta, P.S. From synchronisation to persistent optical turbulence in laser arrays. *Phys. D* **2014**, *286*, 43–58. [CrossRef]
26. Vicente, R.; Mulet, J.; Mirasso, C.R.; Sciamanna, M. Bistable polarization switching in mutually coupled vertical-cavity surface-emitting lasers. *Opt. Lett.* **2006**, *31*, 996–998. [CrossRef] [PubMed]
27. Vaughan, M.; Susanto, H.; Henning, I.; Adams, M. Dynamics of laterally-coupled pairs of spin-VCSELs. *IEEE J. Quantum Electron.* **2020**, *56*, 2400310. [CrossRef]
28. Vaughan, M.; Susanto, H.; Li, N.; Henning, I.; Adams, M. Stability boundaries in laterally-coupled pairs of semiconductor lasers. *Photonics* **2019**, *6*, 74. [CrossRef]
29. Adams, M.; Li, N.; Cemlyn, B.; Susanto, H.; Henning, I. Effects of detuning, gain-guiding, and index antiguiding on the dynamics of two laterally coupled semiconductor lasers. *Phys. Rev. A* **2017**, *95*, 053869. [CrossRef]

30. Martin-Regalado, J.; Balle, S.; San Miguel, M.; Valle, A.; Pesquera, L. Polarization and transverse-mode selection in quantum-well vertical-cavity surface-emitting lasers: Index-and gain-guided devices. *Quantum Semiclass. Opt.* **1997**, *9*, 713–736. [CrossRef]
31. Vaughan, M.; Susanto, H.; Henning, I.; Adams, M. Analysis of evanescently-coupled pairs of spin-polarised vertical-cavity surface-emitting lasers. *arXiv* **2019**, arXiv:1912.06882.
32. Ogawa, K. Simplified theory of the multimode fiber coupler. *Bell Syst. Tech. J.* **1977**, *56*, 729–745. [CrossRef]
33. Marom, E.; Ramer, O.; Ruschin, S. Relation between normal-mode and coupled-mode analyses of parallel waveguides. *IEEE J. Quantum Electron.* **1984**, *20*, 1311–1319. [CrossRef]
34. Dormand, J.R.; Prince, P.J. A family of embedded Runge-Kutta formulae. *J. Comput. Appl. Math* **1980**, *6*, 19–26. [CrossRef]
35. Shampine, L.F.; Reichelt, M.W. The Matlab ODE suite. *SIAM J. Sci. Comput.* **1997**, *18*, 1–22. [CrossRef]
36. Powell, M.J. A FORTRAN subroutine for solving systems of nonlinear algebraic equations. In *Numerical Methods for Nonlinear Algebraic Equations*; Rabinowitz, P., Ed.; Harwood Academic: Reading, UK, 1970
37. Coleman, T.F.; Li, Y. An interior trust region approach for nonlinear minimization subject to bounds. *SIAM J. Optimiz.* **1996**, *6*, 418–445. [CrossRef]
38. Coleman, T.F.; Li, Y. On the convergence of interior-reflective Newton methods for nonlinear minimization subject to bounds. *Math. Program.* **1994**, *67*, 189–224. [CrossRef]

INDEX

M

N

O

P

Q

R

S

9 788197 595769

Printed at Repro India Ltd.